The Evolutionary Epidemiology of Mania and Depression

A Theoretical and Empirical Interpretation of Mood Disorders

To my great business partner
and best friend —
Yau-Dene Chan
all best wishes + many thanks
Gerald A. Cory 3/14/08

THE EVOLUTIONARY EPIDEMIOLOGY OF MANIA AND DEPRESSION

A Theoretical and Empirical Interpretation of Mood Disorders

Daniel R. Wilson
and
Gerald A. Cory, Jr.

With a Foreword by
Farid Dalili

The Edwin Mellen Press
Lewiston•Queenston•Lampeter

Library of Congress Cataloging-in-Publication Data

Wilson, Daniel R.
The evolutionary epidemiology of mania and depression : a theoretical and empirical interpretation of mood disorders / Daniel R. Wilson and Gerald A. Cory, Jr. ; with a foreword by Farid Dalili.
p. cm.
Includes bibliographical references and indexes.
ISBN-13: 978-0-7734-5209-1
ISBN-10: 0-7734-5209-5
I. Title.

hors série.

A CIP catalog record for this book is available from the British Library.

The Edwin Mellen Press
Box 450
Lewiston, New York
USA 14092-0450

The Edwin Mellen Press
Box 67
Queenston, Ontario
CANADA L0S 1L0

The Edwin Mellen Press, Ltd.
Lampeter, Ceredigion, Wales
UNITED KINGDOM SA48 8LT

Printed in the United States of America

**To our mentors, notably,
Paul MacLean, Ernst Mayr and E. O. Wilson.**

TABLE OF CONTENTS

Part III

Bipolar Disorder and Other Neuropsychiatric Syndromes

Part IV

Conclusion

FOREWORD

Concerning this book, Kay Redfield Jamison, PhD, leading scholar on mood disorders and author of *An Unquiet Mind* has written as follows:

This is a fascinating and important book about a topic of immense medical, scientific, and humanistic importance. The authors have put into an overarching theoretical perspective disorders of mood that affect tens of millions of individuals and society itself.

As president of Red Flags Help, a non-profit organization, dedicated to increase the level of depression awareness and related difficulties in California through research and training programs to students, parents, and schools, I welcome the publication of this important work on the evolutionary epidemiology of bipolar, manic-depression.

The authors are uniquely qualified to present the most current scientific research and thinking on this critical syndrome that constitutes an almost incalculable medical and economic burden to individuals, families, and communities, not only in California, but worldwide. Daniel R. Wilson, MD, PhD, Chair of Psychiatry and also Professor of Anthropology at Creighton University, is a pioneer of the evolutionary epidemiological approach to affective disorders, with many scientific publications to his credit. Gerald A. Cory, PhD, a Senior Fellow in Graduate Studies & Research at San Jose State University, is well known for his innovative applications of neuroscience to economics and the social sciences in numerous recent books, articles, and scientific papers.

This book provides an important resource not only to professionals but also to interested individuals, their families, and communities. Beyond its obviously significant contribution to science and understanding, the book also illuminates the hopeful, optimistic case that those burdened with bipolar and other depressive conditions are historically among some of our most creative individuals in the arts, science, and literature. Such understanding not only serves to reduce unwarranted, unhealthy stigma and stress but also promotes helpful attitudes and increases healthy optimism. It further points up and reinforces the urgency of early identification, education, and intervention to mitigate the too often tragic emotional and economic outcomes of ignorance and neglect.

The book presents, not only the most recent research on this vital topic, but also introduces an innovative and powerful new conceptual framework applicable to broader issues of psychiatry, health, and society.

Farid Dalili, Ed. D,
President & CEO,
Red Flags Help,
California

ACKNOWLEDGEMENTS

Daniel R. Wilson, MD, PhD

I wish to thank family and friends who have fostered and sustained my efforts in this work. It is, however, difficult to accurately specify each individual. I regret every inevitable omission. Still, I do wish to particularly thank a number of persons beginning with, Sandra, my wife and life-long best friend and our splendidly adapted daughter, Victoria.

Apart from having chosen my family and ancestors reasonably well, it has been my further good fortune to have had early and sustained exposure to a variety of exceptional teachers. Indeed, many are cited in this book (*i.e.*, some 59 persons who, at one time or another were formally instructors and another 96 with whom I have had opportunity to confer directly in the course of professional or scholarly work). Still, I owe special thanks to Nicholas C. G. N. Mascie-Taylor, Professor and Head of Anthropology at Cambridge, without whose knowledge and practical assistance neither my Ph.D. nor this book would have been feasible to start let alone finish.

Dr. Daniel Pershonok has sustained Socratean support over many, many years and his brilliant translation of papers or concepts from, variously, German, French, Spanish, Swedish, Romanian, Hungarian, Hebrew, Arabic, Greek and Latin sources have greatly enhanced the present work—truly an intellectual *paterfamilias*. Dr. John Price has also become a close friend and mentor. His pioneer writings in evolutionary psychiatry were first mentioned to me when I was ten years old (my father, a GP, upon reviewing John's epochal 1967 item in Lancet, intoned at the dinner table 'finally a psychiatrist who makes sense'). Professors Kay Jamison, Hagop Akiskal, and Alan Schatzberg have also been most helpful and encouraging.

I am indebted to other colleagues from diverse encounters over the years. The late Professor Ernst Mayr was been unstintingly generous in time and attention as well as with highly valuable and quite candid reviews of earlier drafts. Professor E. O. Wilson also has energized my work in an on-going fashion, including his coinage of the wonderfully cogent term 'evolutionary epidemiology' upon reading my earliest draft ideas over two decades ago. Rex Boni of the Russell Trust in New Haven was also a profoundly illuminating influence. So too my co-author, Professor Cory.

Finally, the many patients I have been privileged to come to know through the full range of their human potential and agony in the singular venue of psychiatry have powerfully inspired. Special thoughts go to several patients, friends and relatives who have either not survived or have lost incalculably in their life struggles. I have gained some peace in beginning to draw together evidence that their suffering has a transcendent aspect so, in a way, they remain alive and even passably well.

Gerald A. Cory, Jr., PhD

I owe special thanks to my colleague Dr. Dan Wilson for the opportunity to work with him as a co-author on this important work. In my more recent books *The Consilient Brain: the Bioneurogical Basis of Economics, Society, and Politics* (2004) and *Toward Consilience: the Bioneurological Basis of Behavior, Thought, Experience, and Language* (2000), both published by Kluwer Academic/Plenum, I have had the opportunity to acknowledge the many mentors, scholars, and friends who have contributed to the evolution of my own thought which so closely matches the evolutionary perspective and interests of Dan Wilson. There is no need to repeat those acknowledgements here.

I do, however, wish to note that the production of this book was supported by a grant from the Center for Behavioral Ecology whose associates deserve recognition and appreciation. Among those associates who furnished their efforts and inspiration are Yau-Gene Chan, Mikel Duffy, Manisha Pai, Greg Hobbs, My Thu Bui, Marie Kivley, Meghan Cox and Dale Rowley. I also wish to express sincere thanks to reviewers Stephen Dell, M.D., Don Bradley, Ph.D., and Farid Dalili, Ed.D, who offered many helpful comments and support. Mrs. Patricia Schultz, of Mellen Press was a resourceful, patient, and helpful guide throughout the production process.

1

INTRODUCTION

Darwin's idea of evolution through natural selection has, as an organizing principle, greatly influenced the natural and social sciences over the past century (Wilson, 1975; Masters, 1975; Ruse, 1986; Wilson, 1998). Its consequences have now begun to extend to the arts and letters (Paglia 1990). Even the rarefied domains of philosophy and theology are now beginning to grasp the extraordinary range and power of evolutionary thought (Alexander, 1987; Cory, 2000).

THE NEGLECTED DARWINIAN PERSPECTIVE IN MEDICINE

Yet surprisingly, medicine has so far benefited little from the unique insights of the evolutionary point of view. Until quite recently, psychiatry and related studies have remained almost wholly unenlightened by the evolutionary perspective. This is despite Darwin's (1859, 1868, 1871, 1872) own pioneering efforts to draw human behavior into the domain of evolution and quite apart from Freud's (1917, 1921, 1930, 1987) keen interest in evolutionary ideas as they relate to depth psychology (Sulloway, 1979; Ritvo, 1990).

Meanwhile, seventy years have passed since Karl Jaspers' prediction that the rediscovery of Mendel's laws coupled with Darwinian genetics would soon "bring to medical anthropology generally and psychopathology in particular an exactitude, intensity and unanimity of method from which unprecedented and thoroughly compelling results were imminently expected" (1913: 507). Having ignored such a comprehensive framework, the mainstream of psychiatric research has since continued to move within what Jaspers (1913) so long ago termed "a circle of generalities" only occasionally "punctuated with specific facts." Certainly the sway of rather unreconstructed Freudian concepts endured to inhibit the development of psychiatry longer than Jaspers anticipated.

THE LATELY EMERGENT DARWINIAN INFLUENCE

More recently the corrective force of genetically informed neuroscience has moved psychiatry forward. However heralded this recent so-called biological revolution in psychiatry may be, it seems evolution—that most inclusive science of biology—has yet to comprehensively inform theories or therapeutics within medicine or psychiatry (Barash, 1977; McGuire & Troisi, 1998).

From the biological perspective, psychopathologists are, of course, intensively interested in the neural, humoral, immunological, genetical and all physiological aspects of human behavior and its abnormalities. Likewise, they carefully scrutinize the psychoanalytical, cognitive, interpersonal, social, cultural and other diverse determinants of disordered behavior. But they have not much explored the phylogenetic aspect. Only very lately, Darwinism itself is beginning to capture the more direct attention of physicians, psychiatrists and biomedical researchers.

The advent of the human genome initiative has undoubtedly facilitated this shift of attention. In light of genome research, questions regarding the genetic evolution of disease have become more interesting. Traditional conceptual limitations have become more visible for medicine generally and for both psychoanalysis and biological psychiatry in particular. This is especially so during a time of marked advances in ideas about social biology and genetics. With a nascent Darwinian renewal comes the prospect for a truly integrating model of normal human behavior and its pathology. Thus a systematic theory of psychopathology may be emerging that is neither brainless nor mindless, nor for that matter, sexless (Eisenberg, 1986).

The incorporation of the scientific vigor of the Darwinian framework, however tardy, is a trend to be welcomed by all interested in a complete and systematic conceptualization of medicine and psychiatry. Principles to be abstracted from the taxonomic study of human behavioral forms and their distributions, both normal and abnormal, will generate a deeper understanding of the laws of ecology from which pathological phenomena have originated and evolved. Such principles will eventually reorder theories, teaching, and therapeutics in an approach with more empirical breadth and philosophical vigor than any in the existing schools of psychopathology.

Still, the evolutionary approach to the study of normative behavior is, itself, rather new. Thus, the proposition that the root causes of some pathological behavior are better understood as remnants of past adaptive behaviors is so new that science has done little to contextualize such phenomena. Indeed, there exist no comprehensive texts or courses of study with a specific focus on the positive evolutionary roots of psychopathology. Moreover, basic or clinical research protocols are as yet of limited influence, being both new and small. Yet even as evolutionary psychology acquires "cachet" (Pershonok, 1993), it is ironic that until very recently almost nowhere has there been substantial exploration of the evolved roots of human behavioral pathology, particularly with attention to either genetics or brain sciencc.

Several useful forays, some early, some recent, nevertheless, do stand out. Especially noteworthy are: Darwin (1872); Romanes (1884, 1896); Lloyd Morgan (1909); Lewis (1934); Huxley (1964); Erlenmeyer-Kimling and Paradowski (1966); Price (1972); Gardner (1982); Wenegrat (1984); Bailey (1987); Chance (1988); Gilbert (1989); Glantz and Pearce (1989); MacLean (1990); Badcock (1986, 1990, 1992); Wilson (1993); Williams and Nesse (1991); Stevens and Price (1996, 2000); McGuire and Troisi (1998); Cory and Gardner (2002); Panksepp (2004).

The present work draws much from these efforts. But clearly there is much more to be done. Formal study and research, both theoretical and empirical, is needed to incorporate and systematize within psychiatry the clear conceptual implications of Darwinian principles. The understanding of both normal and pathological behavior will benefit from a thorough examination from an evolutionary vantage point.

The extent and character of mental illness in prehistoric times relative to its expression in contemporary life is, of course, a confounding issue. So, too, the appropriate consideration of mismatch between phylogeny and modern environments. Any significant demographic transition is important to the proper contextualization of the contemporary etiology of certain "diseases". Clearly evolutionists and medical anthropologists have an interest in the matter. But it is also ultimately relevant for metaphysicians, clergy, social scientists, legislators and other professionals. Indeed, an even broader sampling of the general public also has an abiding interest. Many suffer psychiatric maladies themselves or have friends or family who do.

Of particular interest is the question whether any contemporary mental illnesses are biosocial expressions of genes which were advantageous in the course of evolution. The current presumption of medicine is that psychopathology is simply deleterious. Hence, genes transmitting risk are necessarily maladapted. If maladaptive, such genes should be found in the population only at very low levels. Since the work of Darwin, science has known that traits become common, essentially, via promotion of characteristics advantageous to the struggle for existence in the environment of evolutionary adaptation. That is, genes evolve by means of natural selection.

THE INHIBITING LEGACY OF MISAPPLIED DARWINISM

There are further reasons evolution has not yet substantially established itself within psychopathological studies. Most such reasons are nonscientific. They mostly derive from social, political and historical concerns regarding the theory of evolution itself. Scandal still easily attaches to the enterprise of understanding human behavior in terms of evolution. Thus, the introduction of Darwinian ideas to the study of psychopathology has been inhibited, in the first instance, because human evolution remains somehow controversial—at least in the public mind. This is true not just because of the scientifically premature policy ambitions of some social biologists, but often even because of the general principles of evolutionary biology themselves (Campbell, 1975).

The shadows of inhumane social Spencerianism (misnomered social Darwinism) still stretch long. Also, so do the shadows of related eugenical mistakes and pseudoscientific bigotries (e.g., see Beckwith & Alper, 2002). These and other aberrant diversions promulgated over the past century remain, quite justifiably, the source of enduring unease. The Scopes' monkey trial, fascist distortions, and other less perverse but still misguided examples remain within living memory of many. Such ideologically dark and politically clumsy efforts to apply evolution to the human race ought not to be passed over lightly. Any research agenda which at all appears likely to dust off this tragic legacy arouses legitimate fear and genuine loathing.

Darwinism is, then, a residually controversial area of science. Controversy is stirred all the more when the study of human evolution is joined with the central interests of psychopathology. These interests—abnormalities of thought, mood, sex, aggression—themselves are often the focus of considerable popular and even medical derision or unease. To say the least, novel Darwinian studies of such abnormalities in persons already so socially marginalized and misunderstood is work that is unlikely to appeal to the mainstream imagination.

Yet for too often and too long these concerns have been thrown up to—often ideologically and sometimes cynically—"block the way of inquiry" (Peirce, 1899). The scientific study of evolutionary variation is assuredly not license to ascribe value judgments to or otherwise pathologize such variation. Nevertheless, some critics express fear of restrictive social policies which they see as inevitable outcomes from studies of human evolution. For a time such concerns were voiced rather stridently by some who prejudged the agenda to be one of reactionary reductionism (Gould & Lewontin, 1979). These contretemps have subsequently been the basis for a virtual cottage industry of historiographic criticism largely pertaining to the implications of human evolutionism in its social, behavioral and political dimensions (Bowler, 1992).

The anxieties provoked by human social biology have been part and parcel of ongoing and serious discourse in the annals of philosophy and history of science for more than two centuries (Alexander, 1987). In fact, the origins of these debated themes trace back two millennia or more, at least to Aristotle (c. 450 BCE). Most recent arguments, especially those pertaining to human sociobiology, have generated more heat than light. These are now, thankfully, on the wane. Moreover, they are not germane to the topic here at hand.

It should suffice for the present purposes to assure that studies pertaining to human evolution adhere to the highest of ethical standards. Derivative policy recommendations (if any) ought to be few, narrow, and subject to considerable skepticism. It is with this commitment that the authors proceed with the present exploratory effort.

OVERVIEW OF CONTENTS

Emerging within medical anthropology is a subdiscipline that harnesses some of the extraordinary power of the facts of evolution within a manageable medical scope (Wilson, 1992). This subdiscipline, evolutionary epidemiology, is an inherently biomedical synthesis of several basic operations borrowed from evolution and epidemiology, respectively. Evolution and epidemiology are closely related sciences. Each is a type of demographic study. Evolution, (*e-* from, and *volvere-* to turn) concerns the differential transmission of population traits across generations. Epidemiology, (*epi-* that which is among, and *demos-* the population) concerns the array of traits in a given population.

Evolution is itself an inherently epidemiological process. Species transform in evolution principally by selection of those varieties which have practical advantages. Thus, selection hinges on gradations within (*epi-*) populations or societies (*demes*). Evolutionary epidemiology aims to translate trait or gradation prevalence rates, as known to epidemiologists, into frequency rates, as known to geneticists. This synthetic method can thereby deduce which gene systems have a population prevalence so high as to be best explained with reference to a long period of advantageous natural selection in the past environment of evolution.

The principal application here-to-fore has been taken up most prodigiously by microbiologists and infectious disease in the area of host-pathogen co-evolution. However, evolutionary epidemiology also accomplishes two essential things that are the focus of separate parts of this book. That is, it provides:

1. A valuable Darwinian heuristic for an operationalized definition of normality—normality of our evolved sociality is of great interest as both the starting and reference point for psychiatry, as deviations from this

normality may reduce reproductive efficiency and often require psychiatric intervention.

2. Insights as to how genetically-based "disease" may have been healthy in the environment of evolutionary adaptation.

This book is, therefore, an empirical as well as theoretical work in the fashion of Darwin's own "hypothetico-deductive" method. The techniques of evolutionary epidemiology draw upon methodologies and concepts from medicine as well as other natural and social sciences as applied toward a better understanding of both normal and pathological behavior.

Part I applies the evolutionary framework to define health and normality. Normality is the reference point to specify either deviancy or psychopathology. Normality is, however, seldom defined with heuristic precision. Usually it is taken for granted as a type of naturalistic fallacy which assumes the social status quo is the right and proper context. Yet the Darwinian perspective calls for a universal or species-wide and phylogenetic-developmental definition. Thus, evolutionary epidemiology helps provide a framework for the more reliable and valid analysis of *genes, brains, and behavior*—what should be but is, as yet almost never, the central focus of evolutionary psychologists.

Chapter 2 outlines the evolution of human sociality with emphasis on the transition from ancestral vertebrates to mammals to primates and, ultimately, humans. Evolution has produced a human brain that is essentially a social brain. Certain insights into the social brain can only be demonstrated from an evolutionary perspective. Therefore, the understanding of this evolutionary process is necessary to any full appreciation of human social behavior.

Chapter 3 concerns the normal limits of human sociality. An operational definition of such limits is essential to a grasp of normality. To this end the behavioral, neural, and biochemical bases of normality are examined. The human brain evolved to cope with issues of survival of both the individual self and the species. The social brain is essentially a homeostatically regulated neural architecture that evolved to keep social behavior within prescribed limits to assure reproductive success (i.e., 'survival') of self and others. Any complete definition of normality or pathology depends on a proper grasp of this architecture.

Chapter 4 examines the relationship between psychiatry and normality. The evolutionary concept of inclusive fitness is linked to the motivational structures of our evolved neural architecture. It then examines recent progress in defining social competition and social cooperation. Social stress is discussed as a significant aspect of the homeostatic regulatory process that helps lead us to concepts of normality, even optimality.

Part II moves from normality to a focus on the evolutionary legacy of neuropsychiatric disorders with special emphasis on bipolar disorder (BPD) and the manic-depressive spectrum as an exemplary case. **Chapters 5** through **7** set out the foundations of evolutionary epidemiology in greater detail. Here classical principles of Mendelian inheritance, non-Mendelian dynamics, and population genetics are combined with further insights concerning social behavior from the Darwinian evolutionary perspective. Long-established epidemiological evidence relating to frequency and heritability of affective spectrum begs the question as to how such "genetic disease" can be so common. These basic concepts and findings are followed by a discussion of the remarkable advances from molecular genetics and the human genome project which make possible more direct linkage studies.

Chapter 8 focuses upon the case of sickle cell anemia as another exemplary case. This Mendelian medical-genetic "disease" is the benchmark example of how genes that attained high population frequency owing to selective advantages in the environment of evolutionary adaptation still express serious medical pathology. That is, sickle-cell traits conferred resistance to pervasive malaria in that environment but now, in the modern environment, are primarily etiogenic factors in lethal disease. So it is made clear that the clinical and theoretical utility of evolutionary epidemiology applies not only to the physical aspect of medicine but also to the psychological. **Chapter 9** further applies the insights and analytical tools of evolutionary epidemiology to the syndrome of BPD. The prevalence of BPD within the global population reflects (*ala'* sickle cell anemia) clearly a strong selective advantage as a population-genomic polymorphism in the past environment of evolutionary adaptation.

Part III focuses more sharply on bipolar and related affective disorders. **Chapter 10** brings together evidence for defining an evolutionary anthropology of BPD. It examines the roots of manic-depressive behavior in the phylogeny of reptilo-mammalian social competition and the evolution of the neocortical capacity to represent, integrate, and to an extent, balance the competitive circuitries.

Chapter 11 applies the perspective of game theory to manic-depressive behavior using the analogy of hawk-dove game strategies. In doing so, it summarizes recent conceptual advances from the evolutionary perspective, notably social competition theory. In the human ancestral line, dichotomous brain-behavioral substrates of social competition emerged during the transition from early vertebrates and reptiles to mammals. Extreme poles of social competition are manifested in BPD. Such polarities reflect behavioral physiology that was adaptive in our earlier vertebrate history and can be witnessed in some lower vertebrates today. Moreover, these maintained a level of adaptiveness in the transition to reptiles and through mammalian social evolution up to humans. Essentially, this dichotomous strategic spectrum can be represented by two intersecting axes of reptilian self-preservation/mammalian affection and the capacity to escalate or de-escalate both.

Chapter 12, then, proceeds to draw out the basic sociobiology of manic-depressive behavior. The two archetypal opposing strategies from our reptilian and mammalian heritage may be expressed in bipolar extremes or across a varying spectrum. The opposing strategies may further be mitigated and dynamically-balanced by neocortical mediation. The importance of social stress in the homeostatic or allostatic regulation of these interlocking strategies is examined. **Chapter 13** assesses the current state of knowledge of the neural and endocrine correlates of social competition and social cooperation emerging from the interplay of these archetypal phylogenetic strategies.

Chapter 14 focuses on the clinical insights that emerge from evolutionary epidemiological analysis. BPD, as well as other disorders, may not be pathological *per se* but result from a mismatch between prior adaptive phenotypes and the present environmental conditions. Evolutionary analysis suggests that the ratio of adaptation match to mismatch for an individual in a given environment is plausibly a reliable and valid coefficient for defining pathologies, both physical and mental. **Chapter 15** examines several other syndromes that likely arose as adaptive polymorphisms—notably sociopathy—as well as makes a provisional assessment of other neuropsychiatric syndromes to be further clarified by evolutionary epidemiological analysis. It also considers related ethical and clinical issues.

The concluding **Chapter 16** confronts the challenges and opportunities for psychiatry and related—nearly all—disciplines made possible by the systematic application of the evolutionary perspective.

PART I

THE EVOLUTIONARY FRAMEWORK OF HEALTH AND NORMALITY

2

THE EVOLUTION OF HUMAN SOCIALITY

The Darwinian perspective provides a valuable framework for understanding the evolution of human sociality. From the standpoint of evolutionary epidemiology, the brain is a species' characteristic. Its fundamental architecture—like limbs, eyes, ears, and internal organs—is genomically prevalent across the species population. Genetic frequency approaches 100%, with minor variances. The brain is, first and foremost an evolved physiological organ—like all other major organs of the body. The gene-specified neural circuits or architecture constitute the fundamental physiology of the species-wide brain.

Physiologically, the human brain is also a *social* neuromental apparatus. The social brain concept emphasizes both the self-preservational (self-interested) and affectional (other-interested) components necessary to social interaction. The *social brain* is an organizing concept heralded in recent publications—*Textbook of Biological Psychiatry* (Panksepp, 2004), *Foundations in Social Neuroscience* (Cacioppo, et al. 2002) and *Handbook of Affective Sciences* (Davidson, et al. 2003; see also, Cory & Gardner, 2002). Slightly earlier but related volumes include *Descartes' Error: Emotion, Reason, and the Human Brain* (Damasio, 1994), *The Integrative Neurobiology of Affiliation* (Carter, et al. 1997) and *Affective Neuroscience* (Panksepp, 1998). Recent years have thus brought great advances in detailing the many complex and interrelated pathways of interactive social circuitry in the brain.

The genes specifying social circuitry were selected over millions of years of evolutionary history in small kinship groups which required a cooperative interactive dynamic for survival. These dynamic social circuits motivate human social interaction and social exchange at all levels of our lives today.

NEW THINKING ON THE ORIGIN OF NOVEL PHENOTYPES AND SPECIATION

It seems appropriate at this point to digress slightly to discuss important new thinking that is relevant for the emergence of human social architecture. Recently, it has been argued that the origin of novel phenotypes and even speciation depends as much or more on *developmental recombination* and *genetic accommodation* than upon mutation, *per se*.

West-Eberhard (2003, 2005) has integrated a very large body of evidence to suggest phenotypic novelty arises largely via reorganization rather than as the product of altered genes. That is, ancestral phenotypes are first reorganized and then followed by genetic accommodation of change. Moreover, insofar as phenotypic novelty arises from adaptive developmental plasticity, it is not random variant mutation that is actually the main driver of adaptive evolution. Rather, the initial form reflects adaptations accrued over a specific evolutionary historical context. Thus, even as random mutations or novel environmental factors initiate changes in trait frequency, processes of genetic accommodation adapt the threshold or liability by which a novel trait is expressed.

Hence, novelties initiated by environmental factors may drive evolutionary selection more than do novelties initiated by mutation. Moreover, even when actual mutation initiates reorganization—as with genes of major effect on regulation—selection still entails genetic accommodation. That is, genetic change both follows, and is directed by, the reorganized phenotype. Certainly mutation still ultimately provides the genetic variation that allows adaptive evolution, but genetic accommodation need not await mutation.

Developmental recombination occurs in populations as new inputs cause a reorganization of the phenotype. Such inputs may be either a positively selected mutation or a novel environmental stressor. Given the range of individual phenotypic reactivity and variable development, a variety of unique phenotypes for selection and *genetic accommodation* is engendered. If phenotypic variation is genetic, selection leads to accommodation via adaptive change in gene-frequency. If variation is environmentally induced, genetic accommodation can affect either regulation or form. Regulatory adjustment modifies frequency, timing, or circumstances of novel responses; e.g., altered threshold for genetic expression. Adjustments of form optimize features and efficiency of the new trait expressivity.

This understanding depicts adaptive evolution conventionally as phenotypic change linked to selective changes in gene frequency. However, it departs—slightly, but significantly—from a view of adaptive evolution that is exclusively based on mutation-selection. Mutation is a certain source of novelty, but no longer the sole source. With a standing reservoir of genetic variation, genetic accommodation need not rely solely on mutation.

All this is particularly important as research on gene expression is increasingly systematized and accessible. The interests of molecular geneticists can increasingly converge upon those of specialists interested mainly in phenotypes, notably organismic biologists as well as clinical researchers. The intersection is the realm of intermediate processes and organic algorithms, such as endocrinology, neurobiology, and processes of subcellular signaling. These areas, so long estranged from evolutionary biology, emerge as the processes by which environmental, genetic, and phenotypic variation are linked to selection and evolution via mediation of gene expression. This understanding of the developmental nature of evolutionary variation supplies a missing piece of the modern synthesis and especially follows the work of Dobzhansky and Mayr.

Such developmental plasticity may well be a factor in the emergence of various forms of early humankind with increasingly refined social and executive neural architecture.

THE EVOLUTIONARY BACKGROUND

Paul MacLean pioneered the study of neural circuitry substrates comprising the social architecture of the brain. That MacLean was such a path-finder is in keeping with his stature as a leading evolutionary neuroscientist, experimental neuroethologist, and medical psychiatrist who was also the long-time head of the Laboratory of Brain Evolution and Behavior of the National Institutes of Health.

In his 1990 masterwork, *The Triune Brain in Evolution: Role in Paleocerebral Functions*, MacLean tells us that the primary function of the human brain is self-preservation and the preservation of the species* (1990: 17, 23). Although something quite similar may be said of the nervous system of any organism (which, after all, must survive as an individual to reproduce), MacLean formulates not just automatisms or tightly prewired instinctual mechanisms necessary to self-maintenance and procreation, but the evolved architecture of human neuromental circuits upon which family life and social choices are made (1990: 380, 519, 560–61). His concept of brain evolution, appropriately understood and updated, provides the necessary and nearly complete conceptual platform for this undertaking. For a detailed, documented update and critique of MacLean's concepts see Cory (1999, 2002a, 2004).

The R-Complex or 'Self -Preservational Component'

MacLean documents the human brain as evolved in an interconnected, modular three-level structure (Figure 2.1). This structure retains a self-preservational neuromental component reflecting considerable gene-based continuity with our ancestral reptiles that split off from the dinosaur ancestral line

* Here, in keeping with MacLean, we retain the term *preservation of the species* even as this is more precisely a matter of inclusive—not species—fitness.

during the Permian and Triassic periods, some 225–250 million years ago. This is called the *protoreptilian* complex or *R-complex*. The other two main levels were the later elaborated *paleo* (old) mammalian affectional complex, and a most recently evolved higher *neocortex* in the highest centers of the brain.

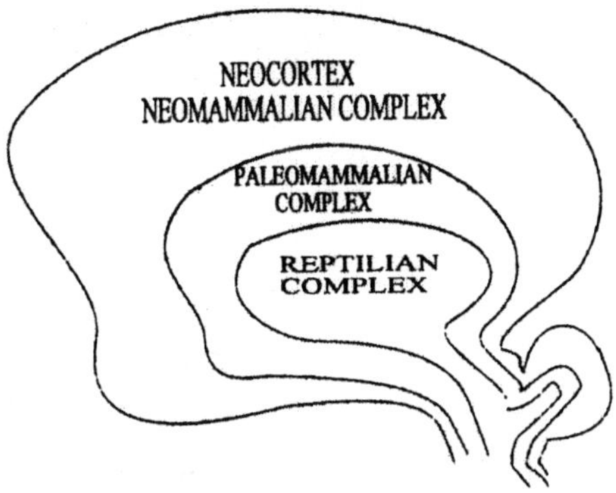

Figure 2.1. The Interconnected, Modular Tri-level Brain (After MacLean). As represented here the three brain divisions do not constitute distinct additions but rather phylogenetically continuous modifications and elaborations of epigenetic homologues.

As brain evolution continued in the branching vertebrate line ancestral to humans, simple vertebrate or protoreptilian brain structure was not replaced, but became modified and elaborated. The protoreptilian elements, thus, provided the substructure and gene-based continuities (homologues) for later brain development while largely retaining basic reptilian character and function. Subsequent early mammalian modifications arose with new emotional neuromental capacities inherent in the paleolimbic and pre-frontal areas.

Unquestionably the human brain, with the neocortical elaborations that followed, reached a zenith. Yet these cortical centers are surprisingly not so relevant to emotionality and behavior so much as they endow domain-specific (language, music, visuo-spatial, etc.) or more associational-analytic talents. While absolutely crucial to what makes the human species so unique, they reveal little of direct interest to the emotional hermeneutics (character and boundaries) of highly social species including humans. Appreciating the qualitative differences of the three interconnected levels is important to understanding the dynamics of human social experience and exchange behavior.

The protoreptilian brain circuits function in humans much as they did in our ancestral vertebrates, to govern fundamentals—daily master routines of life-support (respiration, heartbeat & blood circulation, foraging, defensive behaviors, and reproduction). These functional behaviors necessary to all animate life were sculpted in our ancient ancestral reptiles as well as earlier amphibians and fishes.

Located in what are usually called the brain stem and the midbrain to include midbrain structures at the base of the forebrain (i.e., basal ganglia), this now primal and innermost core of the human brain made up almost the entire and newest neuromental centers in ancestral fishes, amphibians, and reptiles (although not necessarily their modern descendant analog species, because they, too, have undergone further evolution).

The Limbic (Paleomammalian) or Affectional Component

The next stage of our brain evolution and development was largely organized amid a rudimentary mammalian milieu. This MacLean terms the paleo- or "old" mammalian brain with neuromentalities designated collectively as our limbic system. Developing as gene-based continuities from preexisting protoreptilian brain genomics, these new limbic circuits included significant elaboration of such physiological structures as amygdala, hypothalamus, hippocampus, thalamus, thalamocingulate, and orbital-frontal cortices. These modified and elaborated paleo-mammalian early limbic structures facilitated newly essential mammalian innovations absent in our ancestral reptiles such as endothermy (warm-bloodedness), parturition, nursing, infant care, and extended social bonding.

These new characteristics became neuromentally integrated with the earlier evolved life-support functional and behavioral circuitry of the protoreptilian brain to create the more complex life form and pattern characteristic of all mammals.

The Neocortex

The most recent stage of brain modification and elaboration is the neocortex, or isocortex, which MacLean called the neo- or "new" mammalian brain. This great mass of bi-hemispherical brain matter that dominates the skull case of higher primates and humans evolved by elaborating preexisting homologues present in the brains of early vertebrates. The neocortex overgrew and encased the earlier (paleo-) mammalian and protoreptilian neural tissues, but essentially did not replace them. As a consequence of this neocortical evolution and growth, those older brain parts evolved greater complexity and extensive interconnected circuitry with these new tissue structures. All this to subserve behavioral adaptations necessary to primate and humankind's increasingly sophisticated social circumstances.

The unique features of our human brain were refined over a period of several million years in kinship-based foraging societies where sharing or reciprocity was absolutely essential to survival (e.g., see Humphrey, 1976; Isaac, 1978; Knauft, 1994; Erdal & Whiten, 1996; Boehm, 1999). Moreover, this evolutionary neuromental foundation of exchange-economic behavior is of core relevance not only to human psychology and psychopathology but also social sciences, notably market economics of an intrinsically humanely moral type (Cory, 1999, 2002, 2004, 2006ab). Here it is of special importance that such sharing and reciprocity fostered adaptive evolution by which mammalian characteristics of self-preservation and affection constitute the essential motivated expressions of our social brain. Thus, they have become tightly integrated as a neuromental complex that balances conflicting urges arising in behavioral systems.

THE CONFLICT SYSTEMS NEUROBEHAVIORAL (CSN) COMPLEX

The CSN Complex (Figure 2.2) in essence is a systematized, dynamic formulation of the clarified and updated MacLeanian conceptual platform described above. The depiction of self-preservation and affection with their Egoistic and Empathetic derivatives, however, is primarily illustrative and not to suggest definitive lateralization (although there is evidence to suggest that the right hemisphere is typically more instrumental in emotions whereas the left is for more analytical self-preserving behaviors; e.g., Damasio, 1994; Tucker, Luu, & Pribram, 1995; Brownell & Martino, 1998; Henry & Wang, 1998).

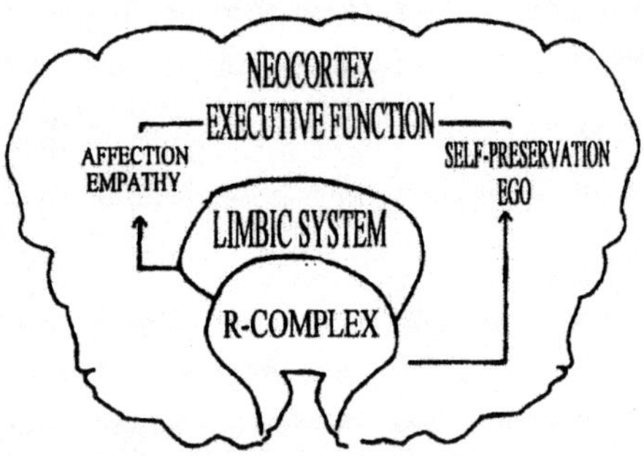

Figure 2.2. The Conflict Systems Neurobehavioral (CSN) Complex. A simplified cutaway representation of the brain showing the behavioral circuits and the derivation of Ego/self-interested and Empathy/other-interested motives and behaviors. Ego and Empathy are capitalized to reflect their use herein as technical terms somewhat distinct from common usage. It should be noted that earlier models (notably, the Freudian Id, Ego, and Superego) postulated conflictual models with three-parts. Freud, however, was unable to tie his model to brain circuitry because brain research had simply not advanced to that point and it has remained ungrounded in neural science (adapted from Cory after MacLean, see Cory 1992, 1999, 2000ab, 2001ab, 2002, 2004, 2006ab).

Heller (et al. 1998), further acknowledges it is well established that particular right hemispheric regions facilitate interpretation and expression of emotional information, but argues that the total experience of emotion is not itself lateralized. Rather, such experience involves dynamic interactions between forward and posterior regions of both hemispheres as well as subcortical (limbic and midbrain) structures. Therefore such complex, highly integrated capacities as Ego and Empathy are better construed as engaging interaction of both hemispheres. Davidson (1995), for example, hypothesizes that the left and right anterior regions of the brain are key components of an affective regulatory system for approach and avoidance behaviors.

Our R-complex and limbic circuits are interconnected neural network circuits that *subjectively* generate and drive specific, and *objectively* observable, behaviors. These core motivational (and emotional) circuits are respectively represented as self-preserving Ego and species-preserving Empathy cognitions of frontal, as well as possibly other, neocortical regions (e.g., see Berridge, 2003; cf. Saarela, 2006; Tankersley, et al. 2007). They serve as dynamic factors of our behavior as they are dynamically-driven and homeostatically regulated by processes of cellular, tissue, and total-bodily metabolism as mediated by *hormonal, neurotransmitter, and neural architecture.*

Ego and Empathy Clarified

As noted earlier, Ego, as used here is not the Ego of Freudian psychoanalysis although there are some similarities. Ego, as indicated in Figure 2.2, is defined as deriving from self-preservational circuitry. It closely matches the standard definition of Egoism in philosophy and as referenced in literary terms such as "self-focused," "self-absorbed," "self-interested," or "narcissistic". Empathy, on the other hand, includes the affectional feelings of sympathic attachment reinforced by cognitions added in the course of neocortical assessment and representation (e.g., see Hoffman, 1981, 2000). Empathy and sympathy, despite their rather distinct meanings in medical-psychiatric settings, are otherwise and especially in recent citations more often posited inclusively (Eisenberg, 1994; Batson, 1991). For this reason and in order to suggest the inclusion of sympathy the term *Empathetic* is used as distinct from the more specific term, *empathic*.

Archetypal Circuits Driven by Behavioral Tension & Stress

Both the R-complex and limbic circuitries are archetypal, inseparable parts of human nature, because each is coded into the species genome by the process of evolution. The degree of genomic control, however, varies with precise mechanisms and also between individuals. Damasio (1994, 1999) uses the terms *preset* and *preorganized* to avoid the implication of an overly-deterministic prewiring or coding in some brain regions sometimes connoted with classical genetical terms such as *canallation.* Terminological nuances aside, phylogenetically older hindbrain and earlier limbic structures necessary for more basic survival are more strongly canallated with neuromental capacities under more genetic and less plastic phenotypic control. Phylogenetically newer tissues such as the neocortex accommodate more environmental experience in the course of phenotypic development.

Behavioral conflict potentially exists, then, simply by the presence of these two large-scale dynamic circuitries in our lives—up and running (canallated or preorganized) even prior to birth. Behavioral tension is often subjectively experienced as frustration, anxiety, or anger, and occurs whenever one of these two fundamental circuits—self-preservation or affection—is activated but meets with some resistance or difficulty that blocks satisfactory expression via neuromental resolution in the CSN Complex. This subjective tension becomes most paralyzing when both systems are activated and seek contending or incompatible responses *within a single situation, i.e., double-bind.* Caught between "I want to" but "I can't"—e.g., "I want to help others (spouse, offspring, friends, et al.), but I can't surrender my needs"—we agonize or ambivalate.

Whether this tension arises through thwarted expression of a single impulse or simultaneous but mutually exclusive urgings of two contending impulses, it leads to worsening behavioral stress whenever it remains unresolved.

The evolutionary process by which the two opposing circuitries motivating self-preservation and affection were combined helped the human lineage adapt in phylogeny new presets for social interaction that allow a wider range of behavioral responses to the environment. When in unresolved conflict, our inborn circuits are a curse only to the degree that we fail to recognize them as a blessing. After all, they are the physiological basis for our highly-advanced sensitivity and homeostatic reaction to the social environment. This very sensitivity generally keeps our interactive behaviors within survival limits and enables us to perceive and appreciate the survival requirements of others. Significantly, if emotionally counter-intuitively, the accompanying behavioral tension (*even the stress!*) is an integral part of this useful function, for it allows us to more immediately evaluate our behavior and the effect it is having on ourselves and others, all as mediated in the CSN Complex.

Behavioral tension serves as an internal emotional compass to guide through the often complicated and treacherous pathways of interpersonal relations. Behavioral stress tells us that we are exceeding safe limits for ourselves and others, and for our larger social, economic, and political structures.

The Frontal Executive

Our executive circuitry is seated largely in our frontal and pre-frontal neocortex (Pribram, 1973, 1994; Fuster, 1997, 1999; Miller & Cummings, 1999; also Goldberg, 2001; Stuss & Knight, 2002). This cortex is a hierarchically structured neural network with a six-layered architecture that *cognitively represents, abstracts, generalizes, mediates*, and to a significant extent, *controls* archetypal inputs it receives from limbic and protoreptilian subcortical centers (also see Berridge, 2003). It makes what may be thought of as our *moral* as well as *rational* choices among our conflicting, impulsive, and irrational or nonrational motivations. This capacity to represent, generalize, and choose—particularly as it is often expressed linguistically—is what differentiates us from even our most closely related primate species (and much of what makes findings in primate behavior, although highly interesting and unquestionably important, insufficient in themselves to fully understand and account for human behavior).

THE MAJOR RANGES OF RECIPROCAL SOCIAL BEHAVIOR

The two master circuits of self-preservation and affection operate as global state variables in a neuromental complex (see Panksepp, 2002; cf. Schulkin, 2002). That is, these two master circuits operate dynamically as mediated by executive neocortex as it assesses inputs received from lower centers all according to *subjectively experienced* and *objectively expressed* behavioral rules, procedures, or algorithms. Such central motive states are likely produced by synchronized firing of neural network circuits that shape our interactive social behavior as phylogenetic presets become behaviorally sensitized (e.g., see Tiitinen, et al. 1993; Cobb, et al. 1995; Traub, et al. 1998; Deth, 2003, 7–10).

The major ranges of the CSN Complex (Figure 2.3) illustrate features of this self-preservational/affectional dynamic as it is homeostatically coupled within normative behavior. In the depiction below, social behavior is divided into three main ranges called, from right to left, Egoistic, Dynamic Balance, and Empathetic. Each range represents a varying mix of self- and other-motivated behaviors "pivoting" in the frontal executive area of our brain. The solid line represents self-preservation-ego and the broken line represents affection-empathy.

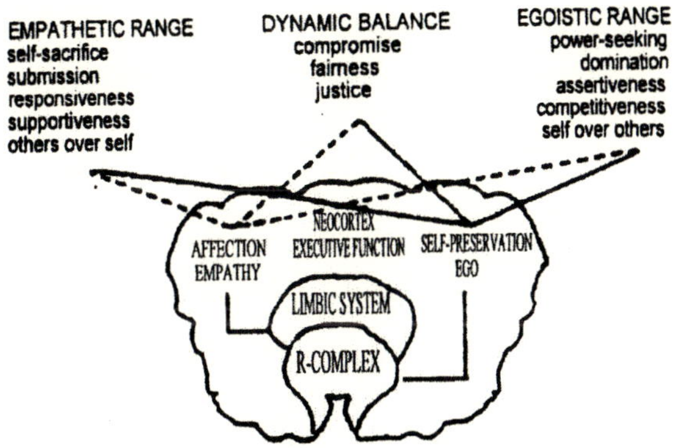

Figure 2.3. The Major Ranges of Behavior. To simplify the graph, the three points are intended to mark the center points of each range, with varying mixes of self-preservation (Ego) and affection (Empathy) on either side of each point. The graph thus indicates not a zero-sum, either/or set of behavioral options but rather a calculus of increasing or decreasing proportions of self- and other-concerned behavior. The figure represents what also may be thought of as central tendencies of interactive social behavior, but is too simplified to represent all the shadings of emotion and motivation (adapted from Cory after MacLean, see Cory 1992, 1999, 2000ab, 2001ab, 2002, 2004, 2006ab).

The Egoistic Range

The Egoistic range indicates behavior dominated by self-preserving intent and actions. Since the two behavioral programs are coupled as an inseparable neuromental complex, Empathy is present here but only in a provisional manner and behavior in this range is self-centered or self-interested. Where Empathy is at its most diminutive along this range, behavior is more Egoic—dominating, power-seeking, or even attacking. When Empathy is more robust along this range, Egoic behavior is moderated and less harsh but still rather controlling, competitive, or assertive. As Empathy gradually increases, the intersection of the two lines of the diagram are drawn toward the range of dynamic balance. Ego behavior will be softened as Empathy is added. But the defining characteristic of the Egoistic, self-preserving range is *self-over-others*. Whether we are blatantly power-seeking or more moderately assertive, in this range we are putting ourselves, our own priorities and feelings, ahead of others.

The Empathetic Range

The Empathetic range represents behavior weighted in favor of affective attachment. Self-preserving behavior persists, but is surpassed by Empathetic tendencies. When Ego is present to a minimal degree along this range, behavior may be Empathetic to extremes of self-sacrifice and submission. When self-preserving behavior is increased along this range, behavior may be Empathetic but moderated somewhat as supportive, responsive, or any of a variety of "others first" behaviors. As the influence of Ego is gradually engaged along this range, Empathetic behavior will approach dynamic balance. Overall, Empathetic range behavior is *others-over-self* or *others first*. Whether we are at the extreme of self-sacrifice or more moderately responsive, we are putting the priorities and feelings of others ahead of our own.

The Dynamic Balance Range

The range of dynamic balance represents homeostasis, or a working balance between our opposing archetypal circuitries. We speak of "working," "rough," or "dynamic" balance because the tug-and-pull between the two circuits continues ceaselessly. The dynamic nature of the circuitry means that "perfect" balance is a theoretical point of interest although unattainable in actual circumstances. Our more balanced behavior tends to be characterized by equality, justice, sharing, and other behaviors which show respect for ourselves and others. In fact, respect for self and others is the keynote of the range of dynamic balance.

Energy or Activation Level—Escalation and De-escalation

The extent to which the circuits of self-preservation and affection are out of balance, or pulling against each other, is a measure of behavioral tension. We experience this behavioral tension both internally and between ourselves and others, in any relationship or interaction. Excessive, unmanaged or unresolved tension engenders behavioral stress along a range of behavior. But the amplitude of escalating or de-escalating energy we invest in the interaction or the relationship is an equally important factor affecting behavioral stress and its manifold physiological sequellae.

Much of this arousal occurs automatically by activation of the HPA (hypothalamic-pituitary-adrenal) axis taken up with greater detail in Chapter 4. Nevertheless, it is also under cortical control to a significant degree. We can put more energy into activities and responses for which we feel enthusiasm—less into others. In highly competitive sports or contests, qualitative differences in energy are easily observed. In intellectual contests, like chess, the energy may be intense, but less obvious.

Such arousal level—or escalation/de-escalation—is applicable across the Ego-Empathy spectrum, giving us two essential axes of social behavior. This is supported by Gray (1987) who has indicated that the brain may moderate behavior for both inhibition and reward-seeking and by Witter and Groenewegen (1992) and Gloor (1997: 325–496) who report a comparator role for the hippocampus, a prominent limbic structure, in the routing of internal and external stimuli in relation to previous affective experience (cf. Iidaka, et. al 2003).

The Ego-Empathy and escalation/de-escalation axes, thus, are two of the most essential axes of social behavior expressed in reciprocal algorithms.

THE RECIPROCAL ALGORITHMS OF SOCIAL BEHAVIOR

Three basic CSN Complex rules, modified somewhat from Cory (1999, 2004), derive from the dynamic interplay of Ego, Empathy, and activation (escalation/de-escalation) level.

Rule 1) *Self-preserving Egoistic behavior creates tension within ourselves and between our selves and others to the degree it lacks Empathy. Tension increases with escalation from low to high activation levels and with moves toward extremes of Ego.*
Subjectively, we can experience tension created by the 'tug' of neglected Empathy along a range from a sympathetic feeling of obligation (to the other) all the way to a more paranoic (if often accurate) expectation that the other might wish to 'even the score'.
Subjectively within others, tension created by our self-interested behavior is experienced along a range from a dolorous or irritated feeling of imposition and hurt all the way to a more paranoic (if often accurate) sense of persecution, accompanied by an urge to 'even the score'.

The dynamic of such behavior is often most clearly seen when children engage in the ethologically important activity of play. Consider, for instance, two children playing on the living-room floor. One hits the other. The second child hits back, responding in kind. Or the children may not hit each other at all. One might instead call the other a bad name. The second child reciprocates, kicking off a round of escalating name-calling. One child may eventually feel unable to even the score and will complain to a parent to intervene. Such interactive play behavior serves to tune and adjust neocortical control of the maturing archetypal opposing circuits in humans as well as other mammalian species. Most of us have experienced such give-and-take as children and have seen it countless times in our own children and grandchildren.

Beyond play, we even see a similar dynamic among adults—in husband-and-wife disputes, bar fights, hockey games, political campaigns, even in the process of lawsuits. The rule operates not only in such highly visible conflict situations, but also in very subtle interactions—in the small behavioral exchanges, the ongoing give-and-take of all interpersonal social exchange relations.

Of the underlying conflictual dynamic of the excitatory-inhibitory motivated neural network architecture, *reactive impulses accrue in healthy proportion to the tension created by Egoistic, self-preserving behavior.*

Rule 2) *Empathetic behavior, because it denies self-preservation or Ego to some degree, also accrues as tension within ourselves and others. This tension increases as we escalate activation levels and as we move toward extremes of Empathy.*

Subjectively within ourselves, tension created by the 'tug' of the neglected self-maintenance (Ego) is experienced as a feeling that 'others owe me one' and a growing need to 'collect our due'. This tension may be experienced as resentment at being exploited, taken for granted, not appreciated, or victimized by other— especially if it continues over time.

Subjectively within others, tension is experienced as a sense of obligation due us.

Reactions that build in ourselves and others are in healthy proportion to the tension created by Empathetic, affectionate behavior. Again, excessive, unmanaged or unresolved tension is experienced as behavioral stress.

When we do things for others—give them things, make personal sacrifices for them—we can feel quite righteous, affectionate, loving. Nevertheless we *do* want a payback. That is the tug of self-preserving circuitry. The tug can be very slight, hardly noticeable at first. But let the giving, the self-sacrifice, go on for a while, unacknowledged or unappreciated (that is, without payback to the Ego) and tension accrues. We may complain that others are taking advantage of us, taking us for granted, victimizing us. Self-preservation circuitry cannot be long short-changed without demanding its due. We may eventually relieve the stress by blowing up at those we have been serving—accusing them of ingratitude or withdrawing our favors and support. Or we may suppress expression of the stress, letting it eat away at our dispositions, damaging our physiological and psychological health.

On the other hand, when we do things for others, they often feel obliged to return the favor in some form to avoid being left with an uneasy sense of debt. Gift-giving notoriously stimulates the receiver to feel the need to reciprocate. A gift received mandates a return. Such obligation was the binding force of primitive family sharing as well as the gift economy of pre-market societies (see Cory, 1999, 2004).

> Rule 3) *Behavior in the range of dynamic balance expresses homeostatic equipoise between Ego and Empathy. Subjectively, within ourselves and others this position of least behavioral tension creates feelings of mutuality and shared respect or 'secure attachment'.*

Most of us find it satisfying to interact with others in equality, with no sense of obligation, superiority or inferiority. When we work together in common humanity, in common cause, we experience behavioral dynamic balance. Certainly, there are many versions of the experience of dynamic balance: the shared pride of parents in helping their children achieve, the joy of athletes in playing well as a team, the satisfaction of co-workers in working together successfully on an important project.

THE RECIPROCAL NATURE OF BEHAVIOR

These algorithms of social behavior operate in the smallest interactions of everyday social life. The dynamic of behavioral tension provides that for every interpersonal act, there is a balancing reciprocal. A self-preserving act requires an Empathetic reciprocal for balance. An Empathetic act, likewise, requires a balancing self-preserving reciprocation. This reciprocity goes back and forth many times even in a short conversation. Without reciprocation, tension builds, stress accumulates, and either confrontation or withdrawal results. If not, and the relationship continues, it becomes a tense and stressful one of inequality or domination/submission, waiting and pressing for the opportunity for adjustment. The behavioral algorithms describe how we get to reciprocity through conflict. They shape the conflict and reciprocity, the give-and-take, at all levels of our interactive, social lives.

Overemphasis on either self-preservation or Empathy, exercise of one archetypal circuit to the exclusion of the other, creates tension and stress in any social configuration—from simple dyadic person-to-person encounters up to and including social exchange interactions among members of the workplace, society at large, social groups, and entire economic and political systems.

VARIABILITY OF THE RECIPROCAL ALGORITHMS

The algorithmic rules of reciprocal behavior operate as neuromental complexes; that is, central tendencies or global state variables of behavior. They also show considerable individual distinctions based upon allelic variability and environmental interaction. They cannot, then, work as precisely as the laws of classical physics or even quantum mechanics since variability and lack of absolute precision is generally true of biological algorithms (e.g., see Maynard Smith, 2002). They are achieved through the process of organic evolution which involves some random processes and natural selection. Sex, gender, developmental, cultural and experiential differences also contribute to variability (Wilson, 1992; Cory, 1999: 42–4).

RECIPROCITY: THE UNIVERSAL NORM

The norm of reciprocity expressing our social neural architecture has long been a major theme in anthropology and sociology (e.g., see Gouldner, 1960; van Baal, 1975) and more recently in economics (e.g., Cory, 1999, 2004; Fehr & Gachter, 2000; Gintis, 2000; Bowles & Gintis, 1998: esp. ch. 17; Eckel & Grossman, 1997). This universal norm observed in all societies, primitive and modern, has been accounted for by or at least compatible with evolutionary theoretical concepts of kin selection, inclusive fitness (Hamilton, 1964), reciprocal altruism (Trivers, 1971a, 1981; Alexander, 1987), and game theory (Axelrod & Hamilton, 1981, Maynard Smith, 1982).

These efforts draw upon so-called gene-centered perspectives that seemingly make the fallacious leap from genes to motivated behavior to construe reciprocity as basically selfish in the *individual* sense. De Waal (1996) attempts to counter this perspective by his interpretation of empathic behavior observed in rhesus monkeys. This departure from the selfishness of gene-centered approaches is welcome, particularly as it looks to the implied motivational mechanisms.

Yet, all these approaches, to include that of de Waal's, are based on the external observation of behavior. They do not take up important hermeneutical considerations nor do they attempt to identify or even speculate upon the neural mechanisms within the organism that necessarily must have been selected in evolution as proximal means to accomplish functions of motivating, maintaining, and rewarding reciprocal behavior as it is so widely observed among humans.

Reciprocity through conflict is achieved in the range of dynamic balance where behavioral tension operating freely *tends* to pull us—as is in keeping with the archetypal opposing circuits of our neural architecture. In dynamic balance, self-preserving behavior and Empathy provide for the emergence of *cooperation and fairness, trust and morality,* in interpersonally social activities. Taking the dynamic balance range to be approaching or approximating the equilibrium of self-preserving behavior and Empathy, as modulated homeostatically by behavioral tension and mediated neocortically, we can derive a formula that expresses this dynamic. Indeed, analyses from the stand-point of evolutionary epidemiology (Wilson, 1986, 1992, 1998) demonstrate that eusociality can be mediated by advanced neuromentalities of affective attachment; all this is germane to new areas for empirical research, particularly psychoneuro-endocrinology as taken up further in Chapter 13.

THE EQUATION OF SOCIALITY

The reciprocal algorithms emerging from the interplay of the opposing reptilo-mammalian circuitries of our social neural architecture have been illustrated graphically by the three major ranges of social behavior (see Figure 2.3). They have also been written in plain English by the section describing their algorithmic interactive dynamic. They can also be stated mathematically in the form of the equation of sociality or social interaction stated as approaching equilibrium or dynamic balance:

$$BT = \frac{Ego}{Emp} \text{ or } \frac{Emp}{Ego} = \pm 1 \text{ (dynamic balance, approx. equilibrium or unity)}$$

In the above formula BT stands for behavioral tension and is a function of the ratio of self-preservation to affection (Ego to Empathy) or vice versa. Because of the physiological homeostatic nature of the dynamic, in which the archetypal circuits act to counter deviations from dynamic equilibrium, either Ego or Empathy can serve as the numerator or denominator to accurately reflect the magnitudes. The degree of convergence or divergence is what is of interest (Cory, 1999, 2004, 2006ab).

This equation gives basic mathematical expression to the interactive social architecture of our evolved social brain structure. As the conflicting circuits of our social architecture approach equilibrium or dynamic balance (represented by the symbolic approximation to unity or dynamic balance, **±1**), behavioral tension/stress are minimized. On the other hand, as the ratios diverge increasingly toward the extremes of Ego or Empathy, behavioral tension increases.

That is, if we have an Empathy magnitude of eight and an Ego magnitude of four, or vice-versa, we have a behavioral tension magnitude of two. At a minimum the neural dynamic serves generally to keep our social behavior, homeostatically, within survival limits which accounts for its Darwinian selection. On the other hand, at the level of optimal functioning, algorithms driven by behavioral tension *tend* to move us toward dynamic balance of Ego and Empathy (*self-* and *other-interest*); that is, balanced reciprocity, or equality.

The formula is deceptively simple, as it expresses what can be quite a variable dynamic equilibrium steady-state that can further ramify in many ways. Thus it is no trivial calculation but actually expresses remarkably complex structural-dynamic relationships. But here we enter into the realm of organic algorithms that is of profound utility in the social sciences, and in contrast to the Newtonian point-reducible force vector algorithms so much more feasible in the exact sciences.

Magnitudes *vs* Precise Calculation

The use of the social brain formula, although expressed in magnitudes, does not inherently exclude precise calculation where data are sufficiently defined and available. As Maynard Smith observed (2002), variability and lack of absolute precision characterize biological algorithms. But the issue of magnitudes versus precise calculation in mathematics goes well back in the history of mathematics.

As French mathematicians Jean Baptiste Say (1855: xxvi) and Antoine Augustin Cournot (1838: 2–3) noted, the application of mathematics to the social sciences—even the physical sciences—is aimed principally at the elaboration of form and structure and to illustrate a connection between causes and consequences.

Cournot, especially, was concerned with the demonstration of functional relationships which could be expressed by the functions of calculus. He was not concerned with numerical calculation and felt some events were resistant to precise numerical calculation except in trivial situations or games of chance. Cournot wrote: "Those skilled in mathematical analysis know that its object is not simply to calculate numbers, but that it is also employed to find relations between magnitudes which cannot be expressed in numbers." (1838: 2–3).

The social brain formula reflects these characteristics. It expresses relationships among neural structures and functions but often—but not always—evades precise numerical calculation.

ABNORMALITIES AND PHYLOGENETIC REGRESSION-PROGRESSION

The CSN Complex is based largely upon neocortical representations of subcortical circuitry that have moved up the hierarchy of neural architecture in the process of human evolution. Such neocortical representations are highly dependent on experience and social learning in the process of phenotypic development. The CSN Complex, then, is essentially a spectrum resonating within the normal ranges of behavior. Although signaled and motivated by homeostatic mechanisms, the spectrum of the Ego-Empathy interactive dynamic relies on a significant degree of prefrontal cortical executive control.

But what happens when this control fails and more primitive subcortical circuits take over our interactive social behavior? Such occasions represent an atavistic reversion to behaviors no longer appropriate to the civilized standards that our higher brain development makes possible.

Kent Bailey (1987) has modeled phylogenetic regression-progression to account for such abnormal behavior when the more recently evolved prefrontal executive breaks down or is overwhelmed by the earlier circuitry. In a recent article he referred to the process as upshifting (progression) and downshifting (regression) the triune brain (2002).

Bailey's insightful analysis brings to light behavior that is *not* normal—that is, behavior out of the range of social behaviors described by the CSN Complex. Bailey's model refers to behavior that may have been normal in the primitive ancestral environment of evolutionary adaptation, but is no longer either normal or adaptive since the evolution of the higher frontal cortex. It represents seriously disruptive social behavior that is released by genetic defect, brain damage, disease, or by extreme social stress.

Bailey applies his analysis to such glaringly abnormal social behavior as displayed by serial killer Ted Bundy and the two youths who perpetrated the Columbine school massacres. Such abnormalities of behavior are important to psychiatry and they point up further the importance of evolutionary epidemiology. They can only be fully understood from an evolutionary perspective.

AGONIA AND HEDONIA IN THE CSN COMPLEX

Michael Chance (1988) developed the concepts of the agonic and hedonic modes by studying the behavior of higher primates and considering our human evolutionary history. Chance's definition of hedonic should be distinguished from what is called hedonic psychology (Kahneman, et al. 1999). The latter is based largely on the elaboration of the pleasure principle of Bentham and Freud and is applied across the entire spectrum of rewarding behaviors, social and non social, indiscriminately.

Instead, Chance held that primate societies and human societies operate in the two modes. In the agonic mode, humans are primarily concerned with self-security, rank hierarchy, and are focused on response to potential threat. The agonic mode evolved in vertebrate reptilian lines well before the appearance of mammals and later primates. The agonic mode is a state of constrained conflict or hostility. A high state of arousal and tension is required to sustain the agonic mode. The high tension/stress level is retained even when agonistic, conflictual behavior is not displayed.

The hedonic mode is, in contrast, affiliative and flexible. It appeared with mammals and was expanded in primates and humans. It leads to conciliation and cooperation with behaviors such as touching, grooming, and kissing which function to reduce tension and stress. In the hedonic mode, the arousal and accompanying tension/stress levels are low. Chance holds that as adults, we humans can be stuck in one mode or the other or move back and forth between the modes.

The similarities of the agonic-hedonic mode concept and the CSN Complex are readily apparent. Both address the same basic phenomena and both are driven by tension at different levels. The differences, however, are notable. The agonic-hedonic lacks a register of amplitude. It also is a dichotomy that tends to polarize behavior. This not only obscures a possible middle ground of moderating homeostasis for dynamically balanced resolutions of behavioral tension, but also does not fully explain the very human extremes of empathic altruism such as the self-sacrifice seen in parents, soldiers, firefighters, and others strongly bonded by affective attachments.

The CSN Complex offers more flexibility and discriminating power while also encompassing the neuroscientific strength *ala'* MacLean. Any human can, within limits, move back and forth across the Ego-Empathy spectrum, in any number of daily social activities. As the two archetypal circuitries tug and pull against each other across the extremes, neocortical architecture mediates tensions that are behaviorally regulated by dynamic, homeostatic mechanisms.

The CSN Complex, of course, summates cooperation emerging with the approach to dynamic balance of Ego and Empathy in behavior. It provides, then, not only for altruism but for the classic 'win-win' behavior that is obscured by the polarity of the agonic-hedonic dichotomy.

The CSN Complex also brings into prominent play the momentously significant characteristic of Empathy so important to our understanding of human social behavior (Ploog, 2002; see also Harris, 2002; MacLean, 1990). The beginnings of Empathy can be seen in the emergence of the capacity for social comparison (placing oneself in the others' shoes and assessing their intentions or strengths). But it really emerges with the premammalian vertebrates and more fully blossoms with the mammalian affectional neuroarchitecture as represented in the neolimbus and neocortex.

The CSN Complex is built directly upon the MacLeanian tri-level platform describing the two primitive archetypal neural circuitries from different periods of our phylogenetic history. From the interplay of these circuitries emerges the bonding affection of human family life that serves the preservation of self and species. The CSN Complex also follows in the traditional ethological approach of Lorenz most fully laid out in Eibl-Eibesfeldt's highly praised *Love and Hate*

(1972). In that work Eibl-Eibesfeldt sees egoistic aggression and altruism as dynamic opposing instincts preprogrammed into human nature. He provides extensive examples from primate and human observations. Primatologist de Waal makes a similar point from the ethological perspective in his *The Reconciled Hierarchy*. De Waal writes that in primates and humans, dominance has an *affective* component which is often overlooked by ethological observers in their preoccupation with competition and aggression (1988: 129).

In sum, MacLean worked to identify the key neural substrates upon which ethological observations such as of Eibl-Eibesfeldt, de Waal and others must necessarily stand. The CSN Complex is an effort to further operationalize these neural substrates in terms that more accurately render social behavioral homeostasis in terms of phylogeny and ontogeny of both individuals and aggregate populations.

An additional advantage of the CSN Complex model is that it is a formulaic equation of homeostatic dynamics expressed as an organic algorithm with wide-ranging applications across many aspects of social behavior (Cory, 1999, 2004). However, as alternative perspectives such as phylogenetic regression and the agonic-hedonic modalities often allow uniquely productive insights, these will be invoked as needed in subsequent chapters.

3

THE NORMAL LIMITS OF SOCIALITY

The prior chapter briefly outlined several basic features in the evolution of human social behavior. This was systematized and depicted graphically across the major ranges of the CSN Complex and described in mathematical terms of both algorithms for reciprocal behavior and an equation of homeostatic sociality. This chapter is devoted to defining the normal limits of sociality in more technical detail as a first and fundamental task of evolutionary epidemiology. Therefore, the state of existing knowledge will be examined for a proper definition of normality as is basic to psychiatry and related basic, translational and therapeutic sciences.

Stated in terms of necessity and sufficiency, the behavioral continuum is "normal" to the degree that it enables Darwinian reproduction as evinced from interactive layers of archetypal neuromental circuitries of our reptilian and mammalian heritage. Thus it is relevant to trace these structures with greater attention to neural architecture, neurotransmitter and endocrine correlates. It is important to emphasize that normality is here approached from the global state perspective of the archetypal opposing circuitries of social neural architecture.

This is a higher level of integration than is usual in extant definitions of normality. More typically such efforts aim at delineation of various dimensional traits of personality and temperament (e.g., Cloninger, 2002). But at a more detailed level of integration, such traits may be considered to reflect, in part, variants of temperament and personality along principle global axes of self-preservation/affection (Ego & Empathy) and escalation/de-escalation.

Moreover, direct evidence that key elements of these structures and functions have been greatly conserved through evolution includes remarkable homologies reaching back to the Precambrian epoch (2000~3000 x10^6 years ago; Doolittle, et al. 1996). Such robust conservation from deep phylogeny provides *the* essential aspect of animate life and the ultimate basis for all behavior, whatever may be the other details of species anatomy and physiology. This absolutely core feature of evolved history of chemical bioregulation repays further consideration in laying the foundation for the ultimate evolution of human social neural architecture.

EVOLUTION OF NEURAL HIERARCHY

Delineation of brain evolution has historically relied on comparative analyses of physioanatomy across nearly all species and a vast compendium of taxonomic similarities and differences has accumulated. But more recently, the development of molecular biology has opened new opportunities for comparative analysis at the much more fundamental level of biochemistry.

The Contributions of Molecular Research

Molecular mechanisms driving neuromental development have become the major focus of brain evolutionary science and fed dramatic insights into fundamental questions. Molecular research casts new light on why neuromental abilities range so broadly across species, why large and energy-demanding brains evolved, and why biomodulators have so greatly diversified (Allman, 1999).

Nature builds more by tinkering with existing developmental plans from past evolutionary solutions, typically adding on a new feature to an extant apparatus. Much has been achieved via the retention and modification of new autologous clones of highly adaptive genes (Wilson, 1992). By illustrative analogy, Allman (1999) recounts his visit to the boiler room of an old power plant. There he noticed a bank of pneumatic tubes adjacent to an array of vacuum tubes near serial types of computer boards, and realized that the plant had never come off line because it was needed for power production. Instead, it was retrofitted as new iterations of technology became available but much of the old systems retained functional roles. The evolution of neural hierarchies reflects analogous phylogenetic tinkering. Thus a brief recap of salient aspects is worthwhile not just as it is illustrative of neuromental evolution but also as the general features of this evolution are directly relevant to understanding the CSN Complex.

All neural hierarchy traces to the matrix-product-matrix loop of DNA-RNA-Protein sequences needed for bioconstruction and reproduction (see Loewenstein, 1999; Cory, 2000). Thus the DNA-product loop is fundamental to structuring the information-processing behavioral circuits of organic life via all manner of bioregulators—chemotransmitters, neurotransmitters, surface and transmembrane receptors, transducers, eVectors, and second messengers—not to delve into peptides and neurohormones.

From this basic starting point, neural hierarchies evolve. In the process, existing neuromental functions are in part subsumed, modified, or represented upward in the hierarchy and new functions emerge. This is evident even in the earliest unicelluate life forms with their intracellular basic bioregulators from which evolved the highly complex series and distributions of receptors and receptor subtypes that make up the central nervous system of higher animals.

Single cells process information from both internal and external sources. The external information from the environment is compared and processed for match or mismatch with the internal information which is the somatic structure or metabolic matrix of the cell. From the successful internal, external match/mismatch processing, the cell incorporates material from its environment. It thus builds and maintains its somatic (bodily) structure and ultimately reproduces a self-copy.

In single cells, external information from the immediate environment is processed by protein circuits (Koch, 1999; Bray, 1995). Protein circuits are the behaving elements of the DNA-RNA-Protein loop. Such processing constitutes the behavior of the cell. The behavior of the cell is inseparable from its somatic matrix or body as the somatic-behavioral link is absolute. The cell can only process what its somatic matrix demands or permits. The unbreakable mind-body linkage is established at the cellular level. That linkage holds in all organisms up to and including human beings—Descartes' mind-body dichotomy never existed.

Origins of Neurotransmission

Scientific support concerning the origin and physioanatomical diversification of bioregulators including neurotransmitters comes from both comparative and newer molecular studies of pre-synaptic and synaptic evolution (Boyd, 1979). Neurotransmitters such as dopamine, norepinephrine, serotonin and acetylcholine originated within unicellulates as primary metabolic molecules that affected chemotaxis—movement toward or away from a chemical gradiant. The functional significance of chemotaxis is to direct cellular behavior at the most primordial level. This can facilitate a number of the most basic cellular functions including exchange of solutes and orientation to sources of light.

Of course a multiplicity of metabolic molecules—ancestral biotransmitters — entrain intra- and extra-cellular responses crucial to chemotaxis. But specific substrates associate with and correspond to specific types of chemotaxia that supply cells with superior access to food, light, and other basic ecological factors and substrates for metabolism. Propitious chemotactic effects contribute to differential success in biosyntheses—a distinct Darwinian advantage in the behavior of even unicellulates.

Origin of Neuroreceptors

Such basic molecules of single cells became transmitters (if not *neuro*transmitters in the strict sense) more than a billion years ago when primitive receptors evolved on cell surface membranes. It is relevant that extra-cellular receptors were phylogenetic precursors of both trans-membrane and intra-cellular receptors. This occurred as these early surface membrane transmitter receptors reacted to variations in the chemophysical conditions of the immediate milieu and released simple biomaterials that influenced these conditions. Here are highly primitive yet identifiable elements of behavior.

These surface membrane functions, thereafter, evolved trans-membrane mechanisms that enhanced ionic homeostasis within the cell. It appears such ion channel regulators first derived from alterations in electrogenic co-transport of amino acids in non-excitable (non-neural) cells but are still quite similar to activities necessary for synaptic transmission as later evolved in neural cells. Eventually, some surface- and trans-membrane receptors were wholly brought into the cell via endocytosis that became phylogenetically fixed in early neuranimate life forms (Shmukler & Buznikov, 1998; Csaba, 1981, 1984).

Thus the very ubiquity and conservation of these substances and processes across the biosphere indicate that chemical communication began first within, then across and among unicellular organisms prior to the emergence of Metazoan (multicellular) life forms at least some billion years ago. Therefore, both uni- and multicellulates share many mechanisms of biosynthetic and functional chemical communication. As mentioned earlier, efficacy of neurotransmission depends on and is controlled by the type and activity of receptors evolved to modulate their activity.

For example, the hormone progesterone so essential for mammalian parturition, and even to modulate cyclicity and mood in women, is evident in even in such primitive fungi as *Trichophyton mentagrophytes* (Schar, et al. 1986). Likewise, the extracellular concentration of neurotransmitters is regulated by high-affinity transporters that arose in primordial evolution.

This serial elaboration of neural evolution has been characterized by major gene duplication events facilitating rapid advances, starting with the primitive receptor system and subsequent elaborations for specialized handling of bioregulators. Chief among these ancient bioregulators are the amine neurotransmitters dopamine (DA) and the closely related norepinephrine (NOR), serotonin (5-HT), acetylcholine (ACH), as well as gamma aminobutyric acid (GABA). In advanced animals, these molecules yet modulate most bioactive psychotropics such as cocaine, amphetamines, and therapeutic antidepressants (Porzgen, et al. 2001).

Dopamine and Norepinephrine

From detailed phylogenetic analyses of amino acid homology of G protein-coupled receptors carried out on species ranging from yeast to humans, it appears that primordial receptors must have evolved more than 800 million years ago. It also appears that local duplications first gave rise to dopamine receptors of the *ADRB-D1/D5-ADRA1-ADRA2* types and that norepinephrine and serotonin receptors arose somewhat later via subsequent duplication and modification of template genes (Pebusque, et al. 1998).

For example, in the cephalochordate-vertebrate transition, the sympathetic nervous system seems to have been quickly derived from the earlier parasympathetic template (Vincent, et al. 1998). Amphioxus, for instance, is a cephalo-chordate approaching the vertebrates. It has a single catecholamine receptor gene that functional and molecular-genetic analyses identify as a monophyletic group subsuming both vertebrate dopamine D1 and beta adrenergic receptor classes. In addition, Amphioxus uses dopamine but not norepinephrine.

By contrast, dopamine D1 and beta adrenergic receptor genes of jawed vertebrates have two distinct but homologous genes as found in the earliest craniates such as lamprey and hagfish (Vincent, et al. 1998). Thus, evolution of greater catecholamine activity throughout the brain occurred as manifold subtypes of G-protein linked receptors emerged as the primordial D1-beta receptor gene was recurrently duplicated and modified to create receptors with unique distributions and/or for entirely new molecules.

Thus, dopamine phylogeny constitutes an ancient neuromodulatory system that has been constantly evolving higher processes including important global state mechanisms for novelty-seeking and reward mechanisms underlying addictive behaviors.

Serotonin

Serotonin is another bioregulator system anciently evolved. It arose in unicellulates and plants well before the appearance even of the simplest prechordate animals at least some 700 million years ago (Azmitia, 2001). Indeed, via its role in the metabolism of its key precursor molecule, tryptophan, serotonin is an essential component in the evolution of life itself.

Tryptophan, an essential indole amino acid, has unique light-capturing properties. Botanic tryptophan compounds absorb light energy for use in burning glucose with oxygen and for the bioconstruction of reducing cofactors. Interactions of tryptophan, oxygen, and reducing cofactors form morphogenic serotonin-like molecules that orient the growth of light-absorbing structures to light sources. A similar serotonergic morphogenesis occurs in animal cells via direct influence on intracellular cytoskeletal development and regulation of proliferation, migration and maturation in a variety of cell types, including endothelial, mast, neuronal, astrocytic, and solid tissue cells.

This differentation is evident in that the three major classes of G protein-coupled serotonin receptors are less than 25% homologous (i.e., 5-HT1, 5-HT2 and 5-HT6). It is further evident in that serotonin 5-HT5, 5-HT7, 5-HTsnail, 5-

HTdro and 5-HT1A receptors diverged as vertebrates differentiated from invertebrates some 600–700 million years ago. Newer receptor subtypes likewise differentiated in the mammalian, primate and human line over the past 120 million years, 50 million years and 6 million years respectively. Moreover, homology analysis also indicates that many G protein-coupled 5-HT receptor subtypes remain to be identified from inverebrates onward, perhaps including humans (Peroutka & Howell, 1994).

Thereby, serotoninergic neurons have come to project all over the brain through at least seven receptor families comprised of at least 14 specific receptors. These all greatly influence most neuromental processes, including intracellular cytoskeletal development up to integrated global state mechanisms of dominance hierarchy in vervet monkeys (high 5-HT) and the risk of violent death by accidents or suicide in humans (low 5-HT).

Of the various serotoninergic receptor systems, of particular interest in the overall discussion here are 5-HT1A and 5-HT2A, as they have oppositive-reciprocal functions in a variety of cellular and behavioral processes. Most pertinently, these include the higher global state mechanisms of reciprocal homeostasis in the CSN Complex. 5-HT1A receptor develops early in the CNS and enhances intracellular stability even as it also directs cell proliferation and differentiation. Clinically, molecules active at the 5-HT1A receptor reduce neuromental activity and serve as anxiolytics.

5-HT2A receptor ontogenesis occurs more slowly and stimulates glycogenolysis (in astrocytes) and increased calcium activity in neurons. These actions can destabilize the cytoskeleton and induce cell proliferation, synaptogenesis, as well as apoptosis (cell death). Clinically in humans, molecules active at the 5-HT2A receptor cause arousal and even hallucinations. These dynamic interactions between the 5-HT1A and 5-HT2A receptors are directly relevant to the CSN Complex and offer important insights as to the etiogenesis of neuromental disease as well as novel strategies for their treatment.

Gamma amino butyric Acid (GABA)

GABA, a neurotransmitter found in widely divergent species from insects to humans, is synthesized from the equally widely distributed excitatory neurotransmitter glutamate. Cumulative evidence supports GABA's role as a major inhibitory neurotransmitter in the vertebrate central nervous system as well as in the central and peripheral nervous systems of invertebrates (Roberts, 1987). Like dopamine and serotonin, GABA's apparent chemical simplicity is made complex by a large number of receptors and post-receptor cellular events (Kleppner & Tobin, 2002; Herbert & Schulkin, 2002). GABA has become a more recent focus of interest as it is found at high concentration in neurons of the striatum, basal ganglia, and also in the nearby hypothalamus. Indeed, GABA is among the most abundant brain neurotransmitters with particular importance in synaptic plasticity (Nakanishi, et al. 1992; Luddens, et al. 1995; McAllister, 1999; Staley, et al. 1995) and neuroendocrine function (Brann & Mahesh, 1994; Trudeau, 1997). In general, GABA functions interactively with glutamate in various structures of the brain.

This quite abbreviated sketch of a few of the principal bioregulators and their basic evolution omits many aspects of a much, much more complex discussion. But from any perspective, it appears that catecholamines, related transmitters, and their various modulators emerged in the course of later vertebro-reptilian evolution as these neurmentalities became highly canallated.

This all eventually led to the emergence of cognitive-emotive capacities such as selective attention, predictive reward and emotional drive, particularly as vertebrates circuits that rely on dopamine and norepinephrine were subsequently recruited to new ends. For example, increased information led to more finely mediated psychomotor functions in bilaterians and later advanced creatures.

Thus, neuroreceptor evolution and arborization is actually more significant to higher neuromentation than are neurotransmitters *per se*. However again, it is all too little appreciated that the main neurotransmitters arose in phylogeny well before neural cells (much less neural networks or systems), or that many major neuromental assemblages were sculpted in deep evolutionary times past.

Ultimately, the CSN Complex template arose as more refined psychomotor adaptive responses were laid down via ever more novel evolutionary elaborations in reptiles, mammals, primates and humans—often with species-specific if phenotypically rather plastic variations (Wilson, 1992). Indeed, alterations of amino acid neurotransmission are etiogenic in several major areas of disease, including neurodegeneration, epilepsy and almost all types of psychopathology (Nakanishi, et al. 1992; Edwards, 1993; During, et al. 1994; Luddens, et al. 1995; During, et al. 2002). Not only is all this of direct relevance for discussions of clinical psychotherapeutics and psychopharmacology taken up in Chapter 13, but it sets the stage for examination of the higher domains in neural hierarchy.

Neurons to Brains

Evolution of the basic neurotransmitters and neuroreceptors in early phylogeny laid the foundation for the emergence and development of neurons, neural networks, and complex brains.

Neurons are essentially like all other cells differing only in that they are specialized for behavioral communication. Specialization occurs in phylogeny as cells become dedicated to perform and amplify distinct cellular functions. Specialized cells group or work together to become the tissues, organs and all higher anatomical structures of the body. Neurons are cells specialized for long distance communication that enables sensory and motoric integration for adaptive responses. The internal biochemical circuits interconnect neural cells to create neural networks that communicate the requirements and conditions of the somatic matrix. Brains—networks of phylo-ontogenic neural subsystems—can be said to have begun once such complexes arose in evolution. Brains process and servo-control the somatic matrices and behavioral repertoires of sophisticated bodies.

THE BASIC VERTEBRATE COMPLEX

Chapter 2 described the basic vertebrate neural complex as a set of circuits dedicated to behavior for self-preservation. Premammalian vertebrates in the human ancestral line displayed limited social life and little if any parental care toward offspring. Fitness was achieved by numbers more than quality; that is, eggs were produced in large quantities sufficient to assure that some survived to reproductive adulthood.

Nature of the circuitry and structures

The basic premammalian vertebrate brain as organized in ancestral amphibians and reptiles consists of the hindbrain, midbrain, cerebellar structures, and a group of structures at the base of the emerging forebrain called the basal ganglia and the striatum. MacLean designated these combined structures as the protoreptilian brain or R-complex and delineated them as the behavioral circuitry for self-preservational functionals, as discussed in the previous chapter. These were basic functions and behaviors that ancestral amphibians and reptiles also had to perform to survive. The brains and behavioral functions of ancestral amphibians and reptiles were limited to these essential elements, albeit with considerable variation reflecting species specialization.

The mid- to forebrain bridging structures of the basal ganglia and striatum are of special interest because they are the more recently developed components of the ancestral reptilian brain. As such they are still largely retained as foundational structures over which the mammalian brain evolved. Although they have undergone elaboration and modification in the mammalian line, the essential preexisting cytoarchitecture, neurotransmitter, and endocrine characteristics of ancestral reptiles, amphibians, and other early vertebrates is preserved.

The basal ganglia were long thought to be devoted almost exclusively to motor control. This early conclusion was based on clinical findings that motor disorders involved dysfunction of these structures. More recent research, however, has revealed involvement of the basal ganglia in a widely distributed system which integrates cognition, motivation, and affect as a basis for appropriate response. Non-motor functions include participation in building up of sets and sequences of behavior for meaningful, goal-directed patterns and repertoires. Thus, the basal ganglia have elements of both mid-and forebrain function.

Findings from reptile experiments indicate the basal ganglia control specific assemblies of innate units of behavior modulated by autonomic neural structures. Such behavioral units are called "species-typical behaviors" (Lorenz, 1950; Tinbergen, 1951; Beer, 1973). The basal ganglia are involved in the expression of such species-typical action patterns in reptiles and primates. Such species typical patterns are focused primarily upon self-preservational behavioral routines and displays associated with social competition.

The extensively interconnected internal circuitry of the basal ganglia is gradually being specified by diverse neuroscientific investigations. It is clear from such dedicated research efforts that multiple reciprocal connections link to more recent mammalian circuits—limbic and neocortical systems in particular. The core and associated structures, are now often regarded by some researchers as a part of a more inclusive basal ganglia system (see Greenberg, 2002 for a comprehensive review).

Architecture, Neurochemistry, and Neuroendocrinology

From the evolutionary perspective as reviewed above, biomolecules and mechanisms are by no means restricted to advanced brain structures but pervade all life. Indeed, as we have seen, phylogenetically, the substances now known principally as neurotransmitters appeared long before the development of the most primitive nervous systems and even before the rise of multi-cellular organisms (Stoka, 1999).

Moreover, neurotransmitter function as such diversified first to form definitive synaptic structural-functional units and then to greatly arborize such synaptic mechanisms in subsequent evolution. Phylogenic elaboration of neurotransmitter functional distributions is a hallmark of higher neuromental evolutionary development. These changes are largely distinct from other quite important evolved functions in non-neural tissues; i.e., alimentation and cardio-respiratory systems (although sophisticated scientific clarification of the interfaces of mind-body are only recently accumulating).

Further evolution of early multi-cellulate life forms to the chordates, though interesting, is not a subject that repays much attention in the present discussion. However, mention must be made of the remarkable explosion of animal forms during the Cambrian period (543–505 mya) when all advanced protostome and deuterostome phyla originated. This suggests rapid evolution of both new general body plans as well as more complex nervous systems (Gabor Miklos, et. al 1994; Prescott & Ibbotson, 1997).

Otherwise, the general chordocephalate structural plan of neuromentation is conserved from chordates on to amphibians, reptiles and mammals. However, with the reptilian innovations and further refinements of the basal ganglia and mid-brain, the foundations for social behavior were emplaced atop the nonsocial central nervous system of chordo-vertebrates (Greenberg, 2002; Marin, et al. 1998).

In the basal ganglia of mammals, the neurotransmitter DA binds to two known classes of dopamine receptors, called D1 and D2. These two receptors are associated with direct and indirect subsystems of the basal ganglia. These subsystems are seen as direct and indirect pathways that compete with each other to cause excitation or net inhibition of activity, respectively (cf. Mink's focused selection and inhibition hypothesis, 1996). That is, a specific motor program is activated while competing programs are broadly inhibited. Motor activity is viewed as the outcome of a balance in activity of these pathways maintained in part by activation of D1 and D2 dopamine receptors. If, for instance, the direct pathway dominated, motor activity might be excessively elevated. On the other hand, if the indirect pathway was relatively more active, the motor activity may be reduced or depressed (Graybiel, 2000).

Dopamine has been thought to have primary reward functions. Such functions have been associated especially with a part of the basal ganglia called the nucleus accumbens (e.g, see Paradiso, et al. 1999; Wise & Bozarth, 1984). Other researchers, however, report that various *stressful* situations evoke dopamine release in the accumbens (Stanwood & Zigmond, 2000; Salamone, 1994; Herve, et al. 1982; Thierry, et al. 1976; Bowers, et al. 1987). The accumbens is also prominently associated with negative emotion in adult humans. Positron emission tomography (PET) of subjects viewing unpleasant pictures revealed increased blood flow primarily in nucleus accumbens. Viewing pleasant pictures, on the other hand, evoked augmented activity in phylogenetically newer cortical limbic areas including prefrontal cortex (Paradiso, et al. 1999; see also Dupue, et al. 2002). The research further suggests that detection and rapid stereotyped response to unpleasant avoidance situations is reasonably coordinated with such phylogenetically older, conserved mechanisms of the basal ganglia as such responses were also necessary to the survival of antecedent vertebrates.

In the light of such research, there is a growing sense that reward functions of DA are not as modal as once thought. In fact the previously interpreted reward function may even be explained by the role of dopamine in synchronizing and amplifying attention (Deth, 2003) and underscoring the significance of stimuli that predict reward. This may, at least in part, explain why non-rewarding or negative behavior can be expressed if it is associated with a dopamine surge (e.g., see Wickelgren, 1998).

THE BASIC MAMMALIAN COMPLEX AND THE NEOCORTEX

Mammals have behaviors that are different from reptiles and other antecedent vertebrates. These behaviors are essentially social. In the explication of the behavioral spectrum in Chapter 2 they were called affectional. They include endothermy or warmbloodedness, nursing, parental-infant care and attachment, and extended social affiliation and bonding as well as various vocal and visual cues that facilitate such social behaviors. Since these characteristic affectional mammalian behaviors were nonexistent or only rudimentary in reptiles and antecedent vertebrates, the substrate for these differences must be sought in the contrasts between mammalian and pre-mammalian vertebrate brain structure or neural architecture (cf. Henry & Wang, 1998).

Thus the central question becomes: How are mammalian brains truly different from the brains of pre-mammalian vertebrates?

An initial difference is that the early mammalian brain showed a great ballooning of neural structures over the prior vertebrate forebrain. These structures were first identified by Paul Broca in 1878, further defined by James Papez (1937) and Paul Yakovlev (1948), and extended even further and more inclusively by MacLean (1952). What Broca called the *grand lobe limbique*, MacLean redesignated the limbic system to acknowledge its extended systemic qualities. Although the parts of the limbic system have been variously defined over the years, currently most scholars agree the system is made of the hypothalamus, amygdala, septum, hippocampus, bed nucleus of stria terminalis, anterior thalamic nuclei, cingulate, and orbital-frontal cortex (see Kandel, et al. 2000: 986–89; Herbert & Schulkin, 2002: 667–69). Some include the nucleus accumbens and other parts of the basal ganglia (e.g., Price, 2002; Morgane, et al. 2005). Since the systems function in an integrated manner, no clear distinction exists between them. Some structures, such as the cingulate, however, seemingly have no representation in pre-mammalian vertebrates (MacLean, 1990: 247).

There are several important points to remember about the limbic system to reduce sources of prior confusion. Firstly, the limbic structures did not arise *de novo*-ly atop the pre-mammalian forebrain. In several cases the limbic structures arose from known pre-existing homologues or related structures in the vertebrate forebrain. It is, thus, not the presence or absence of specific limbic components in the mammalian brain that is notable but, rather, it is their significant elaboration that is of interest. Secondly, the newly elaborated limbic structures are not detached or separate from earlier brain structures. Both are extensively interconnected. Thirdly, the functions of some basic structures have become additionally represented with varying degrees of control shifted upward in the expanded neural hierarchy. Representations move upward but do not replace original circuitry. Fourthly, although there are modularized functions among these neural structures, they are interconnected as a larger system and boundaries among them are not distinct. Thus, the distinction of mid- and forebrain is not definitive.

Mammals again underwent conspicuous evolution of the neocortex or isocortex which was at best very rudimentary in reptiles (e.g., see Godwin &

Crews, 2002: 546; Aboitiz, et al. 2002). The human ancestral line saw the neocortex further evolve to more fully encase the early forebrain and limbic structures. Again the same principles of elaboration and modification of extant structures applied. The newer layered circuits of the neocortex were extensively interconnected among themselves and with the earlier subcortical circuits of the limbic system and the basal forebrain. Some subcortical functions achieved additional representation in the neocortex. Again, representations and control moved at least partially upward in the neural hierarchy but did not replace original circuitry.

The highly developed human cortex is most noteworthy for its diverse and highly evolved capabilities such as abstract thought and concept association as well as domain-specific talents such as language, visuo-spatial interpretation, mathematics, music and so on. Subcortical circuits of the self-preserving protoreptilian complex projected up to become represented as urges and impulses of Egoistic, self-interested preservation. Subcortical circuits of the limbic affectional system became represented as urges and impulses of Empathy, caring for others, or other-interested kin-species preservation. Essentially, a neural architecture emerged that was dedicated to both preservation of self and kin-species—as MacLean so prominently emphasized—through social interaction. The human brain became truly a social brain, and a highly talented one at that.

The Social Brain and Theory of Mind

It is worthy of note that the term *theory of mind,* introduced to ethology by primatologist David Premack (Premack & Woodruff, 1978), is essentially a social brain concept. Premack and associates employed the term while researching whether chimpanzees conceptualized other minds in their fellow primates. It has since been used to account for the deficit of relatedness to others presumed to be central to the neuropsychiatric condition of autism (Frith, 1997, 1989; Baron-

Cohen, 1995; Brothers, 1995; cf. Harris, 2002). It has also been applied to child development, where the standard account is concerned with the child's grasp of the attention, beliefs, and false beliefs of others (Astington, et al. 1988) and to persons with frontal lobe damage such as the much discussed case of Phineas Gage (see Damasio, 1994; for a critique of some current issues, see Grossman, et al. 1997). Any adequate theory of mind must allow, explicitly or implicitly, for a generalized concept of self or Ego as well as a similarly generalized concept of Empathy or other interest. That is, to have a theory of mind of others, there must first be an idea of a mind of self. Such is the essence of a social brain.

Nature of the Circuitry

Underpinning the CSN Complex, neural substrate mechanisms for self-survival (generalized as Ego) project from circuits in the basal ganglia and brain stem (protoreptilian complex) through connections with the amygdala, other limbic structures (early mammalian complex), through to probable cortical representations that add valences of emotion or passion (see Kandel, et al. 1995: 595–612), ultimately to be gated into the frontal cortex by thalamocortical circuitry (e.g., see Sherman & Guillery, 2001; Baars, 1997, 1988; LaBerge, 1995; Crick, 1994).

On the other hand, the mammalian nurturing (affectional) substrate and its associated motivation, may originate in the septal and medial preoptic (MPOA) limbic areas (see Numan & Insel, 2003; Carter & Keverne, 2002; Numan & Sheehan, 1997; Fleming, et al., 1996; Numan, 1994), proceed through hippocampal and amygdaloid circuitry (Bachevalier, 2000; Brothers, 1989) as well as other limbic structures. They are, in turn, gated into the orbital and frontal cortex by neuromodulating thalamocortical circuits (to include the early thalamocingulate cortex). Here conflict with Egoistic inputs is resolved in the pre-frontal and frontal executive of conscious self-awareness and increasingly rational thought. The neuromodulating and gating of *affect* as well as cognition by the thalamocortical circuitry have been greatly under-appreciated (even denied) by many cognitive-behaviorists but are supported by substantial neurological research. Devinsky and Luciano report that the limbic cingulate cortex, a cortical structure closely associated with the limbic thalamus, can be seen as both an *amplifier* and a *filter*, which joins *affect* and *intellect* interconnecting the *emotional and cognitive components of the mind* (1993: 549; cf. Sherman & Guillery, 2001).

Tucker, Luu, and Pribram (1995) speculate that the network architecture of the frontal lobes reflects dual limbic origins of executive function. Specifically, the authors propose that two limbic-cortical pathways apply different motivational biases to direct frontal lobe representation of working memory. They suggest dorsal limbic mechanisms projecting through the cingulate gyrus may be influenced by hedonic evaluations, that is *social attachments*, initiating a mode of holistic and impulsive motor control. On the other hand, they suggest the ventral limbic pathway from amygdala to the orbital frontal cortex may implement a more restricted mode of motor control reflecting adaptive constraints of *self-preservation* (1995: 233–34; see, also Davidson & Irwin, 2002).

The orbital frontal cortex via its connections to the anterior cingulate gyrus, amygdala and other limbic structures seems especially important to the interaction of Ego and Empathy (Eisler & Levine, 2002; Weisfeld, 2002; Schnider & Gutbrod, 1999; Fuster, 1999; Damasio, 1994; Schore, 1994). Such findings are consistent with the CSN Complex in which Ego and Empathy represent conflicting subcortical inputs into the cortical executive apparatus. As noted, several researchers posit a dynamic of conflicting modules, vying for ascendancy in behavior and consciousness (e.g., Edelman & Tononi, 2000; Dennett, 1998; Pinker, 1997; Edelman, 1992; Chance, 1988; Wilson, 1992, 1994, 1998; Gardner & Wilson, 2003).

Architecture, Neurochemistry, and Neuroendocrinology

The affectional circuitry, hosted in great part by the limbic system, or what MacLean also called the paleo(old)-mammalian brain, is distinguished by the range, diversity, and concentration of its chemical transmitters. These chemical agents include neurotransmitters, hormones, and neural modulators of considerable variety. Although many of these transmitters may well remain to be discovered, a good approximation of their circuits and functioning is now possible.

Dopaminergic and serotonergic circuits based in the brain stem and basal ganglia extend through the limbic structures, and into the neocortex. As noted in the discussion of the protoreptilian circuits of basic vertebrate forebrain, dopamine and serotonin likely do not determine the *nature* of behavior but rather are instrumental in the strength, sensitivity, and likelihood of adaptive responses (e.g., see Deth, 2003: 37). The *nature* of the behavior—socially bonding or Egoistic aggressive—is then seemingly governed more by the interplay of other chemical agents. Various neuropeptides, and to an extent, steroids, are important in determining the specific nature of the responses (Herbert & Schulkin, 2002; Carter & Keverne, 2002).

Through connections with the limbic system intensity and emotionality are added to the primarily self-preservational behaviors of the basic premammalian vertebrate brain circuitry. These behaviors, although retaining their original basic function in the earlier circuitry, then, achieve representation up the hierarchy in the more recent limbic circuits.

Importantly, the limbic circuits, and their accompanying modifications of the basal circuitry (the striatum and even the brain stem), introduce the affectional or social bonding characteristics of mammals—characteristics that were absent or only rudimentary in ancestral vertebrates. Research on the nature of these crucially important positive social behaviors, which has been considerably delayed by the unwarranted focus on the more easily identified aggressive behaviors in the past half century, is now proceeding apace.

Generally, it is assumed that affectional social circuits evolved from reproductive circuitry (Crews, 1997; Cory, 1974). Indeed, research bears out that many steroids, peptides, and other transmitters associated with reproduction are also intimately involved in maternal-infant bonding and parental care, including paternal attachment. These basic affectional behaviors are the proximate mechanisms of extended social bonding in adulthood and both depend on the same neural circuits, neurochemicals, and hormonal (endocrine) systems (see Numan & Insel, 2003; Carter & Keverne, 2002).

Given their importance, concise discussion of the various hormones, monoamines, neuropeptides, and amino acids, that participate in this complex affectional architecture seems in order.

Hormones

From the evolutionary perspective, hormones appear early and have significant effects in sex and reproduction among all but the most simple forms of life. In mammals, despite clear cut differences among species, parental bonding behavior generally is not rigidly governed by the specific hormonal events accompanying pregnancy and lactation. Prior exposure to nurturing, alloparental behavior, sensitization processes, and paternal behavior—all widely displayed in mammalian species—testify to lack of rigid hormonal control.

Hormones, however, do have their effects. In nonprimate mammals, such as rats, the estradiol/progesterone ratio is important in promoting the onset and intensity of maternal behavior. In early pregnancy, plasma levels of the steroid hormone estradiol are low and rise through parturition. In contrast, levels of a second key steroid, progesterone, are high in early pregnancy and decline toward parturition. Therefore, around parturition there is a reversal of the estrogen/progesterone ratio leading to a dominance of estradiol. Also accompanying estradiol dominance is the declining level of the pituitary neuropeptide prolactin and increasing levels of placental lactogens. Experimental work, aimed at evaluating these effects, has shown clearly that exposing virgin female rats to high levels of estradiol and prolactin coupled with a withdrawal of progesterone can dramatically stimulate maternal behavior (see the review in Numan & Insel, 2003: 18–19).

Although such clear cut effects are not seen in humans, there is some evidence of a similar hormonal influence. In a study by Alison Fleming and associates (1997) women showing high mother-infant attachment had increasing estradiol/progesterone ratios over pregnancy, whereas mothers showing low infant attachment had decreasing ratios over the same period. Actual estradiol/progesterone ratios at birth, however, were not significant, perhaps, because in humans, despite decline, progesterone levels remain relatively high until after parturition.

The issue of paternal care is less studied and less clear than that of maternal care. Although both the steroid hormone testosterone and the neuropeptide prolactin seem involved in paternal behavior, the extent of contribution is unclear. There is no consistent evidence that high or low circulating testosterone is important to paternal behavior across species. Although some species (e.g., marmosets) show an increase in prolactin when housed with young, there is no evidence in other mammalian species that manipulating prolactin levels changes paternal care.

Concerning humans, a recent survey of hormonal changes in males during the course of their wives' pregnancies showed some decrease in testosterone in the postpartum period. In the early post birth period, increased response to infant cues correlated with low-circulating testosterone. Prolactin, although present, was much lower in the men than in their wives (Storey, et al. 2000). Lower testosterone and cortisol (a stress hormone) in expectant fathers compared to non fathers of similar age has also been recently reported (Wynne-Edwards, 2001). The fact remains that the endocrine correlates of male parenting have not yet been widely studied. Superficially, however, paternal care appears somewhat less dependent on steroid hormonal factors.

Concerning parental care, however, no mammalian species is independent of hormonal influence—humans included. What does seem to vary among species is the base level or degree of parental responsiveness which is displayed in the absence of exposure to the hormonal events of pregnancy, birth, and lactation. Human parental care, then, appears to be linked to such hormonal influences but also built to a considerable extent upon the highly conserved associated core neural architecture through the mediation of cognitive processes made possible by higher cortical development, such as experience and learning.

Monoamines

The chief monoamines implicated in maternal behavior are dopamine, norepinephrine, and serotonin, now highly modified by newer receptor circuitries, to facilitate higher global state mechanisms for more complex social behavior. There is evidence that dopamine is important for motivating or stimulating and amplifying certain maternal behaviors, such as pup retrieval, licking, and nest-building in rats (Stern & Taylor, 1991). Dopamine pathways, especially in the nucleus accumbens (a limbic-striatal structure) with its connections to the MPOA, seem critical to these active components of maternal care. Norepinephrine, on the other hand, has been associated with social cognition, particularly olfactory memory. Studies in a number of mammalian species (e.g., rats, mice, sheep) have shown that release of the monoamine norepinephrine in the olfactory bulb is necessary to learning about individual identity (Ferguson, et al. 2002). The effects of serotonin in the maternal brain seem less pronounced than dopamine or norepinephrine, but seem related to maternal aggression. Overall the multiple interactions of these neurotransmitters with other neuroendocrine functions have yet to be thoroughly investigated and understood.

Amino Acids

As reviewed earlier in this chapter, the amino acid neurotransmitters are the most plentiful and widely distributed in the neural architecture. The two most abundant are Gamma amino-butyric acid (GABA) and glutamate. They are, moreover, reciprocally interactive as GABA is the major inhibitory neurotransmitter in roughly 30 percent of neurons widely dispersed through and across the brain. Glutamate is the major excitatory transmitter in an even higher percentage and with a similar breadth of projection. These neurotransmitters are so widespread that specific functions in maternal or paternal behavior are difficult to identify. It may be assumed that their role is equally widespread and relates to their integral role in neural plasticity which is essential to the long-term changes necessitated by and associated with maternal/paternal care and long term affectional bonding.

Neuropeptides

Neuropeptides are made up of polymerized chains of amino acids. Although built of the same elements as proteins, they lack the length and the distinctively folded three dimensional character associated with full-blown proteins. One neuropeptide important to paternal care, prolactin, has been previously discussed in the section under hormones. Two neuropeptides which are especially important to the affectional system of mammals are oxytocin and vasopressin. These neurohypophyseal peptides are produced in magnocellular neurons of the supraoptic and paraventricular nuclei of the hypothalamus but are stored in and released from the posterior pituitary gland.

Oxytocin is found only in mammals. The phylogenetic origins and development of the family of neuropeptides from which it derives, however, can be traced back to mollusks and insects (see Gainer & Wray, 1992). In lower vertebrates the ancestral peptides (e.g., vasotocin and mesotocin) have long played a key role in fluid homeostasis. Oxytocin, with its emergence in mammals, has however taken on new functions in reproduction as well as in the maintenance of maternal care and pair bonding.

Oxytocin is conspicuously necessary to milk ejection and uterine contraction, but its receptors and central effects are widely distributed in the central nervous system—from the spinal cord, through the brain stem, basal ganglia, limbic system, and frontal cortex. It affects both males and females. In the ventromedial hypothalamus, it is associated with sexual receptivity and the initiation of lordosis behavior; in the paraventricular nucleus, with penile erection; in the ventral tegmental area and medial preoptic area, with the initiation of maternal behaviors; in the midline prefrontal cortex, nucleus accumbens, and connected areas, with pair bonding, social and affiliative behaviors (for a summary, see Zingg, 2002, Table 2).

Vasopressin, the related neuropeptide, emerging from the same ancestral molecules, shares the same family of receptors with oxytocin. There is evidence that in some cases the two neuropeptides can substitute for each other producing similar effects in certain behaviors. Following from the evidence of oxytocin's role in maternal care, the related peptide vasopressin showed similar promise as a mediator of paternal behavior. Although vasopressin does facilitate parental care in some mammalian species, a full understanding of its role and significance awaits further study.

In addition to oxytocin and vasopressin, opioids, corticotropin-releasing hormone (CRH), and other related hormones are known to participate in the function of social bonding (Carter & Keverne, 2002: 300). Altogether these neuropeptides—acting within neural structures with unique cytoarchitecture and pathways and modulated by amines, monoamines, and other hormones—coordinate and control the physiological and behavioral responses of affectionally positive social interaction. Also, there appears to be a degree of built-in redundancy—where some contributing factors fail, others may compensate (e.g., vasopressin for oxytocin). Steroid hormones may not be essential to affectional pair bonding, but may serve to modulate such behaviors (Carter & Keverne, 2002: 331; Herbert & Schulkin, 2002: 679; Simon, 2002: 341).

PARENTAL ATTACHMENT AND THE MAMMALIAN AFFECTIONAL SYSTEM

That the human brain can create internalized models of *self* and *others*, or *Ego* and *Empathy* such as in the CSN Complex—is dependent upon the early attachment to maternal, paternal, and other figures made possible by the evolutionarily highly conserved neural core of mother-infant attachment mechanisms.

Intense interest in the affectional circuitry is not only a strong trend in current neuroscience, it has long been expressed in the extensive literature on the concepts of social bonding and attachment—especially the work done on higher primates and man. Particularly relevant here is the work of psychologist, Harry Harlow on the nature of love and attachment in rhesus/macaque monkeys (1965, 1986). Without relating them to neural substrates, Harlow described five affectional systems in monkeys—maternal, mother-infant, age-mate, heterosexual, and paternal (1986). The CSN Complex, on the other hand, proposes one all-inclusive affectional program capable of functioning as a global state variable.

In the case of humans, René Spitz (1965) and John Bowlby (1969, 1988), who did foundational work in mother-infant attachment behavior are of special interest. All the foregoing, however, reflected field observations, experimental behavioral observations and clinical work. Current affective neuroscience draws upon, clarifies, and elaborates these behavioral concepts by working to ground them in appropriate neural circuitry.

Researchers, Carter and Keverne see the dynamics of reciprocity between the infant and early attachment figures providing the foundation for future social reciprocal interaction of many varieties (2002: 330). Such early positive experience with attachment figures is likely necessary to the full emergence and representation of Empathy in the frontal cortex and to future reciprocal dynamic of Ego and Empathy characteristic of the social brain.

An increasingly emerging scenario sees the nature of response—the mix of affection and self-preservation, Ego and Empathy, in social behavior—resting on a relative mix of numerous, sometimes opposing, sometimes redundant and facilitative neurochemicals and structures. This interactive mixing, mediated by the executive circuitry of the prefrontal cortex, would allow for the spectrum of behavior—the mixes of Ego and Empathy—represented in the CSN Complex as outlined in Chapter 2.

The range of Ego-Empathy response, as appropriate to circumstance, may be considered to set the normal limits of social behavior from the perspective of evolutionary epidemiology, particularly as it optimizes social behavior that fosters proximally adaptive adjustments and, consequently, ultimately enhanced reproductive success. This is a major contribution toward an operational definition of normality that has long eluded social science, medicine and psychiatry.

4

NORMALITY, PSYCHOPATHOLOGY AND BEYOND

An evolutionary epidemiological perspective allows a framework for a definition of normality. In studies of both psychopathology generally and psychiatry in particular, the concept of mental health is defined chiefly by its absence—there is no accepted operational definition of normality. By first defining a concept of normality, psychiatry and related disciplines can better delineate the parameters of pathology and move to an even more constructive model of optimality.

INCLUSIVE FITNESS AND EVOLUTIONARY EPIDEMIOLOGY

Inclusive fitness is a key concept in evolutionary epidemiology. It is, in fact, the key to normality. At minimum it sets the range of survivable adaptive behavior. At maximum it points in the direction of defining optimal behavior or adaptiveness. Some background is necessary to understanding the development of this significant Darwinian concept.

Beginning in the second half of the twentieth century the study of animal behavior took root. The discipline of ethology—the study of the biological basis of animal behavior—made important contributions to the understanding of inclusive fitness. Ethology, had been progressing largely in continental Europe. European ethologists Konrad Lorenz, Niko Tinbergen, and Karl von Frisch won Nobel prizes in 1973 for their pioneering work in the new field. Like Eibl-Eibesfeldt who followed in their tradition (1972, 1989), these ethologists did across-species comparisons and proceeded from an evolutionary perspective. They observed and described behavior. They did not, however, look into the black box of the brain and attempt to speculate on neural mechanisms and neither did they offer specific genetic explanations.

EVOLUTIONARY BIOLOGY TO EVOLUTIONARY PSYCHOLOGY

At the same time evolutionary biology powered by the engine of Darwin's insights (and as strengthened by Mendelian principles), proceeded on a track largely isolated from the mid-century behavioristic mainstream psychology. Despite a few heroic efforts at integration of the biological with the psychological and social sciences, however, the isolation remained essentially intact—often in the face of unfortunate social constructivist ideologies that blocked the path of inquiry.

In the last two decades of the past century, the evolutionary perspective and psychology finally, if hesitatingly, allied in what became known as evolutionary psychology. This advance was propelled by Hamilton's profoundly important solution to the question of genetic altruism that had long nagged Darwinism (Hamilton, 1963, 1964). But like the behavioristic and cognitive psychology before it, much of mainstream evolutionary psychology continues to view the human brain from the outside and as a "black box" (Panksepp & Panksepp, 2000). It then reformulated a good deal of descriptive behavioral-cognitive research for a somewhat more empirical—but almost exclusively proximal—evolutionary psychology. Only recently has interest (if only speculative) moved to consider what could be within—that is, what could be the neuromental apparatus that was passed on in the evolutionary process through our genes.

Still evolutionary psychology continues to pay remarkably little attention to actual brain structure much less to the actual genetics underpinning neuromentation. Initially, elaborate speculations about presumed isolated and dedicated processing modules were offered which dealt with specific environmental challenges that had arisen. Such modules were likened to a Swiss

Army knife, each blade put neatly in place with imaginary precision in the manner of an equally imaginary professional engineer represented by the process of natural selection (Cosmides & Tooby, 1992).

Such improbable speculation proceeded largely unchecked. It was, at times, reminiscent of the pseudoscientific phrenology of the previous century. Phrenology had notoriously sought to identify human characteristics through external examination of lumps on the cranium. The unwritten command driving evolutionary psychology seemed to be "find a behavior, posit a dedicated mental module." The approach was considerably naïve and open to endless proliferation of imaginary neural components with almost no attention paid to epigenetic analyses (e.g., see Wilson 1992, 1994, 1998; Panksepp & Panksepp, 2000; de Waal, 2002; Friston, 2002).

EVOLUTIONARY NEUROETHOLOGY AND EVOLUTIONARY EPIDEMIOLOGY

In the broad panorama of science many tracks proceeded independently at varying levels of reduction and integration—and in isolation often because of our fractionated system of university departments. One of these tracks that moved outside the mainstream of emerging research on neuroscience was the evolutionary neuroethology of MacLean discussed in the opening chapters (see also, Cory & Gardner, 2002).

In a laboratory of the National Institutes of Health, MacLean, nevertheless, in 1952 began and doggedly sustained a research program aimed at understanding evolved behavior through a study of the substrate neural circuitry at various stages of evolution. His emphasis was on certain lizards and lower mammals. MacLean chose lizards because they resembled most closely the ancestral bridge between ancient vertebrates and mammals. His research program focused upon the neural circuitry underlying behavior with emphasis on emotion—especially those nurturing emotions and motivations that lay at the foundation of mammalian family life and that were apparently lacking in ancestral vertebrates.

MacLean's emphasis on the evolution of emotion was far ahead of its time and out of step with the mainstream research in the emergent area of behavioral neuroscience. In keeping with the superficial behavioristic tradition, mainstream neuroscience lacked an evolutionary perspective and focused only on the cognitive mechanics of information-processing. It was *tabula rasa*, redux. It called itself cognitive neuroscience, acknowledging, in its name, the exclusion of emotion. When emotion was occasionally studied it was under the limited, coldly formal information-processing aspect of "learning". When actual emotion was admitted into the chill halls of such science, it was typically a caricature of fear, aggression, and human nastiness in general. At the close of the twentieth century, the study of emotion, based in part on a revival of the long-neglected pioneering work of MacLean, combined with the dramatic efforts of a few dispersed, but bold researchers, seemed at last to gain attention.

The big challenge now seemed to be how the externalized, essentially cold-blooded, gene-centered perspective of evolutionary biology and evolutionary psychology could be reconciled with evolutionary neuroscience and the study of the warm-blooded, nurturing side of humanity. The conceptual bridge came principally from the work of British scholar W. D. Hamilton (to which Triver's reciprocity theory is an important adjunct). In his kin selection model, Hamilton (1964) formalized the issue of how genes for cooperation and altruism could evolve and be etched into the human ancestral genome.

The core of this inclusive fitness model is that it weighed effect of genes not only on the individual who carried them but also on kin individuals who shared the same genes. According to what has come to be called *Hamilton's Rule*, genes coding for cooperation, reciprocity or altruism could evolve if the costs of cooperative behavior to the individual were outweighed by the benefits to related individuals carrying the same genes. Hamilton's rule is expressed varyingly as:

$$rb - c > 0 \quad or \quad c < rb$$

In this formula ***r*** is the coefficient or degree of relatedness or kin, ***b*** is the benefit of the cooperative or altruistic behavior to the kin, and ***c*** is the cost of the behavior to the individual. The equation predicts that as ***r*** increases, benefits will exceed costs and cooperation, reciprocity, or altruism would be favored by natural selection. Hamilton's Rule gives the basic formula for inclusive fitness from the external gene's eye-view. It shows mathematically how it is possible for such traits as altruism and reciprocity to accumulate in the genome.

Neither the concept of inclusive fitness nor the formula, however, can predict the mechanism created by genes to sustain altruism and reciprocity. Accordingly, Hamilton, a biomathematician, did not speculate on the actual nature of such neural mechanisms and structures. He did, however, opine that an adaptive evolutionary epidemiology of mood and its disorders was "entirely sensible" (personal communication to DRW, 1988).

However, the formula of our social neural architecture derived from the CSN Complex, expresses this motivating mechanism and establishes linkage with Hamilton's formula. In other words, the equation expresses the function of the neural architecture that the genes actually *did* produce under the constraints of Hamilton's Rule.

The social brain formula of the CSN Complex can, in fact, be expressed in the same cost/benefit terms that Hamilton uses. Empathy can be considered the cost to the behaving individual, Ego can be benefit to the survival requirements of the recipient. Substituting cost and benefit into the formula, specifies the gene-based operational mechanism for Hamilton's rule (see Cory, 1999: 97–100; 2004). Again, the equation is expressed as approaching equilibrium or dynamic balance:

$$BT = \frac{Benefit}{Cost} \text{ or } \frac{Cost}{Benefit} = \pm 1$$

In keeping with the algorithms of our social neural architecture, *any* empathetic, cooperative, or altruistic behavior may serve to improve the odds of survival of self *plus* others *inclusively* over purely egoistic or self-interested behavior. This is compatible with Hamilton's Rule which says that so long as the benefit to related individuals exceeds the cost to the behaving individual, the genes supporting such behavior are favored by natural selection.

Of course, once the neural mechanism, the social architecture, is in place, empathetic or altruistic behavior may be extended to non-kin—even other life forms and, with emergent religious sentiments, the universe itself—by reciprocity, appropriate education and socialization through the media of language and culture that draw upon, modify, and elaborate the existing neural substrate (e.g., see Cory, 1999, 2004; Buss, 1999; Wilson, 1992, 1998). Not incidentally, this renders the sometimes heated argument between kin- and group-selectionists as spurious at least for species with long-established kinship group living patterns and elaborate reciprocity exchanges, such as humans.

EVOLUTIONARY EPIDEMIOLOGY
NEW APPROACHES, NEW INSIGHTS

A focus on biology in psychiatry and related sciences has a substantial tradition and has provided many useful insights and concepts. Indeed, it dates to Plato with his tripartite explanation of behavior spanning *epithimetikon* (~instinct), *thimoeides* (~emotions), and *logistikon* (~rationality). Most influential through the past century was the pleasure (hedonic) principle emphasized by Freud (1921) as further developed by many others. This has been a mainstay of psychoanalytical as well as a distinct influence on behavioral theory.

More recently, from the ethological perspective, Chance developed the hedonic/agonic conceptual approach reviewed in Chapter 2, which also led him to useful insights (1984, 1988) and, likewise, ideas further developed by others (Price, 2002; Price, et al. 1994; Gardner, 1988, 2002; Gilbert, 1989, 1992; Gilbert & Bailey, 2000; Wilson, 1998). An emphasis on the ubiquitous presence of social hierarchy or dominance/submissive relationships is a hallmark of the ethological approach in animal studies. MacLean pioneered the study of brain circuit substrates of such behaviors. The term neuroethology, which encompasses the study of both brain and behavior from an evolutionary perspective, captures this combined focus well (MacLean, 1990; Cory & Gardner, 2002).

Over the past decades there has been a growing theoretical and clinical literature that proceeds from this conjoined perspective. Neither Freud's approach nor Chance's concept, however, handle the issue of hierarchy adequately. Freud did not tie his thought definitively to evolved neural architecture and was only marginally concerned with hierarchy. Chance proceeded primarily from an externalized ethological perspective and did not make full use of the evidence from evolutionary neuroscience. The CSN Complex and the formula developed from the MacLeanian perspective provide a useful alternative organizing concept for current thinking on hierarchy or dominance/submissive relationships in psychotherapeutic theory. First, a little background.

Price (1967) initiated the present effort to incorporate the concept of hierarchy into human psychotherapy from the ethological perspective by introducing his theory of the dominance hierarchy and the evolution of mental illness. Gardner (1982) carried the effort forward in his ethological approach to the mechanism of manic-depressive disorder. Other scholars took up the call and contributed to more fully developing the social competition hypothesis of depression (see Price, et al. 1994).

An important insight emerged from these efforts. The earlier vertebrate circuitry common to fishes, amniotes, and reptiles (MacLean's protoreptilian brain or R-complex) which scripted self-preserving, essentially asocial behavior, coupled with sometimes life and death highly agonistic contests for territory and dominance, became integrated with the attachment or affectional circuitry of mammals (see also Wilson 1993, 1998, 2002). This combined circuitry permitted the emergence of group living and ongoing social competition—the tug and pull of Ego and Empathy that is modeled in the CSN Complex. Dominance modified by affection leading to cognitively emergent Empathy became possible among higher primates and humans. Dominance, without engagement of affectional mechanisms, could still be a life and death all or nothing contest. Hierarchy, in turn, could now be achieved through alliances that serve and preserve others by engagement of sociophysiologic affectional circuitry. Submission could be accepted and integrated into the social fabric (see de Waal, 1988). Conflict resolution, as we currently practice it in our behavioral, social, and management sciences, thus, became possible.

Of course, behavioral tension, motivating the contests remained. The tug and pull of self-preservation and affection persisted latently or covertly to drive later changes in the hierarchy when the supporting contingencies and/or relative prowess of the contestants changed. Thus, human hierarchies, like those of other mammals and primates (e.g., see de Waal, 1989, 1996; Sapolsky 1992: 315–19; Boehm, 1999) are inherently dynamic because of the latent or potential behavioral tension. In human societies, depending on the degree of social, economic, or political inequalities, they must be held in place by institutions, ideologies, and/or police or military power.

But more generally, the integration of self-preservation with affection, allowed the survival of subordinated others and their inclusion in the social fabric. Management of the stress from the tug and pull of the opposing circuitries was, of course, an ongoing process. For example, managing stress resulting from the contest for mates undoubtedly led to the institution of marriage and helped shape other basic cultural motifs in emergent societies.

The mitigation of extreme agonistic competition by mammalian attachment circuitry and the management of tension and stress within the group, likely spurred more rapid evolution of the higher cortex. The emergent social architecture was a conspicuous success. It made possible the development of complex societies with consequent population growth and eventual explosion. In fact, some prominent observers have been led to question whether the species is not a suicidal success (Tickell, 1993; Wilson, 1993). The institutions of democratic participation and economic upward mobility implicitly recognize behavioral tension underlying the instability of human hierarchy. At minimum they allow institutional vents for tension, promoting social stability amidst social change. At best they lead us toward a minimizing of social inequality.

Birtchnell (1993, 1999), a clinical theorist and practioneer, has developed a relevant neuroethological approach to psychotherapy which, in his own words, "grew out of a preoccupation with the concept of psychological dependence" (1999: xi). His main point was a realization, in his view, that dependence has two parts—getting socially closer or further apart as well as getting lower or higher. Birtchnell posits that we can only relate in four dimensions—being close, distant, upper, or lower are the only ways we can relate (1999: ix).

The hierarchical aspect of Birtchnell's theory is easily captured by the CSN Complex and the basic formula or equation of reciprocal interaction. As a therapist focused on dependence in relating, he can be forgiven for not clearly seeing a fifth possibility in relating; that of dynamic equality (although he does allude to it later in his text, e.g., 1999: 14).

What is more suggestive in light of the formula and the algorithms of reciprocity is that all relating other than dynamic balance or rough equality carries a potentially significant load of behavioral tension. Equilibria of such unbalanced or unequal relationships should always be seen as a temporary adjustment on the way to maturity or psychological independence rather than dependence of any variety. This is important to psychiatry and related disciplines that seek to assist clients to achieve more satisfying, less tension-filled, interpersonal relationships.

EVOLUTIONARY EPIDEMIOLOGY, HOMEOSTASIS, AND SOCIAL STRESS

An understanding of homeostasis and social stress, a significant aspect of evolutionary epidemiology, is necessary in order to fully refine the concept of normality and achieve the therapeutic objective of helping clients to achieve normal, if not optimal, interpersonal functioning. The homeostatic nature of the social brain architecture was touched on the Chapter 2. Homeostasis will now be examined more closely.

As we have seen earlier in Chapter 3, living things from the single cell on have remarkable regulating mechanisms. Without these they would simply die and breakdown into their many component parts. But more importantly, there would be no means for evolution of higher sentient life forms.

In physiology this ability to regulate, to keep all parts working together within set limits, was first described by Claude Bernard, a French physiologist of the 19th century. However, it was Walter B. Cannon, an American physiologist and psychiatrist, who, in 1929, named these regulating processes "homeostasis" (Cannon, 1929, 1932; see the summary article by Lapeyre & Lledo, 1994).

The word Cannon coined comes from two Greek words. *Homeo* means similar and *stasis* means standing. As used by Cannon and other scientists who followed him, homeostasis does not mean something rigidly set, unshakable, or immovable. It describes a condition akin to dynamic equilibria that varies yet at the same time is relatively constant. For the organism it describes a condition that is kept within survival limits. When homeostatic limits are exceeded the mechanism either self-corrects or the organism is dysfunctional. That is, it gets out of sorts, sick, or dies.

Like all other creatures, our human vital processes are homeostatic. Built-in controls keep them generally within healthy and, ultimately, within survival limits. Familiar examples are our body temperature which must be kept within a rather narrow range around 98.6 degrees Fahrenheit. Above that we have a fever. Below that we risk hypothermia. Our body must act to correct both indicated conditions to stay healthy or alive. Another example is our blood pressure which varies rather widely to accommodate our changing levels of physical activity. Nevertheless our blood pressure must remain within limits or we become "dis-eased".

The list of homeostatically controlled physical processes is very long indeed. Moreover, they are characteristically organic algorithmic processes rather than events amenable to more Newtonian specification. In truth, the study of endocrinology is largely the study of homeostasis. Practically every bodily process is regulated within certain varying limits. Even the hormones and neurotransmitters that mediate behavior within our neural architecture are regulated homeostatically. When regulation fails, ours moods, our feelings, and ultimately our behaviors get out of sorts. A huge pharmaceutical industry exists to help us keep our bodily and behavioral processes in relative homeostasis from blood pressure medications, to diet pills, to mood regulators.

The word homeostasis usually refers to the regulation of the internal environment of living organisms. Can we legitimately apply the concept to the workings of our social neural architecture?

Absolutely. All the elements for homeostatic regulation are present—standards, the ability to monitor and evaluate, sensors and effectors. In fact, our autonomic nervous system through its connections with the hypothalamus has primary involvement in behavioral homeostasis. The hypothalamus is closely linked with neural structures in the both the affectional and self-preservational circuitry located in the limbic system and the brain stem (e.g., see Herbert & Schulkin, 2002; Kandel, et. al., 2000: 871–997; Nelson, 2000, esp. pp. 447–94; Goldstein, 1995; Becker, et al. 1992).

As noted previously in Chapter 2, recent research indicates the hippocampus, a prominent limbic structure with wide-ranging cortical and subcortical connections to include the hypothalamus, may be instrumental in routing internal and external stimuli in relation to previous self-preservational and affective experience. This, thereby, contributes to the homeostatic equilibration of Ego and Empathy in our social behavior (e.g., see Witter & Groenewegen, 1992; Gloor, 1997: 325–496; cf. Iidaka, et. al 2003).

To acknowledge the influence of external stimuli, experience, or *learning* on the social circuitry's seemingly rather wide, variable, and modifiable set points and boundaries, some researchers have preferred the broader term *allostatic,* which means *adaptive* and points up the somewhat—yet clearly constrained—modifiable nature of the social circuitry's homeostatic mechanisms (see McEwen, 2003; McEwen & Seeman, 2002; Sterling & Eyer, 1981; cf. Goldstein, 1995). On the issue of adaptability, it is crucial to remember, however, that despite some range of modifiability, there is a genetically set limit beyond which the homeostatic settings of human social architecture cannot be pressed or changed without either compromising or exceeding optimal limits for health or even survival (see Teicher, et al. 2003 for an evolutionary perspective on alternative adaptive pathways which may be selected by early environmental stress).

Our social neural architecture, thus, also depends upon environmental cues for regulation. All the sensory systems of our body feed information into the monitoring neural centers. Many cues—from happy, sad, angry facial expressions, dominant/submissive displays, to vocalizations of tenderness, frustration, surprise, and anger—are genetically pre-wired and fine-tuned through development and experience (e.g., see the discussions in Goldstein, 1995; Sapolsky, 1990a,b; Eibl-Eibesfeldt 1989). The muscles of our body act to carry out adjustments to our social as well as physical environment.

But again, discussion of behaviors that increase fitness logically begins with homeostasis (Herbert & Schulkin, 2002: 659). The stress response traces to the origin of complex life, perhaps some 600 million years ago. All known vertebrates have both the proopio-melanocortin (POMC) molecule that is a precursor of adrencorticotropic (ACTH) as well as opiate-like peptides that are similar to those found even in insects and mollusks. Corticosteroids, characteristic of stress responses, are found in all vertebrates. Such molecules were, from the beginning, involved in defensive, homeostatic responses. They have been remarkably conserved over evolutionary history (see Nesse & Young, 2000).

Social stress results from threats to homeostasis. Although the response mechanisms may vary and to some extent be specific to the nature of the threat (Goldstein, 1995), such responses are experienced by humans more generally or globally as behavioral tension or stress (Selye, 1978). Based on the prevailing evidence, the interplay between our self-preservation and affectional circuitry is regulated by what we experience generically as behavioral tension. When this tension exceeds certain limits it results in the worsening condition of behavioral stress.

It has long been known that environmental stressors produce a generalized stress reaction in vertebrates (Selye, 1936). This generalized reaction involves activation of the hypothalamic-pituitary-adrenal (HPA) axis with release of corticosteroids (Buckingham, 2000) accompanied by other neuropeptides including vasopressin, opioids, prolactin, and cytokines. The composition of the response varies with the nature and intensity of the threat. The interest here is, of course, on social stress for which a considerable research has given rise to an accepted general model. The activation by internal or external stimuli of the HPA axis and its responses signal deviations from homeostasis (Przewlocki, 2002; Blanchard, et al. 2002).

Evidence from animal studies of social competition show that social stress results from social defeat, social instability, social disruption, crowding, and even social isolation (see especially the discussion in Blanchard, et al. 2002). In social contests not only the defeated show evidence of social stress, but also the dominant. Such stresses reflect the fact that the dominant are always threatened in their status. Dominant animals and notably primates are subject to the formation of opposing alliances. The existential fact that dominance is not a once and for all status is an angst that pervades the animal kingdom as well as human society. The agonic mode of Chance, fundamental to the maintenance of dominance, represents a state of constrained hostility in acknowledgement of the persistence of behavioral tension or stress (1988). The CSN Complex captures this well in that either Ego or Empathetic extremes carry increasing levels of behavioral tension and stress.

The fact that all parties to hierarchy, or dominant and submissive relationships, carry social stress is very important to psychiatry and the related healing sciences. The effects of such social stress are far-reaching indeed, which, in turn, constitute a growing proportion of the global burden of disease (Murray & Lopez, 1996). Social stress is a major factor in the etiology of many psychopathologies. Additionally, social stress has deleterious effects on male and female reproduction, heart disease, and degradation of the immune system (e.g., see Wassar & Barash, 1983; McGrady 1984; Hemingway & Marmot, 1999; Nakao, et al. 1998; Patten, 1999; Laviola & Andersen, 2003; Andersen, 2003).

Fuchs and Flügge (2003) report on the damaging effects of chronic social stress on structures of the limbic system. Such effects include changes in the size of brain nuclei, in morphology of neurons, and changes in neurotransmitter and second messenger systems. They note that whereas acute physiological stressors rapidly stimulate the hypothalamus and brain stem (essentially the R-complex self-preserving circuitry), psychological stressors activate higher brain structures to further interpret the perceived danger (see also Teicher, et al. 2003). Of course, extreme social stress is a major factor in phylogenetic regression behavior as set out by Bailey (1987, 2002) and reported in Chapter 2.

SOCIAL STATUS AND SOCIAL STRESS

In humans, social status is of special significance as a producer of social stress. Low status has powerful effects on almost all areas of individual and family life. It influences such major stressors as access to resources, personal safety, and health care (Catalano, 2000; Ranchor & Sanderman, 2000). Research also indicates, significantly, that in addition to the material effects, the fact of low status itself combined with the meaning assigned to such status by the affected individuals in itself adds to stress. Such positional stress is evident in a wide range of social environments from that of schoolchildren, to persons in the commercial workplace, as well as to workers in bureaucratic organizations, and even to international relations. (de Ridder 2000; DeVries, et al. 2003; cf. Sevcikova, et al. 2003; Kukleta, 2001).

IMPLICATIONS: FROM NORMALITY TO OPTIMALITY

The central significance of the evolutionary perspective in psychiatry is readily apparent from the foregoing discussion. Social competition, dominant-submissive relationships, and neuroendocrine cycles, represent adaptive processes sculpted in the evolutionary past. In essence, they are normal, not pathological. The reality is that, in the present day, they are simply prone to much more conflict or mismatch with the optimal values and actual demands of our civilized society.

Such values include health and happiness—optimal functioning for all. Evolutionary epidemiology, by establishing criteria and parameters for normality not only provides perspective and tools for better defining pathology and mitigating stigmatization, but helps properly place the healing sciences in service of optimal human functioning.

The epidemiological analysis proposed here links Hamilton's Rule (and it's externalized gene's eye) with the CSN Complex and formula from evolutionary neuroscience. This, in turn, unites phenomena of social hierarchy with clinical concepts of social competition, depression and mania, and issues of dominance/submission, inequality and equality.

Most significantly, evolutionary epidemiology permits a definition of normality anchored in Darwinian science. In this view, normality is the behavioral range of the archetypal opposing circuitries from our reptilo-mammalian heritage—the Ego/Empathy spectrum—as kept within survival limits by the homeostatic mechanism of behavioral stress. Such a definition frees psychiatric diagnostics from the long confounding issue of social context that concerned such scholars as Thomas Szasz (1974), R. D. Laing (1970), and others. The conflation of *normality* with social context-dependent, normative *desirability* or *adaptability* plagued psychiatric claims to scientific validity and tended to foreclose options for social improvement by default to the status quo. Indeed, the cautionary tale of how Soviet and Nazi totalitarianism co-opted psychiatry to sustain and enforce adherence to coercive ideologies is a well-known tragedy that will hopefully never be repeated.

With the definition of normality and the effect of social context clarified, the path is cleared to define conditions for social and individual *optimality.* That is, how to create a social context promoting individual social interaction with essential harmony or minimal behavioral tension/stress as expressed in the equilibrating social equation of :

$$BT = \frac{Ego}{Emp} \text{ or } \frac{Emp}{Ego} = \frac{Dominance}{Submission} \pm 1$$ ***(dynamic balance or approximate equilibrium or unity)***

This allows psychiatry and psychopathologists, via evolutionary epidemiology, to move into the social sciences more generally. This direct movement is especially with respect to normality as operationalized and better understood on a foundation more broadly valid for research.

In the chapters following, the central issues raised will be examined in more detail from alternative perspectives. Also, the concept and methods of evolutionary epidemiology will be explicated and discussed in the light of recent scientific research.

PART II

THE EVOLUTIONARY LEGACY OF NEUROPSYCHIATRIC EPIDEMIOLOGY

5

CLASSICAL GENETICS OF MANIA AND DEPRESSION

The chapters in Part II will develop further the theory and methods of evolutionary epidemiology. In doing so, they will focus on the specific syndrome of bipolar disorder (BPD) or manic-depression. Mania and depression are, among numerous disorders, interestingly amenable to evolutionary analysis from the perspectives of genetics and population biology.

The heritability of both phenotypic characters and disorders was known even in ancient times. This knowledge is evident in the writings of Hippocrates, Aristotle, and the Talmud—among other sources (Ionesescu & Zellweger, 1983). Such themes also recurred in the Middle Ages well before Renaissance experiments in plant hybridization. It was Maupertius (1746), however, who first clearly identified familial transmission of biological traits in humans with his work on polydactyly. Thereafter, Galton (1865) drew formal scientific attention to the study of twins as means by which genetic features could begin to be distinguished from environmental implants into the phenotype. Meanwhile, the Augustinian Monk Gregor Mendel (1866), working in obscurity, was defining the rules for gene inheritance.

THE MENDELIAN TURNING POINT

A modern turning point came with the essentially simultaneous recognition of the importance of Mendelian concepts by de Vries (1900), Correns (1900), von Tschermak (1900) and Bateson, (1902). Concurrently, Sutton (1902) and Bovari (1902) identified chromosomes as the locus of intergenerational characteristics. Johannsen (1909) coined the term "gene" deriving it from Darwin's own concept of the "pangene". Garrod (1902) described the first epigenetic medical disease mechanism and introduced the useful and still current phrase "inborn error of metabolism".

Nevertheless, human and medical applications of genetics long lagged behind botanical research until the extraordinary blossoming of medical genetics late in the Twentieth Century. This recent application of genetic science to medical issues was first noticeable in perinatal cytogenetics. It is now burgeoning in direct genomic assays of diverse clinical interest. Indeed, the era of genetic surgery is nearly at hand (Ionasescu & Zellweger, 1983; Kelsoe, 1996).

RESISTANCE AND EMERGENCE

Despite such progress in genetics, most psychiatrists and other psychopathologists, up to the most recent decade or so, have maintained psychopathology to be largely the product of social, ontogenetic forces. Although change is underway, the field still maintains a marked aversion to the idea of genetic bases for human behavior (Hyman & Nestler, 1993; Eisenberg, 1995; cf., Cowan, et al., 2002; Gottesman, 2002: xiv; Benjamin, et al. 2002). Only a peripheral group has for long considered heritable vulnerabilities to be etiologically important. Their research has been either largely ignored or regarded as incidental to more creditable developmental explanations. Perhaps, mainstream psychopathologists thought that the nihilism which so often results from psychiatric inquiry could only be avoided if most illnesses were derived from social and developmental circumstances. In any event, the idea that some pathological syndromes—phenotypes really—have an adaptive genetic basis is as counterintuitive to mainstream psychiatry as it is old hat among evolutionary anthropologists. Whatever the reasons, evidence of major genetic origins of and influences upon the etiology of much mental illness is now too compelling to disregard.

MOLECULAR BIOCHEMISTRY

It is commonplace to assume that the recent ascendancy of genetic explanations within psychopathological research is due to advances in laboratory techniques—that is, the analysis of deoxyribonucleic acid via direct recombinant assay. Certainly the complete sequencing of the human genome has aroused great interest. The Human Biology Project to specify the functional operations of the estimated 30,000–40,000 genes will continue well into the new century. Truly, molecular biochemistry holds great promise for psychiatry (Cowan, et al. 2002: 4). Some see it as the beginning of a new era (e.g., Gottesman, 2002: xvi).

Despite recent optimism and efforts, molecular biochemistry, however, has yet to make a comprehensive or definitive contribution to psychopathology. Cowan and associates remind us that with the exception of attention deficit disorder (AD) we have yet to identify any specific gene defects for the major psychiatric disorders (2002: 3–4). Thus, the primary source for reliable and valid data concerning the epigenesis of psychopathology is still to be found in the annals of traditional epidemiological methods of indirect research via twin, adoption, family and association studies. The fanfare attending these newer direct techniques, nevertheless, has had the salutary effect of rekindling broader interest in and acceptance of genes as key agents in the expression of psychopathology. Certainly, the prospect that direct genome studies might soon confirm and surpass the older indirect modes of analysis is most welcome.

GENETIC STRATEGIES FOR PSYCHIATRIC RESEARCH

Even before the advance of direct genomic assays, there has been ample reason to appreciate the role genes must play in the transmission of psychopathology (Essen-Moller, 1985). For the present, the most compelling evidence of genetic factors in mental illness comes from traditional avenues of epidemiology. Evidence has long been accumulating from the indirect assessment of gene action in studies of twins, adoptees, and families. Epidemiology, as a science in its traditional form, has demonstrated that genes are intimately involved in some mental diseases. Additionally, epidemiology has demonstrated genetic details concerning the degree to which diseases run in families, aggregate in lineages, or express other pertinent socio-demographic features such as of sex and age.

Epidemiology has not, however, specified several crucial aspects of the genetic basis of psychopathology. The precise modes of transmission, heterogeneity in phenotypic expression, chromosomal location, and biochemical identity of the gene(s) giving rise to mental illness, with rare exceptions (e.g., Huntington's disease), are not yet known. Indirect epidemiological studies cannot provide answers to these important questions. Rather, they will yield only to direct studies of genetic linkage and molecular biological pathophysiology. Similarly, efforts to reformulate existing psychiatric epidemiological research in light of evolutionary population genetics are now feasible and, indeed, overdue. Such assessments become more feasible and definitive as molecular psychiatry progresses.

THE GENETIC EPIDEMIOLOGY OF THE AFFECTIVE SPECTRUM

At the outset, it may be helpful to define a few key terms widely used in genetics, but likely not well known to the less specialized reader.

Genotype is used in three ways. First, it denotes the entire genetic constitution of an individual. Secondly, it may refer to the alleles (alternative forms of the gene) present at one or more specific loci on a chromosome. Thirdly, it is used to refer to the general type of the species. *Phenotype* is also used in three ways. First, it denotes the entire physiological and biochemical make up of an individual as determined both genetically and environmentally. Secondly, it may designate the individual expressing the phenotype. Thirdly, it may refer to the expression of a single gene or gene pair. *Phenocopy* also has three usages. First, it denotes an environmentally induced phenotype that mimics the phenotype usually produced by a specific genotype. Secondly, it describes an individual expressing such a phenotype. Thirdly, it may designate the simulated trait in a phenocopy. *Concordance* refers to the occurrence of a given trait in both members of a twin pair. It is often used in twin studies that attempt to assess heritability of specific traits. *Penetrance* refers to the frequency of expression of a genotype. If the expression of the trait or genotype is less than 100%, it is said to exhibit *reduced penetrance*. Traits may also be *nonpenetrant* if not expressed in the phenotype. *Proband* denotes an affected individual ascertained independently of his/her relatives in a genetic study. *Autosomal* refers to an autosome, which is any ordinary paired chromosome alike in males and females—distinguished from sex chromosomes which are unlike. For example, there are 22 pairs of autosomes in humans.

Background

As Galton (1884) first noted over a century ago, Identical (monozygotic, or MZ) twins share 100% of genes as transmitted from parents whereas Fraternal (dizygotic, or DZ) twins share, as do any siblings, only 50%. All things being otherwise equal, any differences between identical twins may be attributed to contingent ontogenic effects in the developmental environment. However, environmental influences also begin at, if not before, conception. Likewise, studies of twins by definition involve some sampling bias in that twins have rates of spontaneous abortion, perinatal injury and survival, and other effects that differ from the general population of singletons.

A further methodological limitation of twin studies is that they routinely rely on samples obtained from hospital registers which, again, differ from a more general sampling. Also, extant methodologies assume environmental factors are identical amongst twins; this is less valid an assumption as more subtle phenotypic factors are delineated. Still, twin studies are a most useful preliminary means to distinguish genetic from environmental effects.

Studies of Twins

From early in the past century the genetic predisposition to mental disorders has been rigorously assessed via study of twins (Holzinger, 1929). Thus, a trait which arises from genetic inheritance is more alike (concordant) in identical twins than in fraternals. This ideally factors in the limitations noted above and assumes MZ and DZ twins have equal *relevant* exposure to environmental influence. Estimates of heritability are obtained from contrasts between rates of concordance for traits found among MZ and DZ twins.

Holzinger's (1929) basic quantification of heritability has been thus:

Formula 5. 1: Heritability

$$\textit{Heritability} = \frac{[\%\ MZ\ Concordance - \%\ DZ\ Concordance]}{[100\%\ DZ\ Concordance]}$$

Gottesman and Carey (1983) made further refinements by comparisons of twin data to epidemiology concerning the prevalence of the disease in the general population. This reduces the introduction of statistical errors attributable to heterogeneity and phenocopies. Deviations from 100 percent concordance of MZ twins denotes a corresponding degree of environmental modulation of the phenotype. Such deviations, with respect to single major loci, indicate incomplete or variable penetrance. Also, the proportion of MZ to DZ concordance can indicate the mode of inheritance. A ratio in the two to four range suggests a classical Mendelian single major locus effect (e.g., achondroplastic dwarfism has about four MZ twinnings for each DZ twin pair) whereas generally larger ratios are found in multigene conditions. Significant environmental effects are often noted in association with a ratio of less than two.

The strong genetic basis of bipolar disorder has been evident in studies dating back to 1930 (Gershon, et al. 1976). These earlier studies were reviewed by Tsuang and Faraone (1990). Because many of these studies did not distinguish BPD from unipolar disorders, the concordance and heritability rates they reported applied to a composite mood disorder phenotype. Tsuang and Faraone combined the results of 11 twin studies spanning 1928–1986. Analysis of these studies, made up of 195 MZ pairs and 255 same sex DZ pairs, produced a proband-wise concordance rate of 78% for MZ pairs and 29% for DZ pairs. The estimate of heritability was 63%.

The largest of these studies (Bertelsen, et al. 1977) ascertained 55 MZ and 52 DZ twin pairs from the Danish Twin Register. In each case the proband had been hospitalized for manic depression. The proband-wise concordance rate for MZ twins was 67%, significantly larger than the 20% observed for DZ twins.

A second study conducted by Kendler, et al. (1993) ascertained 486 twin pairs through the Swedish Psychiatric Register and the Swedish Twin Registry. Five of 13 co-twins of a bipolar twin proband were diagnosed for BPD for a concordance rate of 38.5. Only one of 22 DZ pairs was diagnosed BPD for a concordance rate of 4.5%.

However, identical twins may experience more similar environmental conditions than do fraternal twins. Kendler and Eaves (1986) considered this issue in detail only to conclude that differential environmental factors do not account for much of the MZ-DZ variance of concordance for major mental disorders.

Additionally, Cardno, et al. (1999) performed a study of 224 twin pairs ascertained from the Maudsley Twin Register. They reported significantly higher concordance rates for a roughly comparable diagnosis of mania plus hypomania—44% in MZ twins in contrast to 9.1% for DZ twins. The heritability for this diagnosis which may be inclusive of bipolar I and II disorder cases reached a rate of 87%. BPD II is distinguished from the classic BPD I by having episodes of major depression interspersed with hypomanic phases but without full blown mania (Gershon, 1982; Simpson, et al. 1993).

These three more recent studies are of special interest because they were methodologically rigorous and used modern definitions of BPD. Their results, summarized in Table 5.1 below, indicate not only a greater concordance among MZ twins than DZ twins but a significant rate of heritability for BPD.

Table 5. 1: Concordance and Heritability of Largest Recent Twin Studies of BPD

		MZ		*DZ*	
Study		**Pairs[a]**	**C %**	**Pairs**	**C %**	**Heritability**
Bertelsen, ***et al.***	**(1977)**	**34**	**62.0%**	**37**	**8%**	**59%**
Kendler, ***et al.***	**(1993, 1995)**	**13**	**38.5%**	**22**	**4.5%**	**79%**
Cardno, ***et al.***	**(1999)[b]**	**25**	**44.0%**	**33**	**9.1**	**87%**

[a] Nr of pairs with affected bipolar proband
[b] Data for phenotype of "mania plus hypomania" *Adapted from Smoller and Finn 2003*

Studies of Adoption

Studies of identical twins reared apart are needed to further validate this body of twin research on a conclusive empirical basis. If a character is genetic, then offspring should resemble their biological family lines more than their adoptive. Thus the paradigmatic study of adoptees gained interest as a basic means to demonstrate epigenesis underlying familial traits. Later, it was appreciated such studies could also clarify environmental components of phenotypy. Adoption studies involve special problems, not the least of which is that adoptees are, at greater than random rates, taken into ontogenic families which resemble their biological families. It is, likewise, difficult to assess biological parents—particularly fathers—with accurate detail. Such studies also are almost always retrospective in design and rely on uneven medical and bureaucratic records. Also, important sibling data are frequently lost to the processes of adoption.

Several studies have addressed the question of genetic and environmental contributions to mood disorders (e.g., Price, 1968). Unfortunately, however, only two such studies have examined the heritability of BPD using a modern definition.

The first such study, Mendlewitz and Rainer (1977), analyzed illness rates in adoptive and biological parents of 29 bipolar adoptees along with those of 22 unaffected adoptees. They contrasted this study with two additional control sets comprised of parents of 31 bipolar probands who were not adopted and parents of 20 persons who had contracted polio in childhood or adolescence. The polio group was to control for the environmental factors associated with rearing a disabled child. The frequency of affective illness, which included bipolar, unipolar, schizoaffective, and cyclothymic disorders, was 31% in biological parents in contrast to 12% in adoptive parents of bipolar probands. The results, contrasted with the control groups, showed similar risks in biological parents of adopted and nonadopted bipolar probands. Lower risks were observed in the adoptive parents of both bipolar probands and polio probands. Although the small numbers prevented meaningful analysis of BPD alone, the results did support the influence of genetic factors in familial aggregation of affective illness.

A second study reported by Wender, et al. (1986) from a Danish adoption register produced similar conclusions. Rates of illness in biological and adoptive parents of 71 adoptees who had been hospitalized with affective disorders (including 27 with unipolar disorder and 10 with BPD) were contrasted with 71 matched controls. Biological parents of ill adoptees displayed significantly higher frequency of major mood disorders inclusive of unipolar and BPD. Again, the small number of BPD cases prevent meaningful statistical comparison.

In summary, evidence from adoption studies which bears directly on the heritability of BPD is limited. Nevertheless, such evidence is consistent with a genetic influence in transmission of risk for mood disorders inclusively.

Details of adoption, thus, have shown that heritage preceding adoption can predispose a given individual to traits or illnesses that run in family lines. Thus, adoption studies can rule out, as the sole or even main causal factor, developmental factors of either a familial or social type. However, though adoption studies can point to inheritance (or perinatal events) as causative of conditions in offspring, by design these studies cannot elucidate precise genetic issues such as mode of transmission or chromosomal localization of traits. Indeed, adoption studies have a variety of confounding variables. Cadoret (1986) discussed the major methodological criticisms which have been raised and addresses how these factors can be best controlled. Still, with about one percent of persons being formally adopted in Western countries, such studies retain much promise and remain the best means to delineate whether a trait is epigenetic or environmentally acquired.

Adoption studies became feasible as social practices formalized adoptive procedures. Initially, the adoptee studies were used to investigate the relative importance of heredity and environment in the determination of IQ. It was not until the nineteen-sixties that adoption methodology was applied to study the role of heredity in psychopathology. Thus, though predominant interest has been in the use of the technique to define genetic etiological factors in psychopathology, the adoption paradigm is equally able to delineate the role of environmental factors with some precision while controlling for heredity. Subsequently, genetic factors in alcoholism, criminality, personality disorders, antisocial personality, somatization disorder, affective disorder, hyperactivity and schizophrenia have been assessed via the adoption paradigm.

Schulsinger, et al. (1979), via adoption studies, examined suicide—a seemingly valid, reliable marker for affective disorder. While it is, in fact, difficult to establish suicide as a diagnostic category independent of affective disorders, with the use of such a syndromic entity a good degree of specificity is lost (Wender, et al. 1986). However, suicide is especially useful in this study as it is among the least culturally variable of markers for affective illness (Wilson, 1993). In the previously referenced Wender study, biological relatives of 71 affectively disturbed adoptees experienced a disproportionate rate (3.9%) of suicide when compared to adoptive relatives (0.6%) or adoptive relatives of healthy control adoptees (0.3%). The difference in rate of suicide between biological relatives of affectively ill probands versus biological relatives of healthy control probands was statistically significant. This is complemented by Gershon and colleagues' (1987, 1988) family data pertaining to bipolar disorder. In the Gershon study suicide occurred with increasingly clear connection to diagnosed affective illness in recent generations. Cultural and clinical practices were taken to more likely account for this transgenerational increase than a cohort effect. More recent data point to a connection between severity of affective illness—especially bipolar depression and completed suicide (Sapolsky, 1994).

Cadoret (1978, 1979) found similar correlations between adoptees and biological but not adoptive parents. One adoptee study from Sweden failed to link the expression of psychopathology in biological parents and adoptees (Von Knorring, et al. 1983). These findings are less persuasive given that the study was limited to a review of insurance and secondary clinic records. It has also been criticized on statistical grounds (Gershon & Goldin, 1984). Adoption studies—logistically difficult to carry out—are few in number. The entire adoption literature is currently comprised of only 44 adopted, bipolar probands. Moreover, the samples are not drawn in a truly random manner from the general population. In sum, adoption studies merely suggest that risk of BPD is substantially epigenetic rather than primarily acquired in the course of individual development, but they neither confirm nor condemn such a notion.

Studies of Families

Family studies can be used to elucidate the risk shared by blood relatives of probands diagnosed with an illness (Maupertius, 1746; Falconer, 1965). Such studies are typically begun by identifying a patient with an illness (the proband) and then examining consanguine relatives. The rate of illness in the family can then be compared to the rate in the general population. The trait denoted is familial—though not, *ipso facto*, genetic—when the rate of risk is greater than that of the general population.

As with twin studies, the genetic factor can be obscured for traits which also have major environmental components. There is, similarly, the issue of obtaining accurate and complete history. Homologous family environments—transmitting common religion, language, diet and so on—can be confounded with similar genetic transmission (Cavalli-Sforza & Feldman, 1973; Cloninger, 1988, 1995; Cloninger, et al. 1978, 1979ab, 1983; Kidd & Mattysse, 1978; Rao, et al. 1976). Data from families can demonstrate specific modes of transmission in traits that have otherwise been identified as epigenetic.

In family studies of traits with age-related risk, such as most psychiatric disorders, calculations of prevalence rates also must be adjusted to account for the proportion of lifetime risk that each relative has experienced. Still, family risk data can establish patterns of transmission across generations which can then be compared to predictable patterns of genetic inheritance. Genetic transmission is evident when family pedigree distribution patterns strongly resemble specific types of inheritance. Family studies also can identify traits which run together genetically. For example, if alcoholism and bipolar disorder are genetically linked, then relatives of alcoholic probands should show higher rates of bipolar disorder and vice versa.

When individual diseases are thus clearly linked they constitute a spectrum of clinical expression or, in evolutionary terms, phenotypic equivalents. This aspect of family studies is especially relevant to issues of heterogeneity and to technical considerations in direct genomic analysis (*Lod*-scoring). Similarly, bipolar and unipolar illness in relatives of bipolar probands are also of consequence to calculations of *Lod*-scoring, heterogeneity, and related matters. *Lod*-scoring (the logarithm of the odds) evaluates the probability or likelihood that a known genetic marker cosegregates with a disease gene to a greater degree than chance.

Diagnostic Issues: Between Science and Art

In all of these situations the diagnostic presumptions of the clinician may conflict with the specificities of the evolutionist (Wilson, 1994). In particular, Darwinism hinges on genotypes and phenotypes whereas the psychiatric research yet relies on a secondary descriptive scheme by which to classify diagnosis. Psychiatric clinical diagnosis remains simultaneously both more an art than a science and a process of phenotypic presumption rather than specific genotypic clarification (Rice, et al. 1987b).

In the great many studies of subtypes of affective illness within families done in the past four decades, the morbid risk estimates for both relatives of probands and the general population have varied several-fold. These problematic variations are largely attributed to non-genetic factors. That is, much of the range in findings is due to inconsistencies in either diagnostic or research methods while cultural and demographic factors are also contributory.

Comparability and Reliability: Recent Cases

In family studies BPD has most often been linked with related mood and affective disorders. Therefore it is not possible to completely separate the various syndromes with any real precision. Furthermore, the delineation of the two major forms of BPD I and BPD II has confused boundaries somewhat more. An overview of family studies, however, does throw light on the heritability and risk factors of BPD along with other affective disorders. Available studies support the general conclusions that relatives of BPD probands are at considerably increased risk for both BPD and unipolar disorder (UPD). On the other hand, relatives of UPD probands have increased risk for UPD but do not seem to have significantly greater risk for BPD.

In the case of BPD I vs. BPD II, the diagnostic reliability of BPD II is perhaps less validated than that of the former. Recent work, however, has indicated that reliability can be improved through direct interviews of relatives combined with careful diagnostic procedures (Simpson, et al. 2002). Careful family studies using direct interviews of first-degree relatives have helped distinguish BPD II as a closely related but perhaps separate entity. Relatives of BPD II probands have shown elevated risk of BPD II as contrasted with those of BPD I of UPD. Nevertheless, the similarity of familial risk levels across these proband groups suggests these disorders are etiologically linked (Gershon, et al. 1982; Andreason, et al. 1987; Heun & Maier, 1993).

A further development is the breakdown of sharp distinction between the traditional Kraepelian diagnostic categories of manic depression and schizophrenia. The original distinction was based in part upon the belief that the two disorders had a distinct hereditary basis. More recently, observations that the genetic loci of these disorders may overlap have spurred considerable interest in the etiologic boundary between psychotic disorders and BPD (Badner & Gershon,

2002). Results of recent studies have been mixed, some suggesting elevated risk of schizophrenia among relatives of BPD or affective disorder probands (Tsuang, et al. 1980; Taylor, et al. 1993; Valles, et al. 2000), others finding no differential risk of significance (Weissman, et al.1984; Maier, et al. 1993). From the opposite perspective, a meta-analysis of three controlled, direct-interview studies of schizophrenic probands, comprising the Danish Adoption Study, the Iowa 500 Non-500 Family Study, and the Roscommon Family Study, found risk estimates of BPD homogeneous across the studies (Kendler & Gardner, 1997).

The schizoaffective and BPD boundary has also aroused considerable attention but has proved even more difficult to define. This difficulty probably stems in part from the fact that the definition of schizoaffective disorder includes prominent mood symptoms. Nevertheless, results suggest some etiologic overlap in the familial and genetic basis of these disorders. In some studies the risk of BPD among relatives of schizoaffective probands has been equal to or greater than the risk among BPD probands (e.g., Rice, et al. 1987; Maier, et al. 1993). Conversely, other studies have not shown relatives of BPD probands to be at higher risk for schizoaffective disorder (Pauls, et al. 1992; Maier, et al. 1993).

Cardno, et al. (2002) recently examined the overlap of genetic influences on schizophrenia, schizoaffective, and manic disorders by analysis of twins from the Maudsley Twin Register. They reported significant correlations in genetic risk across the three syndromes. The genetic correlations for schizophrenic and manic syndromes were 0.68. For schizoaffective and manic syndromes the correlation was 0.88. Although not without difficulties of interpretation (Kendler, 2002), the results of this analysis provide support for other recent reports of overlapping genetic linkage for BPD and schizophrenia (Berrettini, 2001; Badner & Gershon, 2002).

In summary, the studies indicate a heterogeneous BPD phenotype which in some degree is linked to UPD and psychotic disorders, notably schizoaffective disorder. Further clarification of boundaries between BPD and other affective disorders may depend upon molecular genetic studies to identify specific susceptibility genes.

THE STUDY OF CROSS-CULTURAL VARIATIONS

To many, the study of psychopathology across cultural boundaries seems an exotic activity of marginal scholarly interest. But there is a growing appreciation of the value of cross-cultural research as means to advance a variety of disciplines via analogical inferences (Foley, 1992). Such analogical research can both validate and extend the base of knowledge in psychiatric epistemology which has accrued, heretofore, almost exclusively within the biomedical cultural tradition of the West (Kleinman, 1988). Other historical or cultural traditions are likewise amenable to such analysis (Haldipur, 1989). A thorough review of psychiatric epistemology (such as Jaspers, 1913; Ellenberger, 1958; Hundert, 1989) is, of course, beyond the present scope. But brief mention of certain philosophical issues is necessary to highlight the position of psychiatry within the range of broader intellectual discourse.

Inescapably, psychiatry is an idiomatic genre of phenomenology. After all, psychiatric diagnosis is an attempt to categorize an extreme of the naturally occurring or normal phenomena of mental experience. Disordered mental experiences are to psychopathologists what normative mental experiences are to psychologists—what insects are to entomologists, germs to microbiologists, mountains to stratigraphers, cells to cytologists, stars to astronomers, and so on (McHugh & Slavney, 1986).

In this sense the central problem of psychiatry is identical to that of all taxonomic enterprises: verification. Yet as in any program of phenomenological research, verification of both psychiatric observation and presumption is needed to build a sustainable taxonomy (Kendell, 1975; McHugh & Slavney, 1986). Moreover, a sound taxonomic foundation is necessary to understand those deeper principles which govern the comprehensive operations of dynamical systems (Kant, 1790). This necessity also applies to a comprehensive understanding of psychopathology (Aristotle, circa 450 BCE; James, 1890; Romanes, 1884, 1896; Freud, 1917; Jaspers, 1913; Kleinman, 1988; Wilson, 1987). Part I of the present work has been such an attempt.

The Western Bias: Does it Apply Cross-Culturally?

Unfortunately, the current reliability of psychiatric classificatory systems outstrip their claims of validity (Sapir, 1932, Kleinman, 1988; Hundert, 1989; Good, 1992; Pichot, 1994). This is substantially due to the fact that diagnostic schemae have, for the most part, arisen within the percepts of Western medicine. They, therefore, have both an unclear applicability to other cultures and a strong tendency to the presumptions of logical positivism. Proper cross-cultural studies of mental disorders are needed to appreciate the range of environmental and historical forces which induce phenotypes of psychopathology. In other words, we are back to the issue of social context touched on in Chapter 4. Accurate categories of phenotypic description best derive as the validity of psychopathological categories is negotiated between alternative concepts and experiences noted in unique environs (Kleinman, 1982, 1985, 1987, 1988). Ultimately, the core "universals" of human experience, as suggested in Part I, can thus be teased out from the surface cultural aspects. In an extant phenotype, universals are generally attributed to the organic genotype whereas the cultural aspects are attributed to the environmental component.

Still, there is an increasing appreciation for the as yet unrealized promise of cross-cultural research. At the very least this increasing appreciation complements the great expansion of research which has occurred in the neurosciences that underlie psychopathology. In part, this neuroscience agenda has been made possible by a resurgence in the status of psychiatric epidemiology of a distinctly Western and biomedical type. In turn, this rise has been based on operationalized disease criteria and standardized diagnostic instruments with high levels of inter-rater reliability.

Certainly, sociocultural forces influence phenotypic development of psychopathology. All manner of environmental threats contribute to the clinical presentation of both mental and physical illnesses: stress, alienation, trauma, deprivation, poverty and powerlessness. Moreover, no physiological or genetic etiology has yet been proven as a definite cause of psychopathology. Biological reductionism has once again become the dominant intellectual paradigm in psychiatry more for its likely promise than for its immediate contributions. Still, the pace of research in neuroscientific psychiatry is impressive.

The Potential of Cross-Cultural Study

Cross-cultural study has much to contribute at this juncture, as the study of variant cultural expressions of mental illness is a most important means by which to disentangle the genetic nuances from the environmental components of mental illness. Thus to achieve a more valid psychiatric taxonomy it will be necessary to compare the vast body of Western epidemiological research with studies adapted to and derived from differing populations around the world.

As much was noted long ago by no less an authority than William James who averred such comparative studies were ". . .of the utmost importance for the comprehension of our nature" (1890). Yet for all this theoretical and empirical value there have been surprisingly few cross-cultural studies of psychopathology and even fewer which assess manic-depression, *senso stricto* (Winokur & Clayton, 1986). In practical terms, a necessary and consistent medical anthropology of the manic-depressive syndrome as includes its phenomenology and epidemiology, cannot be said to exist.

There have been some noteworthy efforts. Early on Kraepelin (1904) offered the unqualified observation that both depression and mania were more common among Javanese and Singaporeans than European subjects. Jaspers (1913) and Freud (1930) also emphasized the need to elaborate a cross-cultural phenomenology of mental disorder, albeit in rather divergent tones. Van Wulfften-Palthe (1936) and Oesterreicher (1951) noted a higher rate of depression among ethnic Chinese in Java compared to the natives though neither accounted for the possible confounding role of urban-rural factors.

In one of the few contemporary, operationalized studies, Dunner (et al. 1984) evidenced no difference in the rate of BPD in China and the US. Of particular note, at the international level there is nothing comparable in scale, design or rigor to the benchmark Epidemiological Catchment Area (ECA) study sponsored by the US National Institute of Mental Health (Eaton, et al. 1981; Robins, et al. 1984). It is noteworthy that the World Health Organization did sponsor a systematic comparative analysis of "Depressive Disorders in Different Cultures" (Sartorius, et al. 1983). However, this study is limited in that it pertained only to industrialized cities and failed, quite inexplicably, to specify prevalence rates. Sadly, too, the promising "ECA of the EEA" study initiated by Wilson and Chagnon, which meant to survey rates of major mental illness among the Yamomani, was suppressed by the Venezuelan government the very day fieldwork was to have begun.

Problems in Cross-Cultural Study

What summary is to be made of this important but understudied area? In the first instance, the dearth of cross-cultural studies of psychopathology is puzzling. However, upon further reflection it is perhaps less mystifying. Methodological problems in cross-cultural psychiatry are formidable and it is a notoriously expensive mode of research. These problems of cost and complexity are especially acute as one seeks to move beyond comparisons of homologous populations in industrial nations; i.e., immigrant cohorts or sub-cultural groups within metropolitan societies.

In certain respects, the paucity of cross-cultural studies is arguably commendable as apparently there has been limited perceived need for the allocation of scarce resources toward a simple demonstration of what a consensus of expert opinion already deems obvious. Namely, that there is wide agreement that a few syndromes of psychopathology (including BPD) are core structures of the human genome. Even highly redoubtable social psychiatrists can only conclude the more florid diagnoses—such as manic-depressive illness—are universal features of the human genome.

For example, so Kleinman avers in his serial, meta-analyses of cross-cultural mood disorder studies (1982, 1985, 1987, 1988). Kleinman explicitly grants that these diseases are expressed rather invariantly across cultural boundaries due to the restricted range of phenotypic development imposed by the epigenetic psychobiological processes which cause disease. So it is, widely respected and strong proponents of the sociocultural etiogenesis of mental illness acknowledge manic-depression is a reliable and valid diagnosis with relatively, little idioethnic variance world-wide (Kleinman, 1987, 1988).

This appraisal is entirely consistent with the assessments of researchers with a more neurobiological bent (McKusick, 1990; Goodwin & Jamison, 1990). Moreover, this conclusion has been reinforced recently via an original systematic study (Marzuk & Barchas, 1995). The idea that manic-depression is a rather culturally invariant feature of the human genome is a unique, even remarkable point of agreement in a field often given to contretemps.

To be sure, one might wish for a more robust body of data concerning the degree to which manic-depressive illness has limited phenovariance across cultures. Still, there is an unusually vigorous level of agreement among research psychiatrists that BPD constitutes a universal disease substantially embedded in the human genome and phenotypically prevalent in rather steady rates through the cultures of the world (Goodwin & Jamison, 1990, 2007). It is a genomic as well as sociological universal (Wilson, 1993).

STUDIES OF COHORT AND SECULAR EFFECTS

Populations, cultures and families vary not only across geographical ranges but through temporal sequences as well. Clearly, subsequent generations may have different rates of illness. The epidemiological term *cohort effect* captures just this concept. What is often much less clear is why in the case of a given illness intergenerational rates vary as they do.

Cohort effects are evident in family risk studies of both UPD and BPD (Gershon, et al. 1982, 1985, 1986; Moldin, et al. 1991; Moldin & Gottesman, 1995). The age of onset has been earlier and total lifetime prevalence has been greater in cohorts born since 1940 (Gershon, et al. 1987). Such a cohort effect within one or two generations is not attributable to genomic selection due to differential reproductive success (Gershon, 1983). Rather, these appear to be phenotypic reactions to altered conditions—biological, psychological, sociological—in the environment of adaptive development.

An important but little recognized nuance of such altered conditions may well exist in the sociocultural variations of medicine itself. Genetic anticipation (McInnes, et al. 1993; Petronis & Kennedy, 1995) is an alternative possibility. Indeed, changes in the diagnostic practices of psychiatrists seems a more likely cause (Pope, 1983; Pope & Lipinski, 1978; Pope & Todd, 1990) although an open mind is especially warranted lacking significant research findings in this area.

The overlap of genetic influences on BPD and UPD with schizophrenia and schizoaffective disorder suggested by recent studies points up the need for more direct research to elucidate contributions of genetic pliotropy (different phenotypic forms derived from a single genotype), epistasis (the effects of interactive genes), or convergence (similar phenotypes derived from different genotypes) versus problems of medical taxonomy and psychiatric diagnosis (genuine or spurious co-morbidities). Of central importance is that more than eight decades of research using traditional indirect epidemiological methodologies—twin, adoptee, and family studies—confirm that much of the bipolar spectrum is an evolved population genomic polymorphism with many intrinsic and potentially adaptive features.

Meanwhile though, improved methodologies of quantitative genetics combined with linkage analysis have arrived as the best means to identify specific chromosomal loci and to characterize genomic dynamics more generally. The availability of the human genome sequence and the progress currently being made in the specification of gene function hold great hope for the future. The next chapters will consider these promising directions in greater detail.

6

POPULATION AND QUANTITATIVE GENETICS OF MANIA AND DEPRESSION

The quantitative approaches developed in evolutionary population genetics are highly relevant for evolutionary epidemiology. One clear advantage of evolutionary theory over experimental or field work is that it is essentially logical, even mathematical (Wilson, 1975). Thus, Mendelian and Darwinian laws of genetics become postulates in a largely deductive science reminiscent of algebra (Wilson, 1993). In this way, mathematical and deductive approaches to evolution supersede the more accessible natural history mode of descriptive biology to, at times, achieve more abstract and deeper insights.

For example, in application to evolutionary questions, theory sets parameters of expectation. Since each of the agents of evolutionary change relate to one another in various ways, theory can establish the high and low range of functions which express these relationships. Therefore, a necessary and complete approach to evolution can be achieved only via theory. Experimental and field studies will forever lag behind theoretical deduction due to their inherent logistical complexity. Of course, for these same reasons bad theory is more egregious than is bad experimentation. Moreover, practical experience often runs counter to theoretical insights—after all the sun seems to circle the globe. But before venturing into even basic evolutionary mathematical formulations it is useful to recount several fundamental points.

TYPES OF GENETIC TRANSMISSION

Differences in types of genetic transmission have often led to confusion. Many people seem to intuit that only Mendelian traits are firmly genetic and that continuous traits are somehow softer and less genetic. Many important heritable characters, however, are polygenic or composed of effects of a single major gene locus expressed against a complex backdrop of many small additive features. Such continuously additive traits do not fall neatly into Mendelian matrices of discrete presence or absence of expression. Genetic traits—whether Mendelian, continuous, or mixed—display several consistent features (Neel & Schull, 1954). Gene expressions arise in highly predictable patterns characterized by the following features:

1) Non-blood kin (in-laws) are not affected
2) Blood kin are affected in specific proportions
3) Concordance in relatives is a function of kinship
4) Links exist between gene loci and traits

Ultimately, for the larger purposes of evolutionary epidemiology it little matters if a heritable trait is simple and Mendelian or more complex (Eaves, 1969, 1970). Discrete Mendelian characters are easier to trace, but as genetics has become increasingly sophisticated in both theory and practice, more sophisticated models of transmission have emerged.

QUANTIFYING EVOLUTION

The quantification of evolutionary dynamics is a challenging enterprise. Evolutionary mechanisms, generally, require eons of favorable selections for new mutations to obtain epidemiological prevalence at substantial rates. Further, the threshold rate for specific genes is not precisely known and involves highly technical adjustments for rates of such things as mutation, back-mutation, fertility, penetrance, and dominance. The calculation of threshold rates also assumes broadly stable environmental conditions; that is, an environment of evolutionary adaptedness.

The Hardy-Weinberg Equilibrium and the Prevalence of Genes

A major aim of population genetics is the mathematical clarification of evolutionary dynamics. To this end, the Hardy-Weinberg equilibrium is perhaps the fundamental theorem of population genetics. The equation summates genetic fluctuations across generations in a precise way to specify the net effects of evolution via natural selection (Cavalli-Sforza & Feldman, 1973; Cavalli-Sforza & Bodmer, 1976). Since genes arise by mutation and are selected for or against over time, if a new gene is helpful enough for long enough, it will accumulate. Genes at prevalences well beyond frequency of spontaneous mutation, except in decidedly unusual circumstances, were thus selected into the genome on the basis of positive characteristics expressed in the phenotype. That is, they evolved.

Prevalence and Frequency

In grasping this dynamic, we must understand the essential equivalence of the term *prevalence*, as used by epidemiological geneticists, to the concept of *frequency*, as used by evolutionary geneticists (Wilson, 1992). Given the importance of these terms, it is useful to point out their technical meanings. *Prevalence* refers to the rate of existence of a disease (gene) within a population. *Frequency* similarly refers to the rate of existence of a gene (disease) within a population. Except for the cross-disciplinary distinction, the terms could, in fact, be used interchangeably.

As a rule, prevalence above the mutation rate is likely when a gene occurs in the frequency range of 1/10,000 to 1/1,000 members of a population (Wilson, 1975).

Therefore, in mammalian systems rates much above 0.001% are suggestive of past positive natural selection. Rates beyond 0.01% are *highly* suggestive. Given average mutation rates and neutral fertility effects, it is unlikely that a deleterious gene will attain a frequency of much more than 0.01%. Moreover, to secrete even this random level in the genome might take many generations—up to 10,000. If a simple gene system occurs to such a degree, it was abidingly adaptive—even if no longer so.

On the other hand, deleterious mutations cannot achieve high prevalences spontaneously, nor even via neutral selection (Kimura, 1955, 1968). Consequently, the majority of inherited abnormalities should be rare as a function, more or less, of mutation rate divided by fertility and penetrance. Genetic diseases should be, by definition, of very low prevalence (e.g., achondroplasia). They are often enclosed within sporadic idioethnic lines (e.g., Tay-Sachs disease).

Prevalence of even a deleterious gene will equilibrate over generations at a rate balancing new mutations against the selective pressure. For example, in achondroplasia the equilibrium frequency is one in eight thousand births for what is a classic autosomal dominant condition. Tay-Sachs is seen only rarely and in certain lineages. Some Ashkenazi Jews have it. Laplanders do not (Dobzhansky, 1962). Neither of these conditions evidence past positive selection and, indeed, such positive attributes are not to be predicted in the established tests of evolutionary epidemiology. Though strongly genetic, they are too rare to have been promoted by natural selection. They are counter-exemplary evolutionary noise existing, more or less, at equilibrium rates.

Validity, Reliability, and Frequency

For all the scientific and especially the popular disdain to which psychiatric theory and practice is sometimes subjected, its backbone of epidemiological research is surprisingly strong. Some of the most common, well validated, and reliably diagnosed disorders of the human genome have been established within the field of psychiatry. In passing, it is useful to recall that *validity* and *reliability* are basic, if technical, terms widely used in epidemiology. The former expresses the meaningfulness of a given category of description and the latter expresses the degree to which such descriptions are made by independent observers.

Again, the high prevalence (as population polymorphisms) of such abiding genomic traits points to intrinsic, if overlooked, positive characteristics that caused them to be so avidly retained by long-term and wide-spread natural selection. These genes and their phenotypic expressions can be said to be highly canalized (Wimsatt, 1989).

Thus far, only indirect, phenotypic descriptions of psychopathology have been firmly established. The era of direct, genotypic description is at hand. This era should prove interesting since some diagnostic categories thought to denote discrete diseases may ultimately prove to be mere phenotypic variants of the same etiological genes. Such genomic equivalent through phenotypic 'co-morbidity' can only mask much higher aggregate genetic prevalence rates (Kendler, et al. 1986, 1987, 1993; Hudson and Pope, 1990; Wilson, 1992, 1994, 1995, 1996, 1997). If so, the indices of their twin concordance, prevalence and natural selection will be all the higher.

Indeed, on closer examination it appears that several of the variant genes most common in the entire human genome express psychopathology (e.g., BPD, sociopathy, and obsessive-compulsive disorder). Thus they have been marked Darwinian successes. Moreover, these same diagnoses are universal features of the human genome which evidence little, if any, ontogenetic idiosyncracies of epidemiology due to geographic ethnic, cultural, racial or other influences on the phenotype. Therefore, their Darwinian success was an enduring and universal feature of the evolutionary history of the structure and function of the genome of the entire species.

Still much important psychopathology can already be regarded as the product of natural selection. This is true even without recourse to the greater genetic frequencies which are necessarily involved in concepts of a broader spectrum of extended phenotypes. It is likewise true even without recourse to direct gene analysis. In the era of modern science it is easy to forget that gene characteristics and behavior can be quite usefully scrutinized without direct biochemical technology as the ingenuity of both Darwin and Mendel demonstrated so remarkably. In any case, indirect methods of epidemiology will remain the most useful approach to research in psychiatric research until techniques of direct gene studies are better validated.

MENDELIAN TRAITS, POLYMORPHISMS, AND HARDY-WEINBERG FORMULATIONS

The Hardy-Weinberg formula is a standard point of departure for any discussion of population genetics (Hardy, 1908; Dobzhansky, 1962; Mayr, 1963; Erlenmeyer-Kimling & Jarvik, 1963; Wright, 1968; Provine, 1971; Harrison, et al. 1988). The Hardy-Weinberg formula holds that genotypes can be predicted as conforming to rules of Mendelian genetics and probability. The equation can be applied to more complex questions of gene pool dynamics and, with various emendations, derive insights into the directionality of natural selection (Mayr, 1963). Stable frequencies are found in large, randomly mating populations free of selection, mutation, migration, or drift. The Hardy-Weinberg law (where p is the frequency of allele, M, and q is the frequency for allele, N) is generally expressed:

Formula 6.1: Hardy-Weinberg Equilibrium

$$p2\ (MM):\ 2pq\ (MN):\ q2\ (NN)$$

Original gene frequencies, in theory, are maintained. Thus, in freely mating (panmixic) populations free of selective pressures, all original gene frequencies will be maintained over generations. In application to real populations, however, the formula can be used to assess degrees of demographic or populational change. The Hardy-Weinberg formula was historically among the first means by which to measure natural selection. It has the additional advantage of dealing with that least complicated but very important aspect of genomic analysis—discrete characters. Population variations are discontinuous when a single gene locus (with few polymorphic alleles) accounts for discrete variation. Such genetic effects can be fully understood in purely classical Mendelian terms since segregation is discrete; for instance, tall or short pea plants. An example of such an almost purely Mendelian disease from human medicine is phenylketonuria (Dobzhansky, 1962).

In a larger sense, both evolutionarily or medically, it does not so much matter if these genes are discrete, polymorphic, or polygenic. Discrete traits are not more relevant to evolution nor more genetic than are continuous or complex traits. They are simply more readily analyzed from the point of view of evolution (defined as the intergenerational change of genetic frequencies). Moreover, when frequencies are noted to vary from Hardy-Weinberg equilibria, it is likely the evolutionary agencies of selection, mutation, meiotic drive, gene flow or drift have been operating. Another major source of variance from Hardy-Weinberg predictions occurs with non-random mating.

Also, the classic Hardy-Weinberg formula posits a dimorphic set of alleles at a given locus. Yet loci often encode genes of a polymorphic type having multiple alleles. That is, polymorphisms are present in genomes when two or more forms of a character exist at frequencies greater than that which can be attributed to random mutation alone. Polymorphisms may thus persist in the population when there is a countervailing net effect preventing the elimination of all but one or two variety of alleles. There are three principle types of polymorphisms: transient, balanced, and neutral (Mayr, 1963).

Transient Polymorphisms

Transient polymorphisms are generally those which are simply not yet settled into an ideal equilibrium. Indeed, given fluctuant selective pressures in the evolutionary environment over time, mutation selection balance might never be reached. A transient balance of alleles in a population is rarely a stable feature.

Perhaps the classic example of transient polymorphism is that of the peppered moth (*Biston bistularia*). Obvious fluctuations occurred in these populations during the industrial revolution. Sooty colored moths, formerly quite rare, quickly became more numerous as tree bark became blackened with industrial pollutants (Kettlewell, 1956, 1958, 1961, 1965). Clearly this was linked to the advantages their dark color conferred within an important element of their microenvironment. With a subsequent decline in industrial pollution the selective pressures again changed and the peppered variety returned in areas with lighter bark (Kettlewell, 1961).

Many transient polymorphisms are due to assortative mating but these, in themselves, do not change gene frequency but merely aggregate genes within subgroups of the population. Of course other selective forces also maintain genetic variation in populations, notable types include frequency-dependent, cyclical or counteracting (Fisher, 1930). Indeed, important frequency-dependent behavioral influences on primate reproduction have been verified (Dittus, 1979). Beyond these it is the agencies of evolution (especially selection, mutation, gene flow, and more rarely meiotic drive) which alter population rates of genes over time.

Polymorphism is of special importance with respect to the proper medical analysis of what otherwise appear to be simple, single major locus disease genes. In fact, laboratory studies of isozyme and protein population variability show the genomes of many species, *Homo sapiens* included, carry polymorphic alleles at some 30–50% of genetical loci. Related studies demonstrate that individuals are heterozygotic (hybrid) for some ten percent (Emery & Mueller, 1992). Moreover, these appear to be functional genetic loci with differential selective effects. It is not yet clear to what extent these polymorphisms are the net effect of their genomic balance, transience, or neutrality.

Balanced Polymorphisms and Natural Selection

Perhaps of more immediate medical interest are polymorphisms of either a manifestly balanced or frequency-dependent type. Balanced polymorphisms achieve stability when the ratio of alleles is maintained neither randomly nor fluctuantly, but due to the sustained selective advantage of heterozygotes. Balanced polymorphisms, then, are due to specific genetical mechanisms of a more sustained nature. Generally, balanced polymorphisms arise in nature when the hybrid (heterozygote) state has fitness superior relative to either true breed (homozygote).

Similarly, frequency-dependent polymorphisms confer advantage as a dependent variable proportional to their relative frequency in a population. Typically, such polymorphisms are favored as minority elements against a more modal population. Selective advantages taper off as frequency increases. Therefore, frequency-dependent polymorphisms fluctuate but (absent radical environmental change) do not supplant the type dominating the population.

A clear picture begins to emerge that nearly all species evidence a wide spectrum of allelic variation that is, itself, a necessary and fundamental characteristic of each genome. Such variation is grist for the mills of evolution. Only the extremes of these spectra of genomic variations can be said to constitute disease in the strict sense and, then, only in specific ecological context. A so-called disease gene can quickly become salubrious in shifting ecological conditions, as with the peppered moth. Natural selection along these lines quite likely explains a number of polymorphisms encountered in clinical medicine. Among the first of these to be elucidated were the stable variations of human hemoglobins and immune types. Sickle-cell anemia is the classic human example of which more shall be said later.

Though here the focus is on balanced and frequency dependent polymorphisms, it should be noted that much observed genetic variation is not under the control of selection but the residua of neutral polymorphisms adrift in the population. Newer genetic technologies have revealed there are many such variants that, as "restriction fragment length polymorphisms or 'rflps'," travel through meiosis in conjunction with other genes which are themselves the actual cause of differential selection. This is a presumed mechanism of non-Darwinian, neutral genetic drift.

COMPLEX INHERITANCE

Major genetic effects are clearly amenable to Mendelian analysis (e.g., Hardy-Weinberg formulations and other assessments of largely qualitative traits). Many inherited characteristics, however, are more complex and do not conform straightforwardly to Mendelian principles. Some human characteristics are quantitative or continuous (e.g., height, blood pressure, behavior). Others, such as common diseases, are qualitative or discrete—that is, they are either present or absent—but they do not conform to Mendelian segregation ratios. Although the mechanisms of inheritance in such cases, do not differ essentially from those of purely Mendelian characteristics, they call for some conceptual modifications. Polygenic inheritance of complex traits and multifactorial (multiple genes plus environmental contributions) phenotypic traits, however, generally require a more biometric approach.

Polygenic and Multifactorial Traits

Polygenic traits are those which are due to the combined effects of several genes. It is generally assumed that each locus exerts a slight effect upon the expressed character but only the whole set of genes taken together gives rise to the observed variations of the trait. Likewise, it is often assumed that polygenes express traits in a Gaussian (bell) curve. Actually most polygenic variations appear as stepwise distributions of genetic characters in a population.

However, the phenotypic expression of the genes in the developmental environment does yield a bell curve of trait expression within the population as a whole. This occurs because multiple alleles at each of the separate loci usually confer little effect and these effects often overlap with other loci. Thus, the same phenotypes are often expressed by many different genotypes. So, successive genotypic steps are shaped, through environmental factors, into a smoothly sloping phenotypic population array. The result is a multifactorial trait range.

Still, it is possible that multifactorial traits can express discrete (discontinuous) variations, especially in the case of traits manifest only once a certain threshold of genetic coding is passed. Here, even with a continuous genotypic array, individuals above the threshold manifest a phenotypic trait while those below do not (when trait expression is dependent upon passing a quantitative threshold of genetic effects, e.g., cleft palate (Milerad, et al. 1997; Palomino, et al. 1997). It is also useful to point out that Mendelian traits can, at times, seem rather more quantitative than qualitative. On the whole, however, polygenes are essentially minor loci which, together, interact with the environment to express characters along a continuum.

Methods by which polygenes may be analyzed are different than methods for the evaluation of traits of a more discretely Mendelian type. The continuous distribution of characters expressed by polygenes (especially with multifactorial attributes) does not fall readily into terms adequate for Hardy-Weinberg calculations. Continuous characters are best studied through the application of biometric statistical analysis, e.g., correlation coefficients.

The Challenge of Behavior

Though polygenes are said to be minor with respect to their strength of expression they are not trivial. As this may be especially true in the case of the genetic framework of behavior, it is necessary to more fully consider the modes of inheritance of behavior (Plomin, 1990). Behavior is, after all, an observable and measurable though highly unique form of phenotypic expression. Behavioral genetics has long grappled with the manner in which this most dynamic aspect of nature is nurtured. Behavior is difficult to link tightly to the genome because it is often that part of the organism that exists in a most direct and dynamic equilibrium with the environment (Ryle, 1949).

Behavior is also a challenge in that a great many of its most important dimensions are polygenic rather than Mendelian traits. Rarely, then, is a behavioral character discrete or due to a single major gene. Oligogenic, polygenic, and mixed traits are more common (Rao & Gu, 2002; e.g., see Comings, et al. 2002 on the relationship between dopamine genes and personality traits). Moreover, phenotypes of behavior are more often substantially influenced by environmental and other non-genetic factors than are phenotypes of morphology. This is true with respect to elements of cognition, personality and psychopathology (Piaget, 1932, 1971; Plomin, 1988, 1990; Plomin, et al. 1991; Cloninger, 2002). Likewise, anxiety, sociability, vocational interests, dominance, control and numerous other ethological or physiological domains—ranging from posture or gait to electroencephalographic profile—exhibit phenotypic complexity (Hamburg, 1971; Hamburg, 1974; Gottesman, 1972; Vogel, et al. 1979abc; Ohmark, et al. 1980; Bouchard, et al. 1990; Lykken, et al. 1991; Rushton, 1990; Loehlin, 1992).

There is little doubt that much of interest in the field of behavioral genetics—perhaps especially its subset of human psychology—is best approached via biometric analysis of continuous traits of a complex nature.

Liability-Threshold Models

Liability-threshold models reflect the effort to deal with the challenges of more complex polygenic and multifactorial inheritance. They assume a continuous liability across the general population. Multiple (poly) genes and environmental factors contribute to the liability. Individual surpassing a designated threshold are affected by the trait, whereas those below the threshold are not. In the case of disorders the threshold is set by the lifetime morbid risk of the particular disorder in the general population. Liability-threshold models may have single or multiple thresholds as required.

Model Fitting

Over recent decades a variety of models have been developed for evaluating various multifactorial—genetic and environmental effects. This had made possible biometrical model fitting to test whether and to what extent such multifactorial effects contribute to variation in a quantitative phenotype. For example, biometrical model fitting for data analysis on twins may include models of varying combinations of multifactorial effects. Such effects may include shared and nonshared environmental factors, non-shared environmental influences only, or any other additive factors of interest. Suitable equations are derived and statistical correlations examined. The best models are selected on the basis of goodness of fit and parsimony.

Complex Segregation Analysis

Complex segregation analysis can be employed to identify the likely mode of inheritance of a trait or disorder (Lalouel, et al. 1983). Commonly, this involves the use of a mixed model (Morton & MacLean, 1974). In a mixed model both gene and polygene effects are compared with submodels of single major locus and polygenic inheritance. A significant drawback of such complex analysis is the requirement for large sample sizes to distinguish between models. To date this drawback has greatly limited the utility of this procedure for psychiatric genetics. Nevertheless, such statistical approaches may be useful to explore the possibility of major gene inheritance, prior to launching into the major effort of linkage studies.

Moreover, the mixed model would greatly reduce a recent and quite unfortunate trend among molecular geneticists. Disappointed that they have not often or easily enough replicated major linkage claims, too many molecular geneticists (perhaps especially those working in psychiatric linkage analyses) recently seem to assume that only "complex or continuous genetics" can explain the inconsistencies of linkage analyses to date (see Kennedy, et al. 2003; Mathews & Reus, 2003; Schulze & McMahon, 2003)

This is unfortunate in that it ignores and traduces decades of classical genetic analysis in medicine and psychiatry. But even more unfortunate is that it overlooks a fundamental methodological failure—the reliance of such molecular genetic linkage studies upon a diagnostic schema that is entirely descriptive and overtly not intended to reflect any causal etiologies, notably epigenetic phenotypes. Linkage studies require valid and reliable phenotypes, not consensus diagnostic criteria for clinical use.

In particular, the issue of dimensional versus categorical traits and diagnoses remains an open question. Consequently, what may seem distinct clinical diagnoses may often spring from shared genetic risk, i.e., Bipolar I, Bipolar II and Cyclothymic

Disorders. Indeed, we have seen cases in which a single subject has been identified as the proband in two different National Institute of Mental Health family risk studies—one of schizophrenia and the other of bipolar! With such sampling errors at work, there is no need to speculate on unproven, ancillary notions that these diseases are "highly complex", much less "continuous traits". Instead, it is past time for molecular genetics to acknowledge that some of the major syndromes of psychopathology evolved as population genomic polymorphisms and identify endophenotypes more amenable to their preferred research techniques.

More realistic models have moved human genetics beyond the polemicisms of radical environmentalism or reductive geneticism. Traits and diseases are often expressions of major and minor loci arising in the developmental environs which range from the chromosomal neighborhood to the fetal milieu and on to the external ecosystem (including even mutagenic cosmic rays).

Despite considerable progress in conceptual modeling, the present situation recalls the earlier observations by Sewall Wright, who, in 1968—after some fifty years of rarefied biomathematical research—noted that the phenotype emerged from the genotype in simple terms as follows:

(1) the variation of most traits are polygenic,

(2) universal pleiotropy exists; that is, most alleles can affect many traits,

(3) but each allele has a unique profile of effects,

(4) dominance is contextual and not intrinsic; dominance of an allele can vary with environs or with its profile of different specific effects,

(5) polygenes are not merely additive effects,

(6) homology, phylogenetic or ontogenetic, evinces similar gene effects due to similar developmental needs and

(7) interactions induce discontinuous selective peaks yet continuous trait range.

RELEVANT EVOLUTIONARY THEOREMS

There are a great many principles of evolutionary genetics, some simple and others more difficult. Of these, two are especially relevant to the task of reassessing genetic diseases in light of phylogenetics. They are *prevalence above the mutation rate* and *the range of phenotypic reaction.* Fortunately, these are among the more basic tenets of evolutionary genetics. Certain truths about disease and design derive axiomatically from these two concepts.* Given their importance to evolutionary epidemiological analysis, a brief recap of these two theorems is necessary

Prevalence Above Mutation Rate

Attention to prevalence above the rate of mutation is important because ultimately mutation fuels genomic change. Few mutations are advantageous, but such spontaneous alterations in the genetic material ultimately allow for evolution. Many mutations are evolutionarily neutral given the ubiquity of genomic redundancy. Most, perhaps are deleterious and culled from the genome by natural selection. Thus, the majority of inherited abnormalities should be rare as a function, largely, of mutation rate divided by fertility. But unhealthy mutations do persist in tiny frequencies and at the margins of populations. Consequently, even a deleterious gene will equilibrate over generations at a rate balancing spontaneous mutations corrected for selective force and several other detailed factors.

Therefore, mutations, though marginal in population terms, are nevertheless

---

* Indeed, the present work was substantially improved with the focus on these two tenets as personally communicated to DRW some twenty years ago by Professors E.O. Wilson and Ernst Mayr, respectively.

responsible for much disease morbidity and mortality in individual organisms. Thus, there have evolved numerous mechanisms which maintain the integrity of the genetic material, including variations in the rate of mutation itself (Kimura, 1967; Bridges, et al. 1997). Recently it has been established that the mutation rate per genome per replication in DNA-based taxa is a natural constant near 0.0033 (the rate is much higher for RNA-based organisms; Drake, 1991, 1993). Drake's constant, on a per replication basis, implies regularity of mutation in replication as well as uniformity in the mode and tempo of genetic change.

However, renewed attention has focused on mutation processes in non-dividing cells. Bridges and colleagues (1997) notes replication *qua* replications may be overstudied at the expense of mutational events in resting cells. Indeed, important aspects of DNA synthesis occur apart from replication (Ryan, 1955; Cairnes & Foster, 1991). Thus the controversial notion of entelechy has had a bit of a revival as it is argued some mutations are directed rather than stochastic events. Still, Ripley emphasizes errors resulting from DNA damage typically undergo repair only in the course of replication.

Moreover, significant fine-structure variations exist within this general constant. So, mutation rates can range widely across genomic loci and among individuals in a given population. Further fine-structural distinctions arise between errors due to DNA mal-synthesis and errors due to damaged DNA. Bridges and colleagues (1997) noted further complexity, pointing out that any aspect of DNA metabolism can fail (cf., Petronis. 2003).

Inhomogeneity of mutation rates among cells clearly exists, at least as a fine-structural distinction. Mutator alleles are a likely source of much inhomogeneity. These usually defective genes elude DNA replication or repair. Some 1% of natural isolates of bacteria harbor inhomogeneity. Also transient mutator phenotypes may result from mistranscription of key fidelity genes (Ninio, 1991). The two types of faulty fidelity genes, *transient-phenotypic* and *established-genotypic* mutators may account for the multiple spontaneous mutations which are rapidly accumulating in human genomic research data bases (Ninio, 1991). These, in turn, may be due to mutational hotspots or other selective pressures. In addition to genotypic and phenotypic mutators, simultaneous multiple mutations in different organisms can arise as combined by sex or with complex templated fragment insertions of sequence (Wilson, 1993; Drake, 1991).

Why are mutators relatively common despite the cost associated with production of deleterious mutations? Monnat (1992) notes particular biological attributes of a system may allow mutators to be carried, tolerated and even exploited. That is, high cell loss can be useful (e.g., dermal, immune, or certain tumorous tissues). Monnat further reports many geneticists have readily isolated mutants of interest—namely, naturally occurring mutators in mice consistently arise in the frequency range of 10^{-2} and 10^{-4}. The frequency of recessive mutations due to mutator defects in mismatch-repair genes is about 0.5% in humans, thus the expected frequency of mutator homozygotes is close to 10^{-4}.

In clinical terms, then, rates of specific genetic diseases are quite low. Achondroplasia (dwarfism) is a classic example of autosomal dominant genetic disease without a known Darwinian advantage in humans. The equilibrium frequency is one in eight thousand births for this gene (Dobzhansky, 1962). The disease prevalence is also one in eight thousand. Such very low rates are typical for genetic disorders which are expressive purely of disease phenotypes. At average mutation rates (without assortive mating or fertility effect) a deleterious mutation is unlikely to achieve a prevalence of more than approximately 0.5%. Indeed, to achieve even this random accretion in the genome might take many generations—up to 10,000 (Wilson, 1975). Moreover, the modal mutation rate for genes that shape mammalian neural structures and functions is about one in 10,000 live births (Wilson, personal communication, 1986). Thus, the prevalence of non-adaptive epigenetic traits linked to neuropsychiatric disease expression should not surpass

0.01%. That they do has stunning implications to be demonstrated further on.

Here it must be said that a given allele or gene locus may for some reason be prone to exceptionally high rates of mutation. Such would be quite unusual but, since mutations rates are not known with respect to most traits, high mutability could be a factor in bringing some dysgenic traits to medical attention more than would otherwise be possible. This is perhaps only of academic interest as there is no recognized example of this possibility yet in the annals of human genetics. More typically the genome is sculpted toward a stable and relatively clear standard. Given time and stability, diversity converges upon the normative which converges upon the ideal (MacArthur, 1965).

But heredity is specific to population and environs, specifically the environment of evolutionary adaptation. Environments can and do change or at least resonate about an ecological axis of some variety. Even here, however, normative selection culls the spread of maladaptive mutations and promotes average traits in the genome. Polymorphisms are rarely maintained by mutation alone. As a consequence any phenotype archetypically expressed by the genome in its interactions with the environment was fit in an ecosystem stable enough to have interiorized such potentialities over phylogenetic time. Thus, mutation rates form a threshold of basic Malthusian significance for both evolution and epidemiology.

Range of Phenotypic Reaction

The range of phenotypic reactivity, as well as the rate of mutation, is of basic interest to epidemiology as well as evolution. The phenotype is a compromise or a net product of a variety of forces. These forces include intracellular interactions and more clearly environmental influences, but the developmental pathway is the main arena for epistasis (interaction between different genes). Natural selection will, of course, promote optimal orchestrations of traits because each gene will be favored in systems where it can accrue maximal fitness.

Genetic characteristics of species genomes are far more variable than might be apparent from superficial examination. Waddington (1957) noted how developmental processes could stabilize or otherwise modulate gene expressions. Surprisingly little variation is expressed. In fact, fluctuant genomes can and often do maintain stable phenotypic appearance (Rendel, 1967). Conversely, genomic plasticity sometimes can be stretched toward novel phenotypic expressions (Lund, 1978; Chaudhari & Hahn, 1983; Purves & Lichtman, 1980). There is, then, a range of reaction from which genotypes find expression in phenotypes. In some cases this range is rigid and brittle while in other cases it may be flexible and supple.

Stabilizing selection occurs even in highly stable environments. Variation will exist if for no reason other than inescapable meiotic events needed in sexual reproduction (such as segregation and recombination). Darwinian processes will reduce genomic deviance due to higher selection pressure on such variant phenotypes (Schmalhausen, 1949). These variants have been termed phenodeviants (Lerner, 1954). Many examples of selection against deviant phenotypes have been recorded in the field (Mayr, 1963).

A useful distinction has been made between the two main types of stabilizing selection (Waddington, 1957). A more harsh normalizing selection occurs with the complete removal from the genome of all individual phenodeviant genes. A softer canalizing type promotes epistatic processes which express a standard phenotype despite genomic or ecological variation. Obviously, these two types of stabilizing selection have divergent consequences for patterns of phenotypic adjustment.

Phenotypic reaction and the resulting range of expression are important. After all, heredity transmits to the future while remembering the past. Heredity presupposes environmental characteristics but does not control them. This fact imposes a limit on phenotypic adjustment to ecological change. Beyond this threshold, novel ecological demands force less reproductively effective genetic material out of the genome as pressure of normalizing selection rises.

The phenotype characteristically expresses little of the potential latent in the genotype. Therefore any change in existing environment, or migration to a novel environment, will, when adaptive, cause latent genotypic potential to be expressed. Put differently, it sometimes is possible to reveal otherwise hidden traits through modification of the phenotype under exceptional conditions. That is, different phenotypes may emerge from the same genes in different situations. Fluctuating environs or variations in selective advantage of latent traits, then, are certainly major agents of genomic change. Indeed, evolution often occurs when situational factors bring forth novel or unexpected phenotypes. Quite often this occurs at the periphery of a population, as in migratory segments or those contending with marked change in external conditions.

Clearly, a genotype that can express a range of phenotypes is more likely to adapt to change than a genotype that can express only a rigid phenotype. More will be said of this later on when discussing the sometimes radical phenotypic adjustments demanded of the human genome, especially its sociobiobehavioral components, with the shift from the environment of evolutionary adaptation to the contemporary situation.

7

MOLECULAR GENETICS OF MANIA AND DEPRESSION

In this chapter we turn from indirect classical studies of families, twins, and adoptees, as well as quantitative studies, to the more direct methods of linkage studies facilitated by recent and quite remarkable advances in molecular genetics. As noted in Chapter 6, biometric statistical analysis is never conclusive in establishing mode of inheritance but is useful in identifying potential targets for linkage analysis. Such analyses will allow an understanding of basic pathological processes and eventually should give rise to much in the way of precise therapeutics.

Biological marker and molecular linkage studies are important in assessing the Mendelian character of many psychopathological syndromes, notably bipolar disorder (BPD) and related affective syndromes. The promise of such studies has increased with the sequencing of the human genome. Despite a bumpy series of seeming successes followed by disappointing failures, the promise of those studies grows greater as the delineation of functional attributes progresses. Also, the optimal methodological techniques for this work are subject to considerable ongoing debate.

This research is accruing at so rapid a rate that a text such as this cannot readily maintain current information. Fortunately, there are several readily accessible online data-bases including the US National Library of Medicine *PubMed* (www.ncbi.nlm.nih.gov) and *Mendelian Inheritance Online* (www.ncbi.nlm.nih.gov/entrez/dispomim.cgi?id=125480).

The latter source explains in a helpful interim summary that the bipolar syndrome is a genetically heterogeneous complex trait. It relates in summary that loci for the disorder have been mapped to chromosomes 18p, Xq28 and 21q22. A summary discussion of other suggestive linkages to various chromosomes follows. The mechanism suggested is a mixed model that, as discussed earlier, is a composite of polygenic and heterogeneity effects. That is, the manic-depressive continuum including BPD is an oligogenic trait requiring a combination of moderate to major effects on the cited chromosomes (and perhaps several others). Beyond this important summary point made at the outset, the subject still merits examination a closer level.

THE NATURE OF LINKAGE STUDIES

Linkage studies, in contrast to traditional indirect heritability studies of twins, adoptees, and families, go directly to the genes themselves. Linkage studies analyze the segregation of etiogenic loci within families. Such analysis is possible when an identifiable marker trait lies in close chromosomal proximity to a susceptibility gene. The closer the proximity the less chance there is for separation of the loci at the point of meiotic recombination. The frequency of recombination can be used as an (inverse) index of the distance between two genes. The trait and susceptibility gene need not be causally linked. Rather, they are linked as chromosomal neighbors that molecularly cosegregate. As such they are quite predictably and in tandem passed along to offspring.

Demonstration of linkage between an etiogene and another known gene ("marker") not only proves a genetic basis for the illness in question but also specifies the chromosomal location of the etiogene. Linkage analysis may be the only means to clarify the genetic foundation or mode of inheritance of illnesses where it cannot be clearly inferred from indirect epidemiological studies.

Prior to the era of recombinant genetic technology only some 15% of the genome was amenable to analysis because of insufficient numbers of polymorphic marker loci; such as, blood groups, enzyme variants, electrophoretic protein moieties, antigens, etc., (Botstein, et al. 1980). The development of recombinant genetic technology and its subsequent application to neuropsychiatric syndromes has been rapid and promising. A major milestone was the demonstration of the existence and location of the gene for Huntington's Disease (Gusella, et al. 1983).

With the mapping of the human genome complete, a full array of neuropsychiatric disorders is now undergoing screening for genetic linkage. Unique problems, however, arise in genetic linkage studies of psychopathology (Hyman & Nestler, 1993; Donis-Keller, et al. 1987; Cowan, et al. 2002).

Heterogeneity and Problems of Interpretation

One problem is that, as seen in Chapter 5, heterogeneity may be a factor in the expression of major neuropsychiatric syndromes such as BPD, schizophrenia and Alzheimer's Disease. Heterogeneity, in general, means that there can be more than one cause for what appears to be a single trait or disorder. Genetic heterogeneity means, more specifically, that a disorder or trait may be caused by different allelic mutations.

As a consequence, linkage studies concerning psychopathology are sometimes difficult to assess. For example, linkage may be found in one pedigree and not in another. Likewise there remains the related problem of reconciling clinical diagnosis (a proto-phenotypic entity) to methods of purely genomic analysis (Risch, 1990; Risch & Botstein, 1996ab; Risch & Baron, 1982, 1986).

Therefore, the proper interpretation of subsequent studies which fail to replicate linkage may not necessarily discredit the original findings. They sometimes are most properly taken as evidence of heterogeneity (Hodgkinson, et al. 1987; Kendler 2002). Given diagnostic imprecision and the possibility of heterogeneity, large pedigree series studies are required to directly estimate linkage (Goodwin &

Jamison, 1990, 2007; Cowan, et al. 2002) to the major neuropsychiatric syndromes. Indeed, the field has just begun to focus sufficient attention in this area. Meta-analyses of genome scans of bipolar disorder have found strong susceptibility at a number of loci, notably 13q and 22q (Badner & Gershon, 2002; however, see also, Segurado, et al. 2003) and 6q and 8q (McQueen, et al. 2005).

PUTATIVE RISK FACTOR LINKAGE MARKERS

Epidemiologists have long studied the association between disease traits and control populations. Such studies which link common genetic markers with cases of clearly expressed disease will be needed to fully clarify the expression of epigenetic illnesses. With respect to psychopathological conditions it is likely there is a very long chain of events by which phenotypy is evinced from underlying processes of genetic transmission.

Gershon (et al. 1986) reminds us that, like any biological trait, a marker of psychiatric risk must be:

(1) heritable,
(2) more prevalent in those with illness than those without,
(3) identifiable in ill persons even if the illness is not active, and
(4) more prevalent—cosegregating—in affected family members.

In general practice trait markers of risk are selected via criteria one and two. Criterion two is readily available in compilations of traditional genetic studies of twins, adoptees, and families. Criterion three simply specifies that a marker be a durable trait and not merely a disease state. Criterion four is more complex but has also engendered an array of research. Associations can be established by comparing risk trait frequencies found in affected patients with frequencies found in control populations. This establishes a 2x2 table (Table 7.1) wherein *a*, *b*, *c* and *d* denote the epidemiological incidence of each possible association. From this can be determined the relative risk for an individual to have both the marker trait and the disease in question (and, via *chi*-square analysis, the statistical strength of the association).

TABLE 7.1: Marker Trait

Marker trait		***[+]***	***[–]***
Disease	***[+]***	***a***	***b***
	[–]	***c***	***d***

Associations can be due to several circumstances. Rarely, there may be a true causal connection between the marker trait and disease expression. More often a causal gene may lie quite proximal to the marker trait.

LINKAGE METHODOLOGY

Linkage methodology has long included both parametric and nonparametric approaches. Parametric methodology is the linkage method used in the search for classic Mendelian inheritance by single gene locus or monogene. In such cases the parameters are generally known. The *theta (θ) values* are reflected as the measure of linkage: 0 being perfect linkage; 0.05 being no linkage. The Lod score, on the other hand, provided the logarithm of the odds. A Lod score of 3, which indicated a chance of 1 in 1000 that the observation was false, was considered to confirm linkage.

Stimulated, at least partly, by discouraging results in psychiatry, molecular geneticists developed alternative, nonparametric methods for the elucidation of heterogeneous or multifactorial disorders which were not amenable to straightforward parametric analysis. BPD and other affective disorders fall in this latter more complex group. Nonparametric techniques limited the problems raised by presuming a Mendelian mode of transmission. Instead only shared marker alleles between affected individuals were analyzed. Related methods involved the development of more complex models for the interaction of groups of genes within the same lineage.

DIRECT MOLECULAR LINKAGE STUDIES

The promise of the future seems to rest with direct molecular linkage studies. Traditional indirect measures can, of course, provide leads for studies aimed ultimately at the identification of specific genes and their specific problem nucleotide loci.

Against a backdrop of early inconsistency and the accompanying frustration, researchers developed the more recent direct analytic methods—recombinant genetic analysis and positional cloning. Search for specific etiogenes requires both genetic and physical mapping. Genetic, or linkage, mapping determines the proximity of a disease susceptibility gene to a DNA marker at a known chromosomal locus transmitted to affected members of a family in tandem with the illness.

The development of recombinant DNA technology has enabled genomic mapping to precise chromosomal loci. It is now possible to systematically search the genome for disease genes. Once mapped by genetic methods, a disease locus can be isolated to a more specific chromosomal region, typically of five to ten million base pairs versus some three billion genome-wide base pairs. Thereafter, cloning and sequencing of large regions of DNA will physically map the disease loci as linked to one or more mutations. From this should follow a variety of insights into pathophysiology.

RECENT LINKAGE STUDIES OF BPD

The recent history of BPD linkages studies has been characterized by inconsistency. This history begins with a highly publicized study by Egeland et al. (1987) among the Old Order Amish population. Egeland and colleagues reported linkage to a locus on chromosome 11, specifically at 11p15. Other attempts to identify specific loci followed. Among these was a study by Kelsoe, et al. (1989) that also covered the Old Order Amish pedigree and seemingly confirmed the earlier findings. It produced highly suggestive Lod scores for two markers on chromosome 11. Coincident studies by two other groups in the same time period, however, reported no linkages to chromosome 11. Subsequently, two previously asymptomatic members of the Amish population examined developed serious mood disorders. A re-examination of the data led to considerably reduced Lod scores. The original linkage was subsequently abandoned.

This disappointment was followed by another. A previously reported linkage of BPD to markers on the X-chromosome had been reported by Baron, et al 1987. This linkage had been based on the study of a number of large non-Ashkenazic families in Israel. A follow-on study of other similar pedigrees, however, failed to substantiate the previous findings (Baron, et al. 1993).

A large scale study by the NIMH Genetics Initiative Bipolar Group followed in 1997. The NIMH study included 97 families with data from 232 persons with BPD, 72 with BPD II, 88 with unipolar recurrent depression, and 32 with schizoaffective/bipolar disorder. This study, based upon strict criteria, failed to confirm any of the previously reported linkages. The NIMH study found weak evidence for linkage to chromosome 18p and suggestive evidence for linkages to 6,

10, 7q, 16p, 21q21, 22q and Xq26 (NIMH Genetics Initiative Bipolar Group 1997; Cowan, et al. 2002).

More recently, a large group of collaborators from Stanford, MIT, Johns Hopkins, and Cold Spring Harbor performed a full-genome scan for linkage in 50 families segregating the BPD phenotype (Friddle, et al. 2000). This study used conventional linkage analysis supplemented by a "simultaneous-search algorithm" to get around the problem of heterogeneity within the pedigrees studied. The study failed to identify any single locus or pair of loci to account for the disorder.

Three possible explanations for the failure have been noted: (1) it could reflect the extreme heterogeneity of BPD; (2) if, as indicated by some studies (e.g., Blackwood, et al. 1996), a single dominant gene substantially contributes to the disorder, there must be a number of such major genes segregating in the population; (3) BPD may result from the combined effect of several relatively common disease alleles.

Petronis (2003) reviews the current challenges and notes that, despite the many molecular genetic linkage and association studies of BPD conducted the past two decades, data remain inconsistent and even controversial. He offers that an epigenetic perspective would greatly benefit such molecular studies. Epigenetic factors include alteration of DNA to "orchestrate" genomic activity such as regulation of gene expression. Moreover, epigenetic mechanisms may well account for various "non-Mendelian features of BPD", such as monozygotic (MZ) twin discordance, critical age range of disease susceptibility, clinical sex differences, as well as disease course fluctuations, such as phasing of mania and depression.

A partial summary of the some suggestive findings is reflected in Table 7. 2.

Table 7. 2 Partial Summary of Positive Linkage Findings in BPD

Region	***Authors***
4p16	Blackwood et al 1996; Asherson et al 1998
5q32-35	Sklar et al 2004*; Hong et al 2004*
6p24	Ginns et al 1996; Schultze et al 2004*
6q22	Schultze et al 2004*; Middleton et al 2004*; Pato et al 2004*
8q24	Cichon et al 2001*; Kanyas 2001, Avramopoulos et al 2004*
9q31-q33	Venken, et al. 2005
12q23-24	Craddock et al 1994; Dawson et al 1995a,b, ; Barden et al 1996; Ewald et al. 1998 & 2002*; Jones et al 2002; Curtis et al 2003*, Morisette et al. 1999,
13q14-32	Badenhop et al 2001*
18p11	Berrettini 1994, Stine 1995, Gershon 1996a, Nothen 1999
18q21-23	Stine et al. 1995, Freimer et al, 1996; Chen et al 2000*; Fallin et al 2004*
21q	Straub et al. 1994, Detera-Wadleigh et al. 1996, 1997, Smythe et al. 1997, Gurling 1998, Aita et al. 1999,
Xq24-27. 1	Pekkarinen et al. 1995
Xq28	with various markers, Winokur 1969, Mendlewicz 1979, 1980, 1987, Baron 1977, 1987, 1993, Del Zompo 1984, Berrittini 1990

Adapted and updated from Cowan, et al 2002, Jones, et al. 2002, and Segman & Lerer 2000.

* = new citations added to bibliography from MEDLINE UPDATE . Meta-analysis by McQueen, et al. (2005) support genomewide linkage to BP on 6q and 8q. Lin, et al. (2005) support linkage on 18p and 21q based on assessment of age at onset.

In addition to nuclear DNA linkage studies focusing on probable chromosomal loci, there has been research into the possible role of mitochondrial DNA in BPD. Several studies have indicated a decreased level of energy metabolism involved in BPD as well as other neurodegenerative disorders (e. g., Alzheimer's disease,

Huntington's disease, Parkinson's disease). Such research opens up the possibility of BPD as a metabolic disease in which energy metabolism becomes decreased possibly leading to subtle neuronal damage and cell death (Kato & Kato, 2000; Kato, 2005; Young, 2007).

The search for specific causal or predisposing genetic loci goes on despite the ambiguous, if not disappointing findings to date. In a recent review of genetic linkage and association studies in bipolar disorder, Schulze and McMahon (2003) see a time for optimism (cf., Owen, et al. 2002; Botstein & Risch, 2003).

SUMMARY: THE BIPOLAR SPECTRUM IS QUASI-MENDELIAN

The patterns of inheritance of several major syndromes of psychopathology have begun to yield, via genomic analysis, the means of their phenotypic expression (Marzuk & Barchas, 1995). Indeed, now that the human genome has been sequenced in its entirety, exhaustive searches of chromosomal loci linked to psychopathological conditions are progressing.

Yet even though genetic linkage analytic techniques will eventually specify the Mendelian status of many psychiatric illnesses—perhaps including BPD—there are presently no consistently replicable findings. However, based upon the evidence of traditional indirect methods and despite the inconsistency of direct studies thus far, a few major loci, with subsidiary multifactorial contributions, yet seem more likely than a complex threshold model. Any definitive analysis shall rely on improved diagnostic systems which correlate to epigenesis more accurately, but BPD is clearly not a continuous trait in the human genome.

Thus, the identification of specific genetic loci linked to the phenotypic expression of mental illness remains a challenging problem (McGuffin, et al. 1991, 1993). One arm of the challenge concerns the technical problems arising in the application to cases of psychopathology of laboratory procedures of molecular analysis which were originated in research concerning purely metabolic disorders. Another arm of the challenge is to clarify epistemological vagaries of psychiatric classification and diagnosis in a manner valid for robust molecular research.

For all the promise of the molecular genetic approach, the search for linkage between psychopathology and etiogenes continues. The matter is currently not yet at a definitive stage of investigation. The failure to consistently replicate disparate early findings has fed a variety of efforts to rethink how genes might govern behavior. One broadly accepted idea is that many complex mental traits (such as BPD) are actually composites. As mentioned earlier these several discrete genetic subtypes, each expressing convergent phenotypes albeit from different single major loci, can interact in relatively complex modes (Plomin, et al. 1994). Another promising idea is that some complex mental traits (such as schizophrenia) may be both multigenic and heterogeneous rather than composite. In such a case a single major loci is neither necessary nor sufficient. Rather the trait is due to the additive effects of multiple genes which interact with one another and contribute to the phenotype with widely varying aspects.

For all these ineffabilities, there obtain several important consistencies in research findings. Decades of psychiatric epidemiology based on indirect analyses point to a relatively simple basis of bipolar genetics. This, however, may be misleading. Linkage analysis currently suggests, that like schizophrenia, BPD may be both multigenic and heterogeneous. In a recent review, Berrettini (2003) has summarized evidence for shared susceptibility in BPD and schizophrenia based upon indirect family studies as well as direct linkage studies. He concludes that such studies are consistent with shared susceptibility with several regions of the genome of potential interest. Others argue there is no basis for separation of the two at all (Crowe, 1998).

Berrettini cautions, however, that since no single locus has yet been identified for either disorder or both, the question of shared susceptibility remains a hypothesis to be further tested. This is sensible as, not only have decades of clinical research rather consistently found BPD to be distinct from schizophrenia, but there are clear differences in the overall linkage patterns for the two syndromes. Indeed, a number of quite serious confounding factors of a methodological type are sufficient to account for these inconsistent overlaps in linkage studies.

Nevertheless, the risk of mood disorder in families is a function of the degree of genetic relatedness and greater within pedigrees of the ill than in the general population. The concordance rate of mood disorders in monozygotic twins is at least three times higher than in dizygotic twins. In fact, the inclusive concordance rate of BPD spectrum in monozygotic twins exceeds 90% denoting a relatively mild environmental component for phenotypic expression. Hence, the question of a rather simple genetic basis for BPD is kept open.

This is interesting in light of recent, if tentative, findings of a possible master molecular clock etiomechanism in the hypothalamic suprachiasmatic nucleus (Mitterauer, 2000; Bunney & Bunney, 2000; Benedetti, et al. 2003). This is a gene that encodes protein in the basic helix-loop-helix (bHLH) family of transcription factors. This proteomic function is highly conserved as seen in mouse proteins that regulate circadian activity. Moreover, polymorphisms of homologous proteins in humans are associated with circadian rhythm sleep disorders perhaps the core endophenotypic anomaly of bipolar illness.

The genetic relationship between bipolar and even unipolar mood disorders (much less a variety of other relevant conditions) is not well specified. However, most cases of unipolar depression in pedigrees with high bipolar prevalence may be phenotypic variants of bipolar gene expression. Women are at greater risk of unipolar depression and also transmit bipolar disorder to sons more readily than do fathers. Taken together, this suggests that a major form of the bipolar disorder composite is dominant and X-linked, with possible mother-transmitted mitochondrial involvement. Earlier age of onset correlates with heritability in BPD. Adoption studies, in toto, suggest biological relationships are far more predictive of illness than are sociocultural relationships.

Still, linkage studies are not consistent nor have specific etiogenes yet been identified. Although molecular studies are by no means a mature area of psychiatric research, these will become increasingly useful in the delineation of the genetics and phenotypic expression of mood disorders.

In final summation of current evidence concerning the genetic transmission of BPD, it must be said that the decades of indirect epidemiological analyses of a classical type have generated the most consistent and complementary observations. Even without recourse to the most recent and promising direct genomic studies, it has long been clear that the disease aggregates within kinship groups with exceedingly little transmission via the environment or social vectors (as confirmed in studies of adoptees), and that twin concordance is unquestionably higher in MZ rather than DZ twins.

For subsequent purposes in this work, it is important to note that the bipolar spectrum is: (1) a preponderantly genetic syndrome having few phenocopies and only moderately influenced by environmental factors, (2) of quasi-Mendelian, oligogenic character in which there appear to predominate a few major genetic subtypes consistent with the mixed model of heterogeneity or pleiomorphic expression, that (3) requires naturally valid endophenotypes for reliable molecular linkage research, but again, (4) so common and epigenetic that it must have arisen as a highly evolved population genomic polymorphism.

8

EVOLUTIONARY EPIDEMIOLOGY: THE SICKLE CELL ANEMIA PARADIGM

Disorders that are clinically well-validated, nosologically reliable, epigenetic and common, are amenable to analysis within the framework of evolutionary epidemiology. Indeed, any epigenetic medical or psychiatric disorder that attains prevalence higher than basic theorems of evolutionary genetics predict is of interest. By definition such disorder satisfies the two prime conditions of hypothetico-deductive inference at the heart of evolutionary epidemiology. That is, common genes can be generally inferred to have been selected on the basis of advantageous characteristics. The strengthening of such inferences is a major application of evolutionary epidemiology.

Epidemiology is, as noted, a science of disease which specifies rates (illness prevalences, incidences, distributions, etc.). Evolution is a science of life which specifies changes (gene frequencies, generations, forms, functions, etc.). Evolutionary epidemiology is a deductive synthesis of these two sciences which combines the empirical power of classical observations in genetic epidemiology with the interpretative capacities of neo-Darwinian evolutionary genetics. Novel conclusions are thus derived from primary data integrated from these two disciplines. In particular, prevalence rates of genetical disease are important data points when reformulated for the purpose of analysis in terms of their evolutionary frequencies. As discussed earlier, traits which exceed prevalences beyond frequency of mutation (in Hardy-Weinberg, biometrical calculations and even more complex mixed models) or evidence unusual range of phenotypic reaction are also of special interest.

THE *ALGEBRA* OF EVOLUTIONARY EPIDEMIOLOGICAL ANALYSIS

The integration of evolutionary (population) genetics theory with epidemiology constitutes a new sub-discipline which, in turn, extends basic tools of Darwinian genetics toward an ultimate biological explanation of some diseases as expressions of more normative biological mechanisms. From all this directly follows the hypothesis that much illness is caused by phenotypic friction arising as abiding hominoid genes react with the more recent and brisk rise of human civilization. This appears true because only traits that conferred advantages in the environment of evolutionary adaptation accede to significant rates of genomic prevalence. These deductive methods can be used to test the hypothesis that some diseases have been promoted in the course of natural selection due to intrinsic, if obscure, advantages.

The Relative Irrelevance of Modes of Transmission

The assumption may arise that evolutionary epidemiological analyses can only be used for similarly Mendelian syndromes. This is not the case. Whether pathologic predispositions are coded as Mendelian dominant, recessive or even multigene systems is not in itself crucial to deductive analysis. Traits need only be epigenetic and common to be made amenable to the analytic tools of evolutionary epidemiology.

Clearly, however, Mendelian dominance with strong penetration is the most concise mechanism of putative transmission. It is also the most amenable to deducing that positive selection was a necessary condition for prevalence beyond a certain threshold. Recessive transmission is somewhat more complex and requires higher prevalence to substantiate past positive selective advantage.

More complex polygenic models of transmission are not implausible. However, they do become increasingly cumbersome and less likely relevant in hypothetical positive evolutionary scenarios. The more genes, alleles and complex interactions, the higher the threshold of prevalence needed to establish that positive and non-random selection was responsible. For example, syndromes with complex and weak genetic roots, such as the senile diabetes or dementias, appear to have not accrued through positive evolution.

Several evolutionary scenarios by which newly diseased phenotypes may emerge can be modeled. These include (1) discrete Mendelian traits, whether dominant or heterozygotic, (2) continuous polygenic traits, and (3) more complex mixed, composite or multifactorial systems. These models all presuppose that prevalence rates in epidemiological studies are, with minor adjustments, equivalent to frequency rates for evolutionary calculations.

Therefore, *the frequency threshold which a genetic locus can surpass only via selection* equals the net *prevalence divided by equilibrium* (in the absence of non-Darwinian factors such as drift). Genes which attain such thresholds were not pathological in the environment of evolutionary adaptation unless unusual, non-Darwinian factors prevailed:

Formula 8. 1: Epidemiology as Evidence of Evolution

Epigenetic disease rates divided by equilibrium rates equal selection threshold:

(Pdg)/(Egf) = (Sgt)

(disease gene prevalence)/(gene equilibrium frequency) = (gene selection threshold)

With the redundant signification of genes deleted, this yields:

(Pd)/(Ef) = (St)

(prevalence of disease)/(equilibrium frequency) = (selection threshold)

Or more generally:

$$P/E = S$$

Of course, evolutionary mechanisms typically require eons of favorable selections for new mutations or simple epigenetic systems to obtain epidemiology prevalence at rates much above one-half percent. The precise rate of prevalence involves technical adjustments for rates of such things as mutation, back-mutation, fertility, drift, and assumes broadly stable environmental conditions. Similarly, epidemiological prevalence rates must be adjusted by such factors as the degree of identical twin concordance, phenocopy events and so on. Still, as noted above, if a simple gene system occurs to such a degree it was (though not necessarily, is) abidingly adaptive, *ipso facto* spontaneous maladaptations cannot achieve such prevalence (Wilson & Bossert, 1971; Fisher 1930).

In a classic Darwinian system, evolutionary selective value must be presumed to have existed in phylogeny at the genomic population level wherever genetic alleles of high prevalence co-vary with the manifestation of individual disease. This is the case in the absence of drift due to linkage or similar random, non-Darwinian events. An important deduction follows in consequence of this fact—any common, epigenetic pathology had a likely silver-lining to be best appreciated in the context of ancestral environs. The task of establishing a proper phylogenetic perspective, though daunting, is as central to genomic analysis as are the vantages of molecular or cellular biology.

OLD GENES, NEW ENVIRONMENTS

The deduction follows that such frequency thresholds were surpassed due to the Darwinian selection of advantageous genes over the course of evolution. Features inherent to these syndromes most certainly served, and perhaps still serve, useful functions in human society. This is not, necessarily, to claim that such genes now express entirely healthy phenotypes. Such a claim would be something of a naturalistic fallacy in reverse (Parsons, 1966). The current environments to which young humans are exposed and in which their phenotypes develop are, by and large, different from those in which much relevant human, primate, mammalian and reptilian ancestry achieved increasing optimal and evolutionarily stable strategies (Wilson, 1998).

This underscores the point that the same genes which gave advantages in the environment of evolution might, upon ontogenetic exposure to new biosocial ecology, lead to deleterious phenotypes. Developmental and social forces become important etiological factors by which the ecological cues of the modern environment *nurture* our *nature* in emergent, rather than evolved, ways. The deleterious expressions of such genes, then, are not genetic *qua* genetic but become deleterious because of gene-environment interactions.

Even a disease phenotype which is a 100% penetrant, twin concordant and Mendelian dominant retains a component of environmental expression. This component is a function of the differences between the ideal and the real phenotypes (Wilson, 1975). These two phenotypes, the ideal and the real, arise, respectively, in the environment of evolutionary adaptation (to which the genome and ideal phenotype are, at equilibrium, adapted) and the new environment of ontogenetic adjustment (in which the genome is evoked to express the real phenotype).

THE OBJECTIVES OF REANALYSIS

All this is of importance both for paleoanthropology and the full appreciation of continuing advantages, if any, of these so-called disease genes. Clinically, there must be potential, if latent, benefits of the ideal phenotype residual in genes now more typically expressive of maladapted real phenotypes. There are, of course, further ethical, legal, social and even theological consequences to be explored with respect to these latencies and potential.

But the main objective here is, not to indulge in such related concerns. The objective is rather to explain a method by which standard epidemiological data concerning genetic disease rates can be reanalyzed from an evolutionary point of view. Again, the point of such reanalysis is to establish whether such genetic systems have accumulated over the course of evolution due to the effects of Darwinian natural selection. With the business of *how* specified, attention can be turned in earnest to questions as to *who*, *when*, *where* and *why* all this happened (Mayr, 1963).

With this comprehensive view of genetic processes by which the modern extended phenotype is expressed, certain major epigenetic diseases—most notably mental illnesses—can be better appreciated as atavisms of previously adaptive hominoid biopatterns. An improved grasp of the adaptive essence of mental illnesses may well lead to their de-stigmatization while, also, giving rise to improved methods of rehabilitation.

To reveal any such Darwinian advantages, attention must first focus on how genes now expressing pathology operated in the natural environment of past evolutionary adaptation. Subsequently the phylogenetic how, why, when and what can be considered (Mayr, 1963). All this is difficult to do. Certainly, it is not easy to re-calibrate views formed in the contemporary world to the conditions which prevailed in prehistory. Clinicians, in particular, suffer an ethnocentrism of time. It is quite difficult to imagine phenotypes as they would have emerged in the environment of evolutionary adaptation when one knows little about that environment. Still, this is a crucial step before attention can then focus on whether there any salutary features of these genes are expressed in the current environment. This must be considered in any enlightened clinical practice or public policy (*viz.*, genetic counseling including selective abortion).

Clinicians, in particular often cannot separate evolutionary epidemiological insights from the interesting but separate and far more complex question as to whether genes find adaptive, maladaptive or neutral phenotypic expression in the environs of contemporary development. If genes are rather common and simple then there is little question they were positively selected. This is, however, an entirely different matter than how they are now being selected, much less when, where, why, what and how such selection took place. It seems anthropologists and geneticists commonly are inclined to something of an evolutionary perspective. Physicians rarely are. Having been taught about disease, they cannot readily reorient to the notion that certain syndromes may be more precisely regarded as stories of Darwinian success.

Of course, questions of current selection, if any, are to be answered only by long and painstaking research. Nevertheless, some of the best validated and reliably diagnosed medical and psychiatric disorders may satisfy the requisite conditions of genetic mediation and high prevalence. So, hypotheses about certain disorders can be posited wherever there exists substantive data on the genetics and epidemiology of a given disease. At this juncture the evolutionary epidemiology of sickle-cell anemia is instructive.

A Paradigm in Medicine: Sickle-cell Anemia

Sickle-cell anemia is a benchmark example in which, counterintuitive though it seems, genes expressing a pathological medical syndrome attained high population frequency due to the Darwinian selection of advantages sustained in the environment of their evolutionary adaptation.

Although simple enough to state, such a paradigm has profound consequences. The definition of disease itself becomes context-dependent and subject to significant reappraisals within an evolutionary perspective. A thorough going review of the literature will not be done here but is available elsewhere (for sickle-cell anemia see Kettlewell, 1956, 1958, 1961, 1965; for thalassaemiae see Weatherall & Clegg, 1981). What is notable is how research concerning human hemoglobin polymorphisms and related anemia syndromes is now more widely seen as the first and surely the most well known example of what is here termed evolutionary epidemiology.

Reliable and valid epidemiology when coupled with theorems of population genetics, permits a meaningful reconstruction of the evolutionary scenario by which positive selective forces in the past account for the contemporary distribution of alleles. In all of medicine, the investigations concerning sickle-cell hemoglobinopathy best exemplify the materials and methods by which evolutionary epidemiological analysis proceeds. For the present discussion, it is sufficient to briefly outline relevant research on sickle-cell anemia, *in toto*, to conclude that syndromes of anomalous hemoglobins achieved increased allelic frequency over phylogeny to counterbalance the selective pressure due to the disease burden of malaria.

With the widespread use of modern hematological screening over the past fifty years, the population features of hemoglobin anomalies has become one of the most intensively studied elements of human biochemistry (Livingstone, 1958, 1964, 1967, 1969). By the early 1950's it was clear that a variety of anomalous hemoglobins existed in high frequencies in specific human populations. These high frequencies were found almost exclusively in tropical and sub-tropical climes. Neel (1951, 1954, 1956) first noted how such high frequencies must be due to either high rates of mutation or conditions favoring heterozygotes. Ingram (1956) first demonstrated the molecular etiology of sickle-cell polymorphs.

Allison (1954a, b, c, 1964) gave early evidence such hemoglobinopathies co-varied with population areas in which malaria was endemic. Subsequent investigations (Lambotte-Legrandes, 1958; Delbrouck, 1958; Firschein, 1961; Livingstone, 1958, 1964, 1967, 1969) confirmed that sickle-cell heterozygosity is associated with improved resistance to malaria. In tandem with this hematological research has emerged an appreciation for the Darwinian consequences of this resistance. The high frequency of alleles for anomalous hemoglobins is consistent with a selective advantage to carriers which outweighs the disadvantage to homozygote recessives. It is widely assumed there exists a causal connection, however, the mechanism is not known.

This body of work on sickle-cell anemia is relevant as it underscores how contemporary illnesses may be due to genes which were advantageous in evolutionary selection. This raises the further question of how much human disease genetics may have been promoted by natural selection. The evolutionary perspective—so crucial for the full understanding of the co-evolution of hemoglobin polymorphisms and parasitical diseases endemic in the environment of evolutionary selection—may well be informative in connection with other disorders.

Other diseases genes may come to be seen not as intrinsically pathological but even as expressions of phylogenetic vigor. With this approach, a clearer picture of evolutionary adaptation is possible. A window on the past can be opened. The corpus of medical epidemiology can be readily reformulated as substantive data for reanalysis in evolutionary and paleoanthropological terms. Only recently has this corpus begun to attract the interests of pathologists in a wide variety of clinical fields (Wilson, 1992).

Of course, the vast majority of anomalous alleles in a population are either neutral or deleterious. They are not promoted in genomic frequency but are selected against or, in the case of neutral traits, ignored. It is, therefore, crucial that the incidence of a trait be high and, further, that it be clearly epigenetic to a great degree. It need not, however, be a Mendelian character although such traits are more amenable to evolutionary epidemiological analysis. Only in such circumstances of significant genetic causality and prevalence can Darwinian selection be adduced, particularly that of the heterozygotic type.

Although there are other less common yet possible non-selection causes of sustained polymorphism, heterosis is the primary mechanism in natural systems (Wilson & Bossert, 1971). Certainly a given gene locus may be (for whatever reasons) highly mutable. Phenomenal mutability, though rare, is a possible cause of high frequency for traits otherwise deleterious. Likewise, drift or other non-Darwinian phenomena must be considered.

Heterosis *ala'* sickle-cell anemia is not the sole model. It is merely the first and most simple. Much detailed knowledge was marshalled from genetics, medicine, anthropology and other fields to arrive at a sensible explanation of these anomalies. Unfortunately, the genetics of much biomedical pathology are not known in such detail. This is especially true of many major psychiatric disorders. Nonetheless, more definitive answers as to which so-called disease genes positively evolved appear at hand as the human genome is progressively mapped.

The case of sickle cell anemia demonstrates dramatically the utility of evolutionary epidemiology to the physical aspects of medicine. In the next chapter methods of evolutionary epidemiology will be applied to the socio-behavioral syndrome of manic-depressive illness.

SUMMARY AND CONCLUSIONS

Clearly, evolutionary epidemiology is a simple but powerful instrument of medical and anthropological investigation. This is based on understanding that the term *prevalence*, as used by epidemiological geneticists, is essentially equivalent to the concept of *frequency*, as used by evolutionary geneticists (Wilson, 1992).

These two concepts can be melded in a rigorous fashion to deduce a number of novel insights concerning the ultimate origin of genes now predisposing to phenotypic pathologies of body and mind. Darwinian evolution is, after all, a reflection of past inter-generational increases in the demographic representation of healthy reproductive lines and reductions in unhealthy ones. Darwinism offers a rare longitudinal perspective rather than the cross-sectional vantage more typical of biomedical research.

As noted earlier, Allison's work on sickle-cell anemia is the founding contribution to what is now emerging as a more systematic approach to diverse questions regarding the evolved basis of epidemiologically prevalent genetic illness. But sickle-cell anemia is not a universal morbidity of the human genome. It is, rather, a special gene system which arose in special environments. It is, to some extent, thought of as an idioethnic curiosity of limited medical interest: Africans have it, Laplanders do not. Moreover, it is essentially a disease of somatic, as opposed to behavioral, pathophysiology. Still, sickle-cell anemia provides the conceptual foundation on which might be built a more considerable research program (Cavalli-Sforza & Bodmer, 1976).

Though the full theoretical and practical implications of such evolutionary epidemiological analyses are only now gaining the attention of medical researchers, the work concerning sickle-cell anemia was the first demonstration that carriers of so-called disease genes were endowed with superior Darwinian fitness in certain ecosystems. The prevalence of what had been thought of as simply bad genes was inexplicably high in such ecosystems. Carriers were, it was soon learned, resistant to the ravages of malaria. The malarial organism and the human blood system had co-evolved in the environment of evolutionary adaptation. Sickle-cell anemia, however pathological it is today, was selected into the human genome. More simply and generally, genes common enough to have evolved by means of natural selection can have done so only by virtue of advantages conferred to lineages which carried such genes even if they now express a degree of phenotypic disease.

So it is that techniques of evolutionary epidemiological analysis elucidate whether genes which now express pathological phenotypes nevertheless surpass both their rates of mutation and evolutionary equilibrium and, if so, were promoted by natural selection. In some cases they appear to exceed these expected rates by orders of magnitude. Yet often such genes now seem clearly and rather tightly linked to pathological expression, creating a seeming paradox to our usual expectations based upon our knowledge about the genotypic and phenotypic aspects of genomes in changing environs. As noted earlier, selective parameters shift when an evolutionarily stable genomic strategy experiences a new or changed environment. This is why the *range of phenotypic reaction* emerges as a critical factor in selection as phenotypic capacities are stretched to adapt in the new ecology.

There is, of course, a limit to the amount of genetic stretch in individuals and populations. Less flexible genes are selected against. What was healthy can be made ill. Only such a phenomenon might explain the world-wide excess prevalence of disease genes noted in biomedical pathology. Just as sickled blood cells are no longer needed for resistance to malaria in environs with effective anti-malarial agents, culture can also marginalize the value of other somatic and behavioral gene polymorphisms. It has become harder—in complex, large-scale industrial cultures—to healthfully express some genes which were rather sublime in the more intimate kinship foraging bands and low population density of the paleolithic. The range of phenotypic reaction intrinsic to a genome cannot always or readily adapt to new developmental demands.

Within this framework, much epidemiological data can be reconsidered from an evolutionary genetics standpoint. Such reconsideration will lead to fresh implications about the ancient origins, ongoing processes and current phenotypic modulation of disease genes. There is, then, a vast amount of extant and high quality genetic epidemiological data which will yield new clinical knowledge upon consideration from an explicitly evolutionary point of view. The body of genetic epidemiology accumulated over the past century of medical studies is quite amenable to reformulation into terms appropriate for analysis by the theorems of population genetics.

Of special interest within this considerable body of primary data are the many genomic polymorphisms which evidence significant population prevalence. Any studies which synthesize evolution and epidemiology will assist in the thorough clarification of how natural selection or stochastic processes operated on these variant alleles. Such studies may not only demonstrate some clinical syndromes to be epiphenomena of highly selected phylogenetic forms, but will also, necessarily, stimulate new ideas and insights in paleoanthropology.

In any case, such a program of research will eventually expand medical understanding of the theoretical and clinical consequences of evolutionary epidemiology. Central to this expansion is the fact that several important gene systems now expressing pathological phenotypes had—in a Darwinian sense—intrinsic positive qualities that account for frequency at the high rates observed today. More immediately, several essential tasks are to demonstrate whether a variety of disease genes:

(1) accrued, via natural selection, as advantageous population polymorphisms

(2) retain a normative essence which conferred selective advantage in the past

(3) may respond to novel treatments which evoke healthy if latent phenotypies

or reduce developmental reactions etiologic of pathologic phenotypy.

9

EVOLUTIONARY EPIDEMIOLOGY: "EXCESS" PREVALENCE OF BIPOLAR DISORDER

The clinical and theoretical importance of evolutionary epidemiology extends not only to the physical aspect of medicine as demonstrated in the case of sickle cell anemia, but to the psychological realm as well. As noted previously, the body of psychiatric genetic epidemiology accumulated over the past century is large and cogent. Indeed, this indirect genetic analysis has had a long and productive role in psychiatric research (Goodwin & Jamison, 1990, 2007; Cowan, et al. 2002; Smoller & Finn, 2003). For a variety of reasons this is especially true with respect to bipolar disorder (Winokur & Clayton, 1986).

RECASTING PRIMARY DATA OF DISEASE AND NONDISEASE

This body of primary data is easily recast for reanalysis with the theorems of population genetics. Of special interest are the many genetic polymorphisms with significant population prevalence. Careful examination of these variant alleles may demonstrate that some have been sustained by the operations of natural selection rather than transient or even random events.

Indeed, evolutionary epidemiology may require a refinement in the notion of disease itself. That is, at least as far as such a label might be attached to several of the most common and genetic syndromes in medicine and especially psychiatry. Several important psychiatric conditions (BPD, sociopathy, obsessive-compulsivity, anxiety—even drug abuse and some disorders of personality) are so common and so strongly epigenetic that their epidemiological frequencies surpass even quite conservative thresholds of evolutionary selection.

Epidemiological Catchment Area Study—Toward a Linnaean Taxonomy

A recent, well-designed, pivotal program has generated indirect population genetics data highly (if unintentionally) amenable to reanalysis in the genetic terms of evolutionary epidemiology. The National Institute of Mental Health Epidemiological Catchment Area Study (US-NIMH-ECA) is an excellent example of highly reliable and valid multi-center data on mental disorders derived from a comprehensive, community-based U.S. sample (Eaton, et al. 1981). The program analyzed data derived from structured interviews utilizing highly systematic diagnostic criteria in the best descriptive tradition established by Kraepelin.

A clear example of how evolutionary epidemiological analysis belies any simplistic notions of disease status can be appreciated with the reconsideration of BPD; that is, manic-depression. As reviewed earlier, prevalence has been estimated at 0.6% to 6.0% depending upon the rigor with which the phenotype is defined. With the benchmark ECA it was noted the strictly defined diagnosis has a lifetime risk of 1.2% (Eaton, et al. 1981). More inclusive diagnostic schemes raise the frequency to some 6% (Akiskal, 1977, 1996). The former figure is widely accepted as reliable although the latter estimate may have greater genomic validity.

THE QUASI-MENDELIAN GENETIC CHARACTER OF BIPOLAR DISORDER

As reviewed in Chapter 5, twin concordance rates, adoptee studies and family risk studies have long suggested a strong genetic contribution in the manic-depressive phenotype. Even a highly conservative concordance estimate is that at least 65% of the phenotype of identical twins is due to direct genetic expression; moreover, concordance may well exceed 90%. The consensus view from non-molecular studies has emphasized the essentially Mendelian character of BPD.

Implausibility of Complex Inheritance

Apart from but related to the question of whether BPD is a quasi-Mendelian (or even a Mendelian) trait, there is growing support for the notion that the spectrum of manic-depressive phenotypy is a complex, multifactorial trait. Certainly, it is true that BPD phenotypy constitutes a complex spectrum of phenomenology in a clinical sense. It does not, however, appear to be an additive trait—as is height—that varies continuously in the population. Thus, it is certainly a population polymorphism.

This issue has been thoroughly addressed by indirect analyses of family risk, adoptees and twins reviewed in Chapter 6. Well before the era of direct genomic analyses it was clear the trait breeds true along family lines independent of much environmental effect. At the population genomic level, BPD phenotypy is not evident as an additive factor (Winokur, 1980). It is more consistent with oligogenic—quasi-Mendelian models of transmission than with continuously variable traits (Sham, 1996). This is particularly clear where the phenotypes observed by degrees of relatedness have been mapped to Mendelian ratios (Akiskal, 1995, 1996, 1997).

Further, continuous traits are evident in the population as a Guassian single modal variation, or bell-curve, whereas polymorphisms are at least bimodal (Fisher, 1930). The human genome does not exhibit an obvious bell-curve smoothly linking manic-depressive phenotypy to modal norms. Instead, a strong *j*-curve is noted. The vast majority of individuals and families are free of manic-depressive phenomena. Nevertheless, there are a few lineages which transmit manic-depressive syndromes as a robustly epigenetic trait. These lineages express clinically convergent syndromes (Akiskal, 1995, 1996). Non-familial cases are rare (Cadoret, 1978, 1979).

Darwinian Mechanisms as a Primary Factor

With the inference that manic-depressive phenotypy is broadly Mendelian in nature, questions next arise as to why these genes accrued well beyond putative mutation rate prevalence. That is, what may explain the preponderance of evidence which suggests manic-depressive genotypy is a population polymorphism with selective advantage?

Population genetics offers several possibilities to explain genomic polymorphisms. These begin with the basic Hardy-Weinberg postulate that gene frequencies are stable in the absence of selective forces. The Hardy-Weinberg postulate is useful as a basis to determine aspects of the magnitude and directionality of changes even though assumptions intrinsic to it are not realistic. More realistic extensions of this postulate have been made. The classic selection model and the balanced polymorphism model each include some elements of evolutionary change. The standard selection-mutation equilibrium model assumes mutations come and go in a stochastic manner. This is largely the realm of evolutionary error and inefficiency rather than innovation.

Harmful mutations are removed by natural selection at some equilibrium with the occurrence of new mutations. Most traits of this type are not true polymorphisms as they are rare. The presence of mutant forms near the rate of mutation-selection equilibrium is an important way by which genetic diversity is maintained. Such diverse forms may gain advantage in the event of environmental fluctuations as exaptations, but are otherwise merely unfortunate medical liabilities.

Mutation-selection is the default explanation for almost any medical condition linked to genetic etiology. Often rightly but sometimes quite wrongly, an epigenetic disease is presumed to be a non-adaptive, random mistake until proven otherwise. Generally, epigenetic medical conditions due to this driving force exist in the population at low levels approximating the mutation rate itself. Still, equilibrium explains a wide range of rare medical-genetic ailments which occur near mutation rate prevalence, e.g., many deleterious conditions such as albinism, dwarfism and the like. These syndromes are of narrow biomedical or anthropological interest. They do not cast much light on evolutionary process in any but a stochastic sense.

The balanced polymorphism model posits that differential selective advantages exist between individuals as a function of alternative alleles and zygotic condition. Here certain variant alleles are maintained in the genome at frequencies above mutation rate due to heterozygotic advantage, e.g., sickle-cell anemia. Epigenetic medical conditions due to this driving force can attain prevalence well above mutation rate frequency but, thus far, have been demonstrated for only a few traits. Apart from these examples of Darwinian selection, other deterministic or stochastic forces can account for many polymorphisms.

Neutral Selection

It has been noted that many genetic variations accrue not due to intrinsic Darwinian advantage but because they are proximal to highly advantageous genetic loci. They are, hence, far less likely to be shuffled randomly by the action of polymerases in meiosis (Hyman & Nestler, 1993). Such genetic variants quite simply defy the Mendelian principle of independent segregation.

A gene proximal to an advantageous locus can be swept along in the selective success of the heuristic trait. This accumulation is incidental but can be significant. The accumulation can be particularly significant if the selected locus is highly advantageous, the incidental locus carries little net selective load and the molecular distance separating the two is small. This has been termed *neutral* selection insofar as the incidental locus is typically neither of much benefit or liability (Kimura, 1955, 1968).

Neutral selection is difficult to delineate in the absence of detailed knowledge of molecular physiology. It is, however, less likely a causal factor in the persistence of conditions which are clearly deleterious. Were manic-depression simply a classically Mendelian single major locus effect, it might yet have accrued randomly. It is remotely possible that, were such a deleterious gene adjacent to a locus encoding for a major fetal induction (e.g., the notochord, heart or lungs), it could attain high, even universal genomic prevalence (Hundert, 1990). Currently, there is insufficient evidence to prove or disprove the role, if any, of neutral selection in the maintenance of manic-depression at a high level of world-wide population prevalence. However, no empirical evidence demonstrates that manic-depressive genotypy is sustained by neutral selective events.

Genetic Drift and Founder Effects

Non-random events sometimes produce extra-Darwinian results. The chance assortment of chromosomes at meiosis can be an important factor in trans-generational gene frequency changes. Only a fraction of all possible zygotes become mature adults, therefore, the transmission of many alleles to new generations is random. Further, not all alleles are passed on when diploid sexually reproducing parents have only a small number of offspring.

Such random effects are limited in a large population as these processes will tend toward an average. However, in a small population these effects may be both significant and rapid. Moreover, small populations can, upon adaptive radiation within new environs, account for high prevalences of genes rather independently of selection, per se. The relatively low selective force operative in an unexploited ecosystem coupled with aspects of subsequent inbreeding of the founding stock can induce non-Darwinian fluctuations in genetic frequencies. Binomial sampling errors are the most obvious examples of random genetic drift (Suzuki, et al. 1989).

Likewise in any empirical analysis, interpolation from non-probabilistic genetic samples can misrepresent the total population, sometimes quite significantly. Early medical- anthropological research often relied on exactly these non-random opportunistic or judgment samplings (Konner, 1983). Most early work on cross-cultural psychology suffers from this serious limitation, notably the Human Relation Area Files (Murdock, 1934, 1967, 1972). Such samplings render invalid any statistical analysis based on an assumption of random distributions. By way of example, the universal prevalence rate for hemoglobinopathy could not be accurately derived from samples of blood taken, however conveniently, from volunteers at a sickle-cell clinic.

Moreover, random genetic drift is not limited to species with only few offspring or diploid pro-generation. Windborne plant seeds may land on fertile or less fertile soil independent of their intrinsic traits. Indeed, genetic drift refers to any accidental events that randomly influence allele frequency. For example, earthquake, flood or fire may reduce a population drastically and indiscriminately. Thus a small number of survivors are not likely representative of the original genome. When catastrophe decimates populations the gene pool of surviving generations may differ from the population before the calamity by chance alone.

Examples are noted in a variety of species. Cheetahs have exceedingly little genomic variation due to an apparent reduction their population to only a few animals some several thousand years ago. Similarly, the northern elephant seal was so hunted that by 1890 there were fewer than twenty left. The population now numbers more than 30,000 but there is little genetic variety remaining. The score of surviving animals were more lucky than fit. A special example of genetic drift is the founder effect. It is a type of inbreeding effect in small populations. When a small group branches off a larger population to found a new stock subsequent generations do not reproduce the totality of alleles exactly. There is a randomness which accounts for some component of gene frequency change (Suzuki, et al. 1989).

It is likely that drift caused by such bottlenecks and founder effects may have significantly influenced human evolution in an entirely random manner (Campbell, 1990). Such may account for the striking dearth of blood group B markers among American Indians as is likely due to a very small founding stock which arrived more than 10,000 years ago. Persons with group B blood type either were not among the foundational stock or soon died out.

Certainly, it is wrong to consider natural selection as either the only mode or even the predominant mechanism in evolution. Indeed, one of the most important and controversial issues in population genetics is the degree to which genetic drift and natural selection determine evolutionary change. A key question is whether the immense genetic variety observed in all species is largely inconsequential to survival or reproduction. In this case drift is the main factor. On the other hand, if most gene substitutions do affect fitness, natural selection is the principle factor. Still, much indirect evidence obtains which favors drift as a major mechanism of evolution. Firstly, perhaps most persuasive, is evidence from molecular biochemistry (Harrison, et al. 1988). Studies of evolution at the molecular level have provided strong support for drift as a major mechanism of evolution.

With respect to neurospsychiatric conditions, observed mutations at the level of gene are often neutral and not subject to selection (Hyman & Nestler, 1993). Important to note, however, is the fact that drift cannot be demonstrated in a positive sense. It is impossible to prove the absence of selection. Thus, it is difficult to disprove the possibility of genetic drift, foundation effects or other randomly neutral factors as sustaining elements in the prevalence of manic-depressive genotypy. However, the manic-depressive genetic spectrum has little variation in world-wide prevalence and in no sense has the hallmarks of an idioethnic entity (Kleinman, 1988). It is thus implausible to attribute manic-depressive phenomena, universal features of the human genome, to genetic drift or related processes.

Phenocopies

Phenocopies of manic-depression acquired in the course of ontogenic development may confound genomic analyses. This is a particular problem when the environmentally mediated form is clinically homologous with the classical epigenetic syndrome. Such environmentally engendered clinical equivalents are rare as is evident from studies of twin concordance and adoption outcome. The vast majority of cases of BPD are due to familial genetic risk rather than environmental acquisition (Goodwin & Jamison, 1990, 2007).

In practical terms, manic-depressive phenocopies are limited to a small set of neuromental dysfunctions in the brain. These dysfunctions, in turn, typically arise as organic impairments in centers which modulate affective neurology (Damasio, 1979, 1994; Damasio & Benton, 1979; Damasio & Van Hoesen, 1983; Trimble, et al. 1997). Here there is a predictable complement of acquired etiogenic factors ranging from trauma, toxicity, infection, neoplasia, inflammation and the like (Robins, 1978).

However, in clinical terms organic-affective syndromes are relatively uncommon. Traumatic brain injury, specifically those involving the frontolimbic cortices, is the most frequent source of non-genetic syndromes which are readily mistaken for epigenetic manic-depression (Lishman, 1978; Hales & Yudofsky, 1987). These constitute less than one percent of the total population prevalence of manic-depressive phenomena (Trimble, et al. 1997).

Thus, syndromes acquired in the course of development are not, in evolutionary terms, a significant factor in the population prevalence of manic-depression.

Mutability

Excessive mutability is a further source of high population prevalence for deleterious genetic traits. An extreme of the mutation-selection equilibrium can prevail. Genetic loci differ as to their intrinsic propensity to mutate, largely as a function of chromosomal conformation or size. Some conformations render loci susceptible to disruption by electromagnetic radiation or other mutagenic factors such as oncoviri (Hyman & Nestler, 1993). A genetic locus which is prone to unusually frequent mutation will exist at a relatively high equilibrium rate.

It is, with the paucity of current data as to mutation rates of specific human loci, difficult to completely exclude the possibility that manic-depression is sustained at a high population prevalence due to intrinsically common etiogenic mutation. On the other hand, there is no available evidence that suggests genes etiogenic of manic-depression lie at loci prone to high rates of mutagenesis. Nor does mere mutability help account for certain advantageous endophenotypic features of the syndrome. Here again a non-Darwinian causal factor remains possible, if unlikely.

Fertility

Fertility, a demographic variable of great importance, presents techniques of evolutionary epidemiological analysis with problems similar to the issue of mutability (Fisher, 1928). Some genes are transmitted with considerable waste of reproductive effort. That is, certain traits reproduce in high number but with low efficiency. Thus, there may be relatively high numbers sustained via fertility effects alone in the absence of any selective advantage.

Fertility is not well studied in modern psychiatric populations. Quite remarkably, there are no well-designed studies specific to the issue of fertility in populations of manic-depressives. Essen-Moller (1935) and Hopkinson and Ley (1969) found manic-depressives marry and produce children at rates quite similar to the general population. Others have found reduced rates of marriage and progeny among manic-depressives. Stenstedt (1952) found male manic-depressives to have a 65.8% marriage rate and females a 62.5% rate compared with general population rates of 78.3% and 76%, respectively. Landis and Page (1938) report similar findings. Kallman (1954) reported comparisons in rates of marriage for manic-depressive probands as with their healthy sibs as one control group and general population controls as a third group. The rates of marriage were 72.5%, 92.5% and 87.5%, respectively. Probands were more often childless and also had fewer offspring overall (1.6 versus 2.5). Odegaard (1952) found affectively ill persons had a marriage rates reduced by 10% and fertility rates reduced by more than 20% as well as significantly increased mortality rates. Lewis (1933) found similar reductions also. Further, with respect to mortality, three reports identify a 50% greater death rate after the diagnosis of manic-depression with a life-expectancy reduced by some 15%.

All available data concerning fertility in manic-depressives is either inferred from old studies, studies designed to answer other questions or studies based on inadequate samples (Price, 1996). Any future studies must contend with problems unique to controlling for fertility effects in a syndrome which includes sexual indiscretion as a core feature. Such indiscretions—affairs, prostitution, incest, clandestine adoption, abortion and the like—make accurate appraisal of fertility quite difficult. Likewise, the issue of increased, if indirect fertility, as may be due to collateral reproduction (via putative kinship selection evident as *normal* proband-kin were the most fit in Kallman's 1952 study immediately above) has not been studied and poses considerable logistical difficulties. Perhaps systematic analyses of fertility in manic-depressive lineages are in progress but reliable and valid reported results are not available from well-designed, specific studies. It is, however, noteworthy that no reports of significantly enhanced fertility among manic-depressives exist.

Assortative Mating

The non-random matings of similar parental stocks is not, strictly speaking, an extra-Darwinian mechanism. It is instead simply a departure from the Mendelian and Hardy-Weinberg postulates of random breeding. Associative mating often poses interesting correlations with the distribution of normal traits in the population (Mascie-Taylor & Boldsen, 1988). However, clinicians sometimes offer assortative mating toward an explanation for the high prevalence of certain disease traits (Rosenthal, 1974; Rice, et al. 1978). In fact, the non-random breeding by similar types does not alter population frequency rates of genes, per se (Fisher, 1918, 1928, 1930).

More precisely, assortative mating can account for variations from expected ratios for some Mendelian characteristics, e.g., recessive homozygotes. This can be further complicated by sampling errors. However, assortative mating does not alter the totality of genomic alleles. Rather, as more recessives are sequestered amongst themselves, there are fewer recessives phenotypically obscured by dominants. Thus the population is partitioned or stratified to some phenotypic degree but the overall genotypic ratios are retained. When blue-eyed parents breed preferentially, there are more evidently blue-eyed children. But there are not more genes for blue eyes in the population as a whole. Recessive traits can thus achieve greater degrees of phenotypic expression without increasing in genomic proportion via assortative mating.

Affective disorders, including BPD, subsume considerable assortative mating (Goodwin & Jamison, 1990). This is a significant clinical and research fact. However it is not one which has not been exhaustively researched at this juncture. Clearly, both direct and indirect epidemiological studies must carefully control for assortative matings to maintain valid statistical power (Hyman & Nestler, 1993). Likewise, clinicians must often contend with complex pedigree dynamics with manifold risks (Jamison, 1996). However, assortative mating cannot account for the prevalence of a genetic trait well beyond its mutation rate threshold. Therefore, it is not a factor in the frequency above mutation rate for genes in the manic-depressive spectrum.

There are reasons, then, to turn from non-Darwinian causal explanations to natural and sexual selection in a search for the factors which have sustained the high frequency of genes expressive of the manic-depressive phenotypic spectrum. It is possible and, indeed, likely these genes accumulated in the environment of evolutionary adaptation due in no small part to selective advantage. This possibility can be approached by evolutionary epidemiological inference.

Likelihood of an Oligogenic Pseudodominant Effect

As noted earlier, current evidence concerning the epigenesis of BPD supports a quasi-Mendelian cause and does not support etiology due either to additive or continuous traits. In particular, decades of indirect epidemiological analyses have generated consistent and complementary observations concerning manic-depression. It is a disease which aggregates in kinship lines, for which social transmission is uncommon (as confirmed in studies of adoptees) and for which twin concordance is higher in monozygotic than dizygotic twins. More recently, the search for genes linked to bipolar disorder has moved beyond these historical, indirect methods of epidemiological analysis to methods of direct molecular analysis. Inconclusive and frustrating, these more direct methods have yet to resolve a satisfying view of the epigenesis and phenotypic spectrum of the BPD syndrome.

In a recent review, Tsuang and Faraone (2003) set the challenges for resolving the difficulties encountered thus far in finding genes for BPD. To confirm etiological heterogeneity over homogeneity would require separating bipolar probands into at least two classes having different etiologies and different pathological signatures. Since homogeneity, or a single cause, is the more parsimonious assumption, Occam's razor would require that it be the null hypothesis. To establish either homogeneity or heterogeneity implies the capacity to fully describe one or more mechanism of etiology. Although there are suggestive clues, such description is presently beyond our capability. Confounding the problem further is the fact that different etiologies can lead to convergent phenotypes.

Estimated Index of Selection for Manic-Depression

With these "first principles" clearly set out, attention can usefully turn to forms of demonstration in which the value of syllogistic reasoning is contingent upon the reliability of its constituent premises (Bacon, 1620; Descartes, 1641). With this, alternatives to received notions can emerge in the face of compelling inferences.

The manic-depressive phenotype is strongly genetic in a fundamentally Mendelian sense and very common at the population level. By the most conservative estimates its degree of twin concordance of at least 65% with prevalence is at least 0.5% of the human genome world-wide. *Taken together, these facts lead to the highly conservative deduction that bipolar disorder is some three hundred times more common than might be calculated from Hardy-Weinberg equilibrium predictions alone*:

CALCULATION 9.1: Minimal Estimate of Quasi-Mendelian Selection:

$$(P/E = S) = [0.325]/[0.001] = 325$$

(where 'P' is prevalence [adjusted by concordance, oligozygosity, *etc.*, O.5 x 65%)] and where 'E' is the frequency [known or estimated by Hardy-Weinberg law] then 'S' is the coefficient of natural selection). N.B.: Oligozygosity [Z] reduces the selection index by a factor of S/Z where Z is the total constituent loci.

Moreover, with the higher prevalence figures emerging from the ECA (1.2%) and with an extended clinical spectrum (and the higher rates of twin concordance this such a spectrum denotes, i.e., 90%), a more realistic estimate is that the manic-depressive phenotype is beyond one thousand times more common than it would be in the absence of the sustained operations of natural selection:

CALCULATION 9.2: Moderate Estimate of Quasi-Mendelian Selection:

(P/E = S) = [1.2]/[0.001] = 1200

If a broader range of psychiatric conditions are, indeed, phenotypic variants of manic-depressive genotypy then frequency may be as high as six percent. This, with a high concordance rate, would sustain a genetic prevalence some six thousand times mutation rate levels alone:

CALCULATION 9.3: Maximal Estimate of Quasi-Mendelian Selection:

(P/E = S) + [6]/[0.001] = 6000

Certainly, the epigenetic transmission of manic-depression may occur in a non-Mendelian mode. Still, the frequency of the genetic factor appears to surpass that attainable by stochastic events alone. This is particularly persuasive given the complete absence of evidence that the syndrome is a continuous genetic trait in the human population. In any case, these genes so greatly exceed their predicted rate prevalence as to perhaps be best explained with reference to their promotion in the operations of natural selection sustained in the environment of evolutionary adaptation (Wilson, 1992).

SUMMARY

This chapter has examined the issue of the excess prevalence of the manic-depressive syndrome in the light of theorems from population genetics. Apart from examples of the apparent influence of primary Darwinism mechanisms of mutation and selection, we have examined the possible roles of other deterministic or stochastic processes that may possibly account, in part, for the excess prevalence of BPD polymorphisms. Such processes included neutral selection, genetic drift and founder effects, phenocopies, mutability, fertility, and assortative mating.

At this point on available evidence, bipolar disorder may yet be most plausibly considered:

(1) a preponderantly genetic syndrome having few phenocopies and only moderately influenced by environmental factors and,

(2) of substantially Mendelian or quasi Mendelian character in which heterogeneity remains a confounding factor.

Most manic-depressive phenotypy is not due to polygenic transmission, highly heterogeneous expression, or phenocopy. Instead as few as three fixed loci appear to be inherited together as a core and functionally dominant pseudomonogene (McKusick, 1990).

PART III

BIPOLAR DISORDER AND OTHER NEUROPSYCHIATRIC SYNDROMES

10

TOWARD AN EVOLUTIONARY ANTHROPOLOGY OF BIPOLAR "DISORDER"

In this chapter we return from our foray into the technical aspects of genetics and population biology to the theme of Part I. The MacLeanian perspective developed in Chapter 1-4 provides the foundation for a definitive evolutionary epidemiology. It also provides a context by which much of psychopathology and evolutionary anthropology can be merged to create the formal subdiscipline of evolutionary epidemiology.

The insights gained from the neuroethological perspective, examining both brain and behavior, may be articulated at varying levels of integration and reduction. That is, at one level our individually-applied psychotherapy may reductively examine the detail of personal circumstance. At another level we may stand back from specific detail and examine principles at higher levels of integration.

Multiple levels are useful for full appreciation not only of the etiology and symptomatology of bipolar disorder (BPD), but especially its many advantageous phenotypic traits. This also applies to some disorders. The Ego-Empathy, escalation/de-escalation nexus is especially useful for higher level of integration. It serves as an umbrella beneath which increasingly detailed reductive analysis can be traced yet remain incorporated within larger species-wide and even higher frameworks. Much of the research reviewed in this and following chapters serves to illuminate alternative viewpoints and levels of integration.

RECONTEXTUALING PSYCHOPATHOLOGY AND THE NEUROETHOLOGICAL PERSPECTIVE

Proceeding from the neuroethological perspective, we will examine the recent literature on social competition and mood with special relevance to BPD. Social competition is, of course, fundamental to the CSN Complex (after all, the full name is the *conflict systems neurobehavioral complex*) and mood is captured, at least in part, by the escalation/de-escalation features of the model. This chapter and those following will attempt to review and pull together much of the recent relevant work.

The present effort is in keeping with reappraisals of many other strands of psychopathology. A general theme is to appreciate these as variations from more phylogenetically stable neuromentalities. Another more specific theme is that these only now yield more pathophenotypic expression in the contemporary environments of individual development (Wilson, 1988; Gilbert, 1989; Stevens & Price, 1996; McGuire & Troisi, 1998).

Evolution has been advanced in the explication of an increasing range of psychopathology and distress, including mood disorders (Bowlby, 1980; Beck, 1974a, b, 1983, 1987; Nesse, 1990; Gilbert, 1992; Price, 1967, 1972; Price, et al. 1994); social anxiety (Beck, et al. 1985; Ohman, 1986; Trower & Gilbert, 1989); phobias (Marks, 1987; Marks & Nesse, 1994); neuroses (Slavin & Kriegman, 1992), diverse anxiety syndromes (Marks, 1987); substance abuse (Wilson, 1992); violence (Bailey, 1987); both child abuse and homicide (Daly & Wilson, 1988; Wilson, 1992), as well as pedophilia and rape (Ellis, et al.1992).

It is also noteworthy that this recontextualizing has been taken up by writers from various schools, among them: cognitive-emotive (Beck, et al. 1985), psychobiological (Glantz & Pearce, 1989), and psychoanalytic (Nesse, 1990; Badcock, 1992; Slavin & Kriegman, 1992). Likewise, it has been argued that much psychopathology reflects adaptive responses within specific social contexts rather than abnormality, per se (Wakefield, 1992). Ultimately, evolution has been recommended as the necessary basis for wholesale revision of both the taxonomy and theory of psychopathology (Wilson, 1992, 1993, 1994; Lilionfeld & Marino, 1995).

A parallel theme, gathering strength in the past generation, is the perceived need to develop more integrated, biopsychosocial models of psychopathology which detail how social competition influences psychological and biological processes and which reflect evolved social behavior, (Spuhler, 1959; Chance, 1967, 1980, 1988; Chagnon, 1968, 1988; Tiger, 1975; Durham, 1976; Eisenberg, 1986; Hinde, 1989; Cronin, 1991; Cacioppo & Berntson, 1992; Gilbert, et al. 1995). With this, the arena of sociosexual competition becomes a crucial component by which sociobehavioral phylogeny and ontogeny can be traced in terms of both the normative and aberrant (Ellis, 1926; Ford, 1945; Ford & Beach, 1951; Ellis & Abarbanel, 1961; Caspari, 1972; Symons, 1979; Clutton-Brock, 1988).

BPD AND THE EVOLUTION OF SOCIAL COMPETITION

The phenotypic spectrum of manic-depressive behavior, particularly, has demonstrable evolutionary roots in the phylogeny of reptilo-mammalian social competition (MacLean, 1990; Price, 1993). Indeed, the very evolution of the manic-depressive behavioral repertoire depends greatly on the fact that archetypal socially competitive behavior is biologically ritualized (Price & Sloman, 1984).

The Importance of Mood

Mood is a significant component in social competition. The raising of mood—and all this entails—provides the capacity to acquire socially allocated resources. On the other hand, the lowering of mood induces the incapacity for such acquisition. Significantly, however, lowering of mood also serves to protect the yielder from physical attack, injury or death.

Such capacities and incapacities modulate social competition and foster adjustment to the new equilibria of the mammalian milieu. Thus, mood substantially co-varies with social rank and behavior in species with stratified societal structures. By extension, mood co-varies rather directly with ultimate reproductive success (Nesse, 1991; Stevens & Price, 1996).

This is not surprising, of course, as the brain underlies both behavior and mood. It is, after all, that organ which adjusts the organism to its environment in the most total sense including, specifically, the social dimension.

Social Self-evaluation

Social accommodation has been accomplished in phylogeny through increasingly complex arrays of evaluative capacity. Thus, abnormalities of self-assessment are an underlying feature of mental disease. Phenotypes which evoke evolved responses (e.g., anxiety, depression, paranoia) rather directly index the individual's social appraisal and—rightly or not—net self-esteem (Gilbert, 1992; Wilson, 1996; Price, 1996).

Put differently, critical brain structures and mental processes have evolved to analyze social interactions and attitudes for possible advantageous or even survival-dependent algorithms (MacLean, 1990; Cosmides & Tooby, 1992). Beck (1974ab) further emphasizes how these calculations may be influenced by and derived from events both in the world at large as well as the internal dynamics of an individual. Such functions are surely modified by sociocultural processes but there is a greater aspect that is species-typical and phylogenetically based (Weitz, 1977, 1979).

Further, the degree of safety provided or reduced varies with dimensions, individuals, and cultures. Therefore, the neural apparatus to allocate mood for both self-evaluation and interaction with others is a key and delicate aspect of psychopathological phenotypy.

This may be a useful avenue to further investigate the evolutionary mechanisms of dysfunction that arise in psychopathology, particularly those due to mismatch in novel environs. This is in part because the transmission to the self of environmental, largely social, information can significantly influence individual development (Eccles, 1951; Popper & Eccles, 1977; Hundert, 1989; Gilbert, 1992). Pathophysiologies in the neural modulation of self-assessment via mood thus have been linked, both theoretically and clinically, to manic-depressive behavior (Price, 1995, 1996; Gilbert, 1992).

Social feedback closely correlates with evolutionary consequences—food, care, protection, status and reproduction. Evolved phenomena of biosocial care, support, protection, safety, status, prestige and the avoidance of isolation, defeat, humiliation, attack become increasing complex in the *scala natura* (MacLean, 1990; Stevens & Price, 1996; cf: Thompson, 1942). Still, significant neural changes often ensue when biosocial goals are impeded. Indeed, disruptions are all the more likely given the increasing complexity in the arrangement of neural and mental structures—it is easier to tune a Volkswagen than a Ferrari, but high performance often requires high maintenance.

Social Communication for Approach and Avoidance

In the evolution of complex biosocial interaction, social communication becomes especially relevant. The dimensions of threat, intimacy, social and public distancing are among the most central concerns of psychopathology. Gardner (1988) emphasizes how much psychopathology is a disorder of communication. He argues all intraspecific communication is reduced to two basic messages as conspecifics *approach* (link with) or *avoid* (space from) one another. This emphasis of an essential, inevitable and usefully dichotomous anthropomorphism echoes other primary (Asquith, 1984) and interpretative reports (Stevens & Price, 1996).

Thus psychopathology both results from and expresses details of social spacing (in territorial as well as social species). Kegan (1982) sensibly relates this to the need of a growing child to balance drives toward autonomy against basic dependency concerns. Spacing is a form of proximity-controlling behavior, the dimensions of which determine under what conditions interactions register with respect to conspecifics as encroachment, violation, invasion or containment. For example, Beck (1983: 274) delineates two basic types of depression: the autonomous depressive who spaces himself and the "sociotropic" depressive who yearns for intimacy. Such observations are suggestive of the self-preservation/affectional dynamic and reminiscent of Angyal's (1958) concepts of autonomy and homonomy (relatedness), Buber's (1958) I and Thou, Tillich's (1950) courage to be and courage to be a part, as well as other duality concepts.

More generally, psychopathology involves attributions or comparisons of how one is appraised or entreated by others (e.g., love, abandonment, jealousy, etc.). As such, psychopathological conditions either subserve evolutionary strategies or otherwise have (possibly negative) evolutionary results. With such evolutionary significance, neuromental capacities have grown to better assess interpersonal motivations and behavior and so adjust individuals to the social milieu (Gardner 1988, 1995).

The competent interpretation of the behavior of conspecifics—affective or intellective inferences from facial expression, posture, intention movements, and so on—confers strong evolutionarily advantages (Darwin, 1872; Romanes, 1884, 1896; Chance, 1974; Wilson, 1975; Damasio & Van Hoesen, 1983; Damasio, 1994).

Moreover, one's ability to conceal intention and thereby to shroud personal intentions with respect to biosocial goals can be a source of advantage (Machiavelli, 1531; Trivers, 1985; Byrne & Whiten, 1988; Cosmides & Tooby, 1992). Further, self-deception prevents intentions and motivations from entering full consciousness; that is, the liar who believes himself is more convincing to others (Trivers, 1985).

Social Influence

A most potent advantage in intraspecies competition is the display of social influence—the power to influence others (Chance 1988; Hawkes, 1990). This occurs largely through the control of social attention (Gilbert, 1988, 1992). Communications which fail to influence others are poor social tactics. As noted earlier, in humans social attention display is culturally embellished but remains rooted in deeper neuromental phylogenies (MacLean, 1990).

These phylogenies are evident in innate algorithms of social reasoning which have been likened to accountancy ledger sheets (Tooby & Cosmides, 1990a, b). Unbalanced accounting in such social exchange accounts, manifested at the level of global state algorithms in the Ego/Empathy complex, is expressed as behavioral tension or social stress. Cory (1999, 2004) has argued that such interpersonal tension and stress provide the motive power of social exchange—from sharing among primitive hominid kinship groups to the gift economy of non market societies—extending even to the market transactions of modern commercial societies.

From the reductive standpoint, Gilbert (1984, 1988, 1992) makes the clinical-theoretical point that this accountancy model provides a basis for an improved model of psychopathology. We can use it, thusly, to examine the unbalanced or balanced ledgers—that is, balanced or unbalanced reciprocity—among individuals. At a higher integrative level, of course, all social competition, social communication, and social exchange are facets of the same intrinsic human reciprocal sociality that begins with mother-infant affectional exchange (Carter & Keverne 2002). Glantz and Pearce (1989) likewise note how neuroses are phylogenetically more rational than they may appear in individual ontogenic terms alone. Moreover, such an approach takes the view that social behavior has evolved over millions of years and, further, that it can be best understood in terms of inclusive fitness and cost-benefit algorithms which direct evolutionarily stable patterns of affective-behavioral response (Maynard Smith, 1972).

Phenotypic Expression and the Changing Environment

Of course no phylogenetic advance is forever adaptive. Phenotypic expression is dependent on the environment of ontogeny. This is inevitably fluid. Naturally selected benefits can quickly become liabilities with environmental change. Industrial melanism of peppered moths is a classic example of range of phenotypic reaction (cf., Ch. 6). Yet the environment of evolutionary adaptation is still, generally, the appropriate context for phenotypic development.

However, the more an ontogenic situation varies from the environment of evolutionary adaptation the more likely there will be higher selection index with more developmental failures. This is true of both psyche and soma, but perhaps more crucial to behavioral repertoires than physical formation. As Gilbert (1992) notes, the context-inappropriate expressions of phylogenetically interiorized developmental trends constitutes an ultimate cause of both normative phenomena, such as migratory or hibernation states, as well as pathological conditions, such as neuroses.

It remains unclear how far the acculturation of social behaviors can stray from the parameters laid down phylogenetically (Hallipike, 1984; Fox, 1986; Bailey,

1996). Culture is itself an emergent aspect of phenotypic development which is now different from conditions in the environment of evolutionary adaptation (Spuhler, 1959). In brief example, however, people in literate cultures may gather status based on the fluency of their writing rather than via raw physical power. Consciousness and self-consciousness allow greater modulations of response patterning in the course of development as nature is nurtured (Hundert, 1989). Individuals can make conscious, even free, decisions, as to social behavior. Thus, learning history, societal attitudes, and cultural context can shape phenotypy perhaps beyond limits of accrued selectionism (Paglia, 1990, 1994).

In this vein Fox (1986) has noted:

> ... sociobiology speaks of organisms "maximising reproductive success" . . . They will strive to accrue resources of all kinds—wealth, power, access to sex, and normally reproductive success (inclusive fitness) will follow . . . They are not in this argument motivated to maximize reproductive success *per se*, but those things that will, in the normal course of events, lead to it. (p. 193).

Fox then raises the crucial issue of how individuals successful in the contemporary cultural arena may not be so in reproductive terms. Thus the basic biosocial goals are not directly reducible to inclusive fitness, but only those activities which correlate with such fitness in species-typical environs.

Gilbert (1992) carries this forward in clinical terms noting the dearth of patients with a chief complaint of unmaximized genetic potential. Instead, patients are preoccupied with proximal aims, e.g., the avoidance of humiliation or defeat, the accumulation of power or nurturance from others—all expressions of the Ego-Empathy complex spectrum set out by the authors in Part I. Serious impediments to these goals are, of course, typical sources of stress which, in turn, can cause disturbances in neuromental functioning. Disturbances produced by the stressful thwarting of biosocial goals can directly reduce survival through increased disease, debility and predation (Murray & Lopez, 1996).

It appears that impaired satisfaction of biosocial goals is the force driving much human pathology and, especially, psychopathology. Indeed, the thwarting of biosocial goals generally, not reproduction specifically, is proximally etiogenic of neuromental disorder. Otherwise, infertility would be among the most common and debilitating of life events which, it must be said empirically, it is not (Paykel, 1978, 1982).

PHYLOGENETIC SUCCESSION IN THE BRAIN: CHANNELS FOR AGGRESSION AND AFFILIATION

The brain evolved as a system with the primal purpose of making survival sense of the environment and, thereafter, increasingly to facilitate adaptive behavior largely via interpersonal signals. In higher elaborations the brain has become a meaning-making organ (Jaspers, 1913; Papez, 1937; Sperry, 1964; Sarnat & Netsky, 1981; Gilbert, 1988; MacLean, 1990; Hundert, 1990; Mithen, 1996). In other respects, the brain is an algorithmic information processor which integrates and imbues meaning upon stimuli. This integration and interpretation has been directed by selection, including inclusive fitness dynamics. Therefore, the basic neuromental apparatus of the brain recapitulates competencies.

It is useful to note the brain and its nervous system traces its phylogenetic origins back at least some 500 million years in the dorsal ectodermal tissues of marine organisms and perhaps to the much earlier emergence of bilaterians (Reichert, 2005). This basic design is retained in humans and is the template for most animate life forms. Approximately 300 million years ago terrestrial reptiles derived from the reptilian epoch that much of the more deeply canalized biobehavioral and proto-social neuromentalities were evolved. These have to do fundamentally with strategies for the control of territory. This can be appreciated better if one takes behavior in the perspective of how greatly it has evolved by natural selection. Such a perspective makes relevant empirical studies of animals as well as the application of theoretical elements of behavioral modeling (Price, 1992).

THE EGOISTIC, SELF-PRESEVATIONAL SOCIAL REPERTOIRE OF REPTILES

For some 300 million years most vertebrates have been manipulating each other's behavior, self-esteem and mood by means of signals (Parker, 1984). So, then the study of our common ancestry with contemporary reptiles can illuminate important aspects of human behavior quite directly (Carpenter, 1978). Reptiles evidence little affiliative, parental or pair-bonding behavior. Apart from mating, their entire social repertoire largely reduces to the evaluation and control of self-esteem.

Most work on reptiles has focused on lizards. The Rainbow lizard, *Agama agama* is especially well studied (Harris, 1964). Male lizards emit four basic signals. Two of these are overt self-expressions: "I am strong" or "I am weak." As most reptiles are deaf, these social signals can be plainly seen as in gait, posture and skin color. The other two types of signal convey the results of assessments between rivals: "I am stronger than you" or "I am weaker than you."

The form of these messages varies between species. The stronger often ritualistically butts heads or hits sideways with the tail, whereas the weaker often blushes or flees. In whatever form, most vertebrate brains thus have an analogue of self-esteem with some capacity to be raised or lowered by signals from conspecifics. This has been identified in virtually all terrestrial vertebrates including humans (Eibl-Eibesfeldt, 1975).

In dealing with such signals we rely on an apparatus which is phylogenetically quite old and deeply rooted in those elements of our genome which organize basic behavior. Of course, these primitive signals have become greatly elaborated as mammalian and primate neural capacities subsequently evolved (Bucy & Kluver, 1955). Such elaborations have come to overlay not replace earlier features. Somber reptilian algorithms reside beneath the warmer mantle of mammalian parentalism, pair-bonding, or affiliation (Eisenberg, 1981; MacLean, 1985, 1990).

Therefore, a thorough conceptualization of human socio-behavioral phenomena accounts not only for reptilian origins but ongoing retentions with modifications as well. These retentions are modified due to a need for their subsequent integration with novel mammalian and primate capacities of the later limbic and neo-cortical structures. Innovative capacities are needed to channel aggression toward higher-level sociality with increasing nuance.

INTERSPECIFIC VS. INTRASPECIFIC AGGRESSION

Aggression is behavior wherein one animal assaults or attacks another. Aggression is of two types: interspecific and intraspecific. Interspecific aggression is that conflict which arises between members of different species; e.g., predation wherein the loser is consumed by the winner and aggression for defensive purposes or toward competitors for resources such as food, water or mates. Interspecific aggression is a component of sustenance behavior for feeding and maintenance; it rarely derives from emotions such as anger.

Intraspecific aggression is more directly significant in Darwinian terms as members of the same species attack. Intraspecific aggression is evident in all vertebrates. Fishes nip and clench jaws with conspecifics, reptiles wrestle tails, birds harry one another, rats paw and bite, bulls, sheep, and goats butt heads.

Intraspecific aggression evolved because conspecifics are often in direct competition for resources (food, mates, and habitats). It is channeled largely by the assessment of relative risks and potential benefits in possible aggressive encounters. It is true that some animals will fight to the death for resources. This is the case almost exclusively when defeat is tantamount to genetic suicide, e.g., male sea elephants will be killed before ceding a breeding harem. More commonly there is a sub-lethal show of force after which the lesser foe retreats; e.g., male elk engage their antlers until one becomes fatigued. Such tactical retreat may be rewarded in later mating seasons.

Significantly, most intraspecific aggression does not engender bodily harm, e.g., antlers are shed at the end of rut and do not serve as weapons with which to stab or maim rivals. Evolutionary mechanisms restrain the morbidity and mortality of intraspecific aggression. Such conflict is wasteful of reproductive potential. Territoriality is a key mode by which intraspecific aggression is epigenetically minimized. Thereby, conflicts are limited mostly to border skirmishes. Another epigenetic means of limiting wasteful conflict within species is with constraints on violence. This is made possible through the ritualization of agonistic behavior. By way of illustration, contesting vipers wrestle bodily but do not use their fangs while apes deter rivals by shaking trees or gesturing dramatically without often causing bodily harm.

MacLean (1949, 1985, 1990), elaborating on Papez' earlier idea (1937), provided what has become the classic account of reptilian behavior. This account is classic as both a general explanation for the channeling of intraspecific aggression and with special relevance to inclusive fitness among other selection factors essential to later mammalian developments. Ritualized power display is the hallmark reptilian repertoire of intraspecific aggression. These competitive displays determine the allocation of resources among conspecifics.

Dominance is the mechanism for hegemony over preferred nests, foods, sexual partners, and all manner of other opportunities for survival and differential reproductive success. Thus was stabilized in evolution the marking and enforcement of territorial boundaries, ritualized challenge and threat displays, protection of mates and offspring, etc.

Again, territoriality is a proximal means by which animals control a variety of resources. These resources in turn relate to ultimately reproductive success. Typically males compete for territories by performing ritualized agonistic behaviors as tests of strength. Losers may cede territory or be forced to less desirable niches with fewer or poorer food sources, nesting sites, and mating opportunities. Aggressively morbid, even mortal, battles can occur but these are limited as they are quite wasteful of reproductive effort. Instead, territory is enforced principally via the ritualized display of likely competitive advantage.

Such ritualized agonistic behavior is advantageous because likely winners avoid occasional chance debilitation while the presumptive loser evades thorough physical defeat. Moreover, a wounded or weary victor may not well-mobilize for the next challenger or may become prey for other species. Clear indicators of acceptance of defeat that terminate agonic encounters before injury occurs is in the interest of both contestants. For example, upon submission cichlid fish retract fins; dogs expose soft underbellies; and gulls present their unguarded necks. Such signals of acceptance of defeat halt further aggression.

These various and phylogenetically old behavioral repertoires can manifest, more often atavistically than adaptively, in creatures with more advanced neuromental capabilities, including humans. This phylogenetic regression is developed fully by Bailey (1987, 2002) and discussed in Chapter 2 of this volume. Again, phylogenetically old adaptations are typically modified or added to before they are superseded in evolution but they are still retained to be operationally released in certain circumstances (e.g., see also Thompson, 1942). This is evident in both somatic and behavioral dimensions. The release of motoric activity mediated by the extra-pyramidal tract (e.g., Parkinson's Disease) is an example of the somatic type and Morris (1977) elucidates diverse variations of the behavioral type.

As discussed previously in Chapter 2, MacLean (1985, 1990), posits that the human brain can be divided into three basic assemblies, leaving aside the simplest vegetative aspects. The oldest component is constituted of neuromentalities largely fixed in the course of reptilian evolution. These capacities derive from functions assembled largely in the structures of the hypothalamus and midbrain and is termed the R-complex. Over this midbrain array the early mammalian (limbic) and later mammalian (neocortical) lobes grew.

The R-complex is the locus of behavioral algorithms directed to the needs of reptilian reproductive strategies, a mostly individualistic and competitive array of biosocial repertoires. The R-complex also has direct and powerful efferents to centers of attention and arousal which it can recruit in a variety of circumstances. As such, the R-complex is the likely source of many atavistic behaviors which higher animals, including humans, typically evidence when primitive threats are perceived. There is a strong reptilian residue in the interpretative mechanisms by which mammals, primates, and hominoids organize their behavior (Brown & Schafer, 1888; Goltz, 1892; Sherrington, 1906; Cannon, 1939; Dusser de Barenne, 1920; Kluver & Bucy, 1937; Bard & Mountcastle, 1948; Hayek, 1952; Pribram, 1958; Bucy & Kluver, 1955; Brady, 1958; Harlow, 1958, 1965). The clarification of this residue is an aim of the paleopathology of behavior quite as Bailey (1987: 63) notes:

> We must acknowledge our species possesses the neural and motivational-emotional proclivities . . . of our reptilian ancestors, and, thus our drives, feelings, fantasies and thoughts are thoroughly conditioned by emanations from the R-complex. Reptilian carry-overs provide automatic, compulsive urgency to much of human behaviour, where free-will steps aside and persons act as they must, often despising themselves . . . for their

prejudices, hatreds, compulsions, conformity, deceptiveness and guile.

The hypothalamic, midbrain and paleocortical assemblies of the R-complex are intimately involved in autonomic adaptive behavior of both parasympathetic and sympathetic types. The R-complex is, likewise, implicated in the establishment of some learned behaviors and related phenomena of reinforcement.

Certainly the neural advancements which characterize mammalian evolution, i.e., the elaborated limbic lobes and neocortex, have served to inhibit both the structures and functions of this reptilian residue. Thus evolution was spurred by inclusive fitness strategies increasingly consequential to parallel growth of mammalian sociality, such as live birth, kindred bands, etc.

Functional tensions can arise between the R-complex and upper cortical operations. Typically, cognitive capacities effectively modulate more primitive reptilian drives. The R-complex can, however, prevail either adaptively or, more often, maladaptively as discussed in Chapter 2. The maladaptive release of reptilian repertoires is usefully termed phylogenetic regression (Bailey, 1987, 2002). Also, this notion was appreciated even a century ago as in the work of the founder of forensic psychiatry, Krafft-Ebbing (1880), who emphatically detailed the "atavistic and instinctive nature" of many crimes of perversion, notably those with a sexual dimension.

THE NURTURING, EMPATHETIC REPERTOIRE OF MAMMALS

Mammalian innovations evolved with live birth and thermoregulation modulated by the combined and increasingly complex resources of the limbic system as well as the pituitary-adrenal axis. Maternal behavior is, of course, one of the key aspects of mammalian evolution, involving parturition, nest-building, suckling, and all manner of protection of the young. From the mother-offspring bond grew the vast array of mammalian sociality and kinship dynamics (Wilson, 1994; Carter & Keverne, 2002).

Part and parcel of these innovations was the emergence of new behavioral repertoires to elicit or mete out care in mammalian social kindred bands. Apart from providing inhibitory regulation of the R-complex and rudimentary prosocial tendencies, the limbic system facilitates playfulness among the young. Until the cognitive-emotive foundations for prosociality emerge circa age two, youngsters remain concrete and highly egocentric (Piaget, 1932, 1971; Kagan, 1982, 1984).

Play—a sophisticated form of co-operative behavior—cannot be formulated in the absence of an awareness of both self and others, that is, Ego and Empathy. Such play greatly shapes the neural networking laid down during brain myelinization. It, therefore, also unfolds amid critical phases of development in which sociocultural imprinting must occur (Lorenz, 1981; Sarnat & Netsky, 1981; MacLean, 1990). There is, further, evidence that paleocortical structures of the limbic system have potent effects on complex perception and memory. It is with such mature neuromental capacities that developing organisms come to balance aggressive individualistic competitiveness with more social behavior, such as the reciprocal interplay of the CSN Complex and features of the hedonic social mode.

Clearly, inhibitory mechanisms are not perfectly formed in phenotypic development. Certainly sibling rivalry, as for parental attention, does arise—as do the persistent, if less raw, drives for social dominance (Badcock, 1992). Still, biosocial maturation among mammals elicits phenocompetencies which regulate overt behavior. Awareness of the social consequences of action can substantially direct and modify the development of behavior—whether aggressive, care-eliciting or care-giving. Ontogenic influences of both a positive and negative type sculpt the adult pattern of social motivations, anxieties and stratagems.

NEOCORTICAL CAPACITIES OF RATIONALITY AND DYNAMIC BALANCE

The final component of the triune brain is the neo-mammalian cortex—the executive circuitry—with an enormous and well-differentiated capacity for symbolic representation, planning, foresight and linguistic communication (Penfield, 1938, Pribram, 1958; Deacon, 1983; MacLean, 1990; Damasio & Van Hoesen, 1983; Damasio, 1994; Fuster 1997). Also crucially, the neocortex led to the essentially egalitarian social behavior so characteristic of the early foraging kinship bands of our evolutionary environment of adaptation (e.g., see Boehm 1999; Knauft 1991, 1994) The cognitive, rational capacities of particularly the frontal cortex allowed for the dynamic balancing of the two opposing circuits from our early mammalian and reptilian heritage that is captured in the dynamic balance range of the CSN Complex (Cory 1999, 2002, 2004,2006ab; see also Chapter 2 this book). The ability to dynamically balance the opposing subcortical circuitries constituted a highly significant adaptation in human evolution and is a source of significant mismatch in the civilized societies of today.

The Issue of Generalism vs. Locationism in Neocortical Function

As a matter of historical interest, the crucial role of neocortex in cognition and specific domains of rationality has been discussed for nearly two centuries. In the early 1800s debates passed between anatomists (e.g., the phrenologists Gall and Spurtzheim) and the physiologists (e.g., Flourens). At issue was whether the cerebral mantle operated as a single entity or was constituted of specific sub-regions—generalism vs. locationism.

While data were long subsumed under one banner at the expense of the other, more recently something of an integrated view has emerged with the notion that some cognitive considerations are of a rather general type involving dispersed brain functions (e.g., Berns & Nemeroff 2003: 78), whereas others are highly specialized in specific areas. In some cases the anatomical specificity of cortical function is being mapped with exceptional and surprising detail (Damasio, 1994).

A classic example of specificity is hemiquadranto-achromatopsia as occurs with tiny strokes in the blood supply of a small sector of the visual association cortex. This renders patients unable to appreciate the color of a quadrant of the visual field even as they can see the full field with clarity and, in three-fourths, with color. There appear to be myriad other perceptual components in the cortex, most of which may be more relevant to neurology, in the narrow sense, than to psychiatry or psychology, in the broader sense (Weiss, 1950; Brain, 1960; Damasio, 1994).

Adaptive-Maladaptive Features of the Neocortex

In any event, an important feature of the higher cortex is its capacity to filter, modify, and represent limbic and lower drives through processes of rational, intellective analysis. A closely related feature is the transmission of highly synthesized input from the neocortex to the limbic system where it is imbued with an emotional dimension (Denny-Brown, 1951; Damasio & Van Hoesen, 1983; Damasio, 1994). Thus, Papez' (1937) original concept of a anatomico-functional loop connecting the hippocampus, fornix, mammilary bodies, anterior thalamic nuclei, cingulate gyri, cingulum, hippocampus is borne out and further enriched by recent demonstrations that considerable processed input arrives from the neocortex itself, as well (Damasio, 1979; Damasio & Van Hoesen, 1983; Brady, 1958).

As noted earlier, while these capacities are crucial to sociality, much less acculturated behavior—atavisms—can arise from latent phylogenetic assemblies. This is not merely possible but, indeed likely, at least under certain circumstances. Clearly, the hierarchies of neuromentation are neither discrete in themselves nor entirely predictable in that there can be a vying for supremacy between the functional layers—even in their neocortical representations as reflected in the CSN Complex.

Indeed, Gilbert (1989) proposes that a number of psychiatric disorders maladaptively either directly express—or indirectly repress—amplification of phylogenetically old behaviors. He cites pathological jealousy disorder as an example. Pathologically jealous individuals have much difficulty monitoring their mates. They are hypervigilant with considerable attention given over to the search for evidence of cheating or checking on the movements, activities or associates of mates. Such behavior has obvious relevance to ultimate reproductive success. Likewise, and as we shall soon consider further, the matter of phylogenetic regression in specific reference to manic-depressive escalation and de-escalation of social competition is of particular interest in the present discussion.

In summary, this chapter moved us from evolutionary epidemiology toward an evolutionary anthropology of manic-depressive behavior. This excursion was based upon a neuroethological perspective and the literature on social competition as it sought to capture the evolution of sociality from ancestral reptiles to mammals and ultimately to humans. Our social interactions, our social competition, even our social exchange activity, were seen to be based on the interaction of conserved circuitries from our reptilo-mammalian heritage. In both the normal and exceptional functioning of these social neuromentalities, the emergence of BPD symptomatology might readily be evinced. In the next chapter, we will consider the contributions of game theory in further delineating the evolutionary epidemiology of BPD with relevance, of course, to other affective disorders. But perhaps most revelatory, a more sophisticated appreciation of the advantageous endophenotypes that spring from the abundance of affective etiogenes in the human genome is also highly relevant. With this, we enter the realm of what might be deemed the *neuropathologies of talent*. to be discussed more fully in Chapter 14.

11

GAME THEORY AND BIPOLAR DISORDER

In this chapter we turn to game theory to further explore the way in which evolution appears to have sculpted sociobehavioral structures and processes. Game theory, with its mathematical approach to behavioral interplay, makes a special contribution in deriving phenotypic insights from standard epidemiological analyses. In its bare bones formalization, game theory sharpens our focus on essentials. It further acts as a corrective to more speculative tendencies, including the "just so" dismissals that often greet evolutionary ideas.

In applying the techniques of game theory we also draw upon relevant evolutionary aspects of social psychology and behavioral neurology. Evolved brain structures and mental processes direct social behavior toward the ultimate enhancement of differential reproductive success. Following our game theoretical analysis, discussion is, then, extended toward a broader sociobiological perspective.

GAME THEORY AND EVOLUTIONARILY STABLE STRATEGIES

This survey is not intended as a definitive review of each component area. It is intended, rather, as a basis for the further development of a preliminary evolutionary genetic and neuropsychological model for expressing both normative and manic-depressive behavior with a high degree of reliability and validity. Game theory is especially useful for this purpose. Such analysis is quite compatible with and has long informed central sociobiological themes of ethology and psychiatry.

Evolutionary game theory offers preliminary models of how proximal social behavior links—via neuromentalities evolved at serial stages of phylogeny—to ultimate sociobiological issues, notably reproductive success. Such fundamental issues require neural capacity for two alternative strategies: escalation, *Hawk*; or de-escalation, *Dove*. Furthermore, these two strategies derive from deeply canalized and successive layers of neuromentality (Price, 1996; Wilson, 1998, 2003). Game theory, with its formal mathematical operations. not only renders such a model particularly robust and practical, but it also points the way toward more direct, clinically relevant investigations ranging from ultimate epigenesis to proximal pathogenesis. For the present discussion, game theoretical mathematical models of neuromental sociophysiology become useful for fuller demonstration that bipolar disorder (BPD) especially its full spectrum with many advantageous phenotypes, is a quasi-Mendelian balanced polymorphism of the human genome.

Criticisms of Game Theory

The application of game theoretical concepts to issues in evolutionary biology has been criticized by some population geneticists who contend crucial aspects of reproductive biology are neglected; e.g., the dynamics of sexual selection, modes of inheritance, and other phenomena of genetic selection. However, recent research has begun to reconcile selective dynamics with evolutionary game theory, specifically general mating systems, polygenic traits and situations of frequency-dependent selection (Bomze, et al. 1983; Hammerstein, 1996; Weissing, 1996). Nevertheless, polymorphic populations may not be reducible to game theoretical models and, if they are, it is not yet clear how stabilization is attained at the molecular level of gene substitution (Weissing, 1996).

For instance, it is increasingly evident that behavioral strategies by which the dynamics of social rank are modulated became fixed in the environment of evolutionary adaptation. That is, in the course of phylogeny, resources adaptive in the proximal sense were allocated by these genomically interiorized and increasingly fixed neuromentalities. Plausibly, such allocations of proximal resources correlated with the ultimate reproductive success of manic-depressive genotypy (Ghiselin, 1967). This population polymorphism of BPD is not, then, a disease in the strict sense. Advantageous characteristics have been identified and certain disease features of manic-depression appear to be abnormal phenotypes expressed in reaction to mismatches between the current environment and that of evolutionary adaptation.

KEY CONCEPTS AND BASIC APPLICATIONS OF THE MATHEMATICS OF GAME THEORY

Game theory is that branch of mathematics which analyses conflicts of interest. Game theory originated in the study of checkers, cards and the like. The primary aim is to clarify, in specified conditions, optimal choices or those which result in a desired outcome. Game theory has become a serious means of systematically understanding important conflicts and, despite its origins and indeed its name, has influenced diverse fields (i.e., sociology, economics, politics, and military science).

Formal study of game theory was initiated by the French mathematician Émile Borel in his several papers on the stochastic dynamics in games of chance and theories of play (1953abc). However, a comprehensive and exact framework for game theory was first propounded in the 1920s by the Hungaro-American mathematician John von Neumann (1928). Although game theory originated as a rather abstract mathematical technique it had a major empirical influence on World War II military doctrines. Logistics, submarine and air warfare strategies derived directly from postulates of game theory. After the war, game theory developed rapidly in an array of social sciences.

An important distinction is made between *normal* and *extensive* games. A game in extensive form has a list of rules governing the activities and preferences of each player. Simple parlor and card games, such as checkers or rummy, are actually examples of extensive game forms. An extraordinary range of strategies are obtained in even these simplest of extensive games. Thus, mathematicians have reduced more extensive forms of games to normal conditions which allow computations can be carried out completely. In a normal game, all outcomes for every combination of strategies can be specified for each player at any point. Thus, normal games do not depend on player subjectivities—since these, although not absent, have been completely formalized.

A game has *perfect information* when all moves are known by or obvious to the players (checkers or chess). In games of *imperfect information*, players have only partial and different data (poker). A *strategy* is each player's algorithm of choices optimal at every stage of a given game. A strategy is a means of optimizing outcome given all contingencies. Game theory discriminates games by the number of players and the conditions of play. The most basic *one-person games* involve only the interest of a single player, e.g., solitaire. Singular games, lacking a genuine conflict of interest, are essentially probabilistic summations. From both a game-theory and behavioral perspective these lack interest as no adversaries contend.

Plural games, of which dual games are the prototypical type, include checkers, chess, and many card games. Dual, or two person games, have been closely analyzed by game theorists and the choices and payoffs can be specified. In certain respects, *n*-person or *plural games* resemble dual games. However, the interactions of three or more players (or teams) can induce unique conflicts having more complex dynamics, such as coalitions, cooperation and collusion—as, for example, in bridge.

A *zero-sum game* is one in which the total payoff at the end of the game is zero, that is, the amount won by one player equals the amount lost by the loser. Total resources are stable in zero-sum games—no production or destruction of value occurs. Importantly, all *plural non-zero-sum* games reduce to an *n + 1 zero-sum game* and further reduce to the special case of the *dual* zero-sum game (von Neumann & Morgenstern, 1944). Moreover, it has been further proven all dual zero-sum games are governed by maximal-minimal, or minimax, strategy (von Neumann 1928). Consequently, dual player zero-sum games constitute a core area of game theory mathematics.

The broad utility of game theory accounts for its steadily growing influence beyond the first efforts of von Neumann and Morgenstern (1944) to use it in the analysis of economic behavior (especially conflicts in maximizing profits or distribution of goods and services). In the social sciences, *n*-person game theory has interesting uses in the study a range of economic, political and societal forces. Examples include the equitable division of property (as in inheritance) as well as the summation of power in legislative procedures. Procedures of majority rule and individual decision making can be modeled in game theory, as well. Similarly, epidemiologists use game theory; e.g., to optimize immunization schemes and test protocols for vaccines or drugs. Military strategists model battles using game theory, as outcome is victory or defeat in non-zero-sum games.

The application of game theory to human events has been criticized as a dehumanizing and dangerous oversimplification (Gould & Lewontin, 1979). Certainly, human dynamics can be more complicated than game theory might predict. It is, however, an advance in the analysis of complex dynamic systems, including human behavior.

EVOLUTIONARY APPLICATIONS OF THE MATHEMATICS OF GAME THEORY

More recently the issue of evolutionary psychology—notably aspects of aggressive and subordinate behavior—has been recast in the specific mathematical terms of game theory. Maynard Smith, in collaboration with Price and Parker, clarified the idea of the *evolutionarily stable strategy* or *ESS* (Maynard Smith, 1972, 1974, 1976, 1982; Maynard Smith & Price, 1973; Maynard Smith & Parker, 1976). Their concept of the evolutionarily stable strategy developed from ideas adumbrated earlier (pertaining to evolution: Hamilton, 1963, 1964a, b, 1967, 1971, 1975; Axelrod & Hamilton, 1981; MacArthur, 1965; Darwin, 1859, 1871; as pertaining to game theory: von Neumann, 1928).

In this unique application of game theory, a *strategy* is posited as a phylogenetically interiorized neurobehavioral repertoire (e.g., attack and pursue if prey flees but run away if prey retaliates). Therefore, evolutionary game theory is not a merely abstract model of behavior—manifold innate biosocial patterns appear to have evolved within the basic dichotomy of evolutionarily stabilized strategies of aggression or subordination. This strategic dichotomy is a highly conserved and genetically programmed neuromentality, with important embellishments at various taxonomic levels (Gardner, 1988).

For example, both reptiles and birds establish rank hierarchies in connection to territorial breeding hegemony. Typically, the holder of territory has resource advantage, including reproductive opportunities (Archer, 1988). These advantages have been noted in primates but, being subject to greater corticolimbic control, they are expressed with more plasticity than in lower animals (Dunbar, 1988). Still, there are typical interactions which have become stabilized in the course of evolution. Including several of direct relevance to which we now turn.

Ritualized Agonistic Behavior (RAB)

One such stereotypic interaction, *Ritualized Agonistic Behavior* (RAB), became stabilized when competition for control of territory was at issue in phylogeny (Lorenz, 1950, 1981). Contestants stiffen and square off standing upright, often with secondary bobbing and weaving (Archer, 1988). This is a primitive power display in which eye contact is often crucial as a signal of fearlessness (Gilbert, 1992). The neuromusculature for such displays is largely conserved in the midbrain of reptiles and on through humans. Indeed the goose-step of soldiers and athletes is ethologically derived from repertoires evolved in earlier vertebrates (MacLean, 1974, 1985, 1987). Thus, RAB is an evolutionarily stabilized neuromentality of the reptilian midbrain subserving instinctive, antagonistic encounters (Stevens & Price, 1996). It is the most primitive unconsciously *instinctive* neuromental module of the agonic social mode (Chance, 1988) or the egoistic, self-preservational circuitry of the CSN complex (Cory 1974, 1999; 2002, 2004, 2006ab; also Ch. 2, this volume).

Maynard Smith (1982) sees the ritualized display of RAB as a constituent part of an *assessor strategy*. This evolutionarily stabilized assessor strategy allows individuals (or again, *players* in the parlance of game theory) to determine their power relative to a contestant. An assessment which predicts relative strength can sustain success. On the other hand, the timely assessment of relative weakness may minimize loss or injury. The ritualization of these assessments constituted a necessary phylogenetic precursor to later evolutionarily stable strategies. These, in turn, are increasingly mediated via interpersonal displays of an increasingly rational-emotive type of cognition and behavior (Pavlov, 1929; Hawkes, 1990; Ellis, et al. 1992). As noted earlier, in the ability to assess or compare may be seen the beginnings of the emergence of empathy.

Resource Holding Power (RHP)

A second typical interaction which has become stabilized in the course of evolution is resource holding power (RHP). As Parker (1974, 1984) noted, the very display itself often serves as a ritualized and symbolic signal of RHP. Thus, sentient creatures have a strong reproductive interest in the phylogenetic elaboration of accurate evaluations of relative power, both their own assets as well as the assets of rivals. Such evaluation can lead an animal to engage in contests of either high return or low risk and the avoidance of the converse. The evolutionarily stabilized neurobehavioral apparatus for RHP is conserved from late ancestral reptiles through humans. It also appears in birds. Thus, the brain and behavioral components of RHP arose with even the earliest limbic structures to subserve proto-conscious, *emotive* agonistic encounters (Stevens & Price, 1996). Such circuits are the foundational basis for the agonic social mode (Chance, 1988) or the self-preserving egoistic programming (Cory 1974, 1999).

Animals perceiving substantial advantage or crucial stakes are more motivated to signal dominance and more likely to act out their interests via conflict insofar as the relative risk is assessed as worthwhile (Gilbert, 1992). While the precise modes of such social comparisons are unknown, they appear to be expressions of innate but context-dependent algorithms (Hinde, 1974, 1987, 1989; Cory 1974, 1999). These algorithms can be modulated by a variety of sociophysiological factors such as past experience of success or failure, age, hormone and neurotransmitter state (Archer, 1988; Gilbert, 1992). Also, while mechanisms of social comparison become decidedly more complex in higher animals, a strong biogenic continuity of these earlier modes exists even in the human line (Festinger, 1954; Swallow & Kuiper, 1988; Wood, 1989; Abrams, et al. 1990; Gilbert, 1992; Cory, 1974, 1999).

SAHP and the Importance of Empathy

Evolutionary additions have been made to these earlier systems which originally mediated social control and attentional structure via displays of threat and aggression. As noted, in the primitive capacity for social comparison necessary to

RHP, may be seen the dawning of the potential for empathy. This RHP stereotypy has been modified—but not replaced—by affectional circuitry making possible social *attractiveness* mediated by empathy. Status and regard, elicited from others voluntarily, are outcomes of continuous reciprocal social relations and are empathy dependent (cf. Cory 1999, 2004; Buss, 1988, 1991, 1992, 1994; Buss & Plomin, 1984). This latter refinement of RHP, based on mammalian affectional circuitry, has been defined in terms of *social attention holding power*, or SAHP (Gilbert, 1989, 1992).

Despite the one way emphasis of the definition, SAHP is necessarily a two-way street. The SAHP apparatus arose from the mammalian affectional circuitry to include the thalamocingulate gateway into the later neocortical structures. Such structures functionally support the more overtly conscious affective as well as cognitive aspects of neuromentality in the social mammals and, particularly, the primate order and other eusocial species (MacLean, 1949, 1974, 1985, 1990; Gilbert, 1992; Devinsky & Luciano, 1993; Stevens & Price, 1996). Social attractiveness requires empathy. Others require empathy to judge one as attractive. To be attractive one must also display a degree of empathy in return. SAHP, then, is a two-way interactive, reciprocal process. Such other-oriented circuits are a necessary condition of both the hedonic social mode and the empathetic behaviors of the CSN complex.

This motivation to attain positive attention from others as well as to give it—that is, to interact empathetically—serves as a proximal and powerful mode of resource allocation to include sharing among social animals

Hawk-Dove Evolutionarily Stable Strategies

The phylogenetic succession of RAB, RHP, and SAHP is elegantly subsumed in a specific game theoretical analysis called, *Hawk and Dove* (Maynard Smith, 1972, 1974, 1976, 1982; Maynard Smith & Parker, 1976; Maynard Smith & Price, 1973). The concept of divergent yet evolutionarily stabilized strategies was applied to the social governance of competition and cooperation. A strategy can be said to have been stabilized in evolution when it is generally adopted by a population without recourse to a superior neurobehavioral algorithm (Maynard Smith, 1974).

An Idealized Formal Paradigm

An ideal species in which only two contrary options for social interaction was postulated and then analyzed within the parameters of game theory (Maynard Smith, 1974). These divergent strategies were dichotomized—allegorically not ethologically—as Hawk or Dove. In this simplifying paradigm, all individuals are either hawks or doves. None can recognize the nature of others nor recall the outcome of prior encounters. Hawks fight without restraint until seriously injured. Doves posture and may intimidate but do not engage in physical aggression. When two doves strive against one another they display competitively until one backs off. Also, doves quickly evade attacks from hawks. Meanwhile, when two hawks engage they fight until one is either dead or seriously wounded. Hawks will, of course, defeat doves in most encounters.

Dovish behavior is optimal in the narrow frame of each player. Nevertheless, a complete prevalence of doves does not evolve. Under the formal assumptions, a complete prevalence of doves would both invite and greatly reward hawkishness with runaway success. Yet hawks do not prevail either. This is because at some point doves become too rare for aggression to reap sure rewards and the assessment of cost becomes increasingly uncertain. Further, the population does not oscillate between hawk or dove ascendance. Indeed, thorough game theoretical analysis demonstrates that neither Hawk nor Dove is itself an ESS. Instead, a stable ratio between the two obtains (Maynard Smith, 1974; Maynard Smith & Parker, 1976).

This is indeed a simple, highly constrained and formal theoretical approach. Nevertheless it may underlie increasingly elaborate and naturalistic paradigms. At a basic level, it has the virtue of formalizing a crude understanding of rank, hierarchy, and related social dynamics, including genomic polymorphisms (Maynard Smith & Price 1973).

The Artificial Constraints of the Hawk-Dove Game Model

As noted earlier, the Hawk-Dove dichotomous model is highly formal, simple, and artificial. As applied to humans, at best it simulates reptilian agonistic behaviors of dominance and submission with no alternatives. It forces individuals into stereotyped polar extremes of behavior with no options or middle ground. It is also completely externalized and without insight as to the motivation mechanisms—the neural circuitry, neurotransmitters or endocrine systems—required to sustain such polar strategies.

Additionally, it operates, reptilian-like, on a one time dyadic encounter, not allowing for the continuous social engagement or accumulated memory of mammals. Like much formal modeling, it has, at best, a tenuous connection with reality. The CSN Complex, on the other hand, more fully addresses these shortfalls. Ego and Empathy are neocortical representations of limbic affectional and reptilian self-preservational circuitry. The tug-and-pull of these opposing circuitries is driven by behavioral tension/stress and maintained homeostatically within survival limits. Nevertheless, the CSN Complex—inherent to us all—can account for the hawk-dove dichotomy more realistically even at the extremes. But it does much more. It provides for a spectrum of mixed behaviors or options reflecting varying degrees of competition and cooperation and, importantly, their ultimate dynamic balance or equilibrium.

The Phylogenetic Dichotomy of Competition and Cooperation

As set out in Chapter 2, there are two great archetypal systems governing conflict in mammalian social life. Although they do so differently, the models of both Chance and Cory deal with these systems.

In Chance's terms one has to do with issues of territory, power, rank, and status. These are dealt with in an essentially agonic mode. The other has to do with attachment, affiliation, cooperation, and solicitude. These are dealt with in an essentially hedonic mode.

The CSN Complex, in contrast, avoids the ideas of modes in that the two archetypal systems interact within each player and between each to produce the behavior of the moment. We may be egoistic/agonic or empathetic/hedonic across a varied spectrum as appropriate to the situation. Additionally, we may achieve a dynamic balance of the systems as predicted by the game model. The CSN Complex has an additional advantage of being expressed concisely as a game theoretical mathematic equation reflecting its homeostatic nature.

Each of the archetypal systems can operate adaptively but both can engender pathology as well. This possibility is more clearly indicated in the CSN Complex where tension and stress increase with movement to the extremes of either ego or empathy. As stress accumulates, the effects can become pathological. The agonic/hedonic model does not effectively capture the homeostatic nature of the social circuitry nor its dimensional range across a spectrum.

The basic evolutionary theme presented thus far is surprisingly simple to recap even though this phylogeny unfolded over some three-hundred million years. Our earliest vertebrate ancestors competed on an individual, asocial basis. In fact, many vertebrate species still do. Classical reptilian behavior and its functional neural circuits then became stabilized in dyadic contests. Proximal control of territory ultimately led to other resources such as food and mates.

As mammalian affectional circuitry evolved, however, it made group living possible. With sharing of territory and resources dyadic competition became less discrete and more continuous. It was redirected toward the proximal acquisition of rank and status. High rank, although it did not exclude sharing with others, ultimately led to a larger share of resources necessary for reproductive success. Since all individuals had similarly opposing circuitry, the contests over rank and hierarchy were not achieved once and for all. Motivated by behavioral tension, residual in both winners and losers, hierarchy was always subject to challenge and therefore inherently dynamic and often unstable.

BASIC SOCIOBIOLOGICAL CONSIDERATIONS

Evolution at its very core concerns the study of selections between genetic alternatives within reproductive groups (Mayr, 1950, 1961, 1963, 1972, 1977, 1988, 1992, 1997; Barash, 1977). In social animals these alternatives are often behavioral, although physical form is not without continuing relevance. Therefore, the study of evolution in social animals can usefully focus on tactics and strategies advantageous in the acquisition of proximal resources which, ultimately, enhance reproductive success.

The force of these integrating theories and complementary data sets was marshalled in *Sociobiology: The New Synthesis* (Wilson, 1975). Its last and openly speculative chapter on the evolution of human behavior caused the book to be challenged early on. This, in turn, led to strong initial controversy concerning evolutionary biopsychology in general and human sociobiology in particular. Some critics perceived a recrudescence of rigid biogenetic determinism (cf., Bowler, 1992). Although the criticism has abated, the term sociobiology retains an air of skepticism in some respects.

Sociobiology explicitly extends the consequences of natural selection to the social behavior of animals, including humans. Subsequently, evolutionary social genetics has generated robust research. Diverse elements of human behavior have been reconceptualized under this rubric; for example, non-verbal communication and language acquisition, mate choice, incest avoidance and certain personality traits. Sociobiology has been most successful in analyzing the evolved behavior patterns which defy understanding in terms of the survival of individuals alone, particularly those of the social insects. It has also fed recent advances in understanding the evolved basis of much human behavior.

The term sociobiology is recent but the problems it addresses are not. Indeed, Darwin long struggled with the question of altruism (Desmond & Moore, 1991). He was perplexed by the willingness of some individuals to assist others, even at the direct reduction of reproductive success (the caring for young by sterile worker ants, the sacrificial defense of young by older animals, etc.). Less overt forms of altruism were also problematic. Even grooming a conspecific would reduce at least marginally the groomer's reproductive potential. The issue of altruism likewise has been a redoubt of sorts for the analysis of nature as something other than metaphorical "red in tooth and claw". Indeed, it is emerging as a putative foundation for the phylogeny of moral virtue (Ruse, 1986; Ridley, 1996).

How selection could favor self-sacrifice was unclear to Darwin. The genes expressing altruism would seemingly be de-selected insofar as they reduced reproductive success in a direct fashion. Still, Darwin thought altruism could be subsumed under natural selection and his efforts to bring altruism into the fold fully anticipated the concept of kinship selection. He did not, however, formalize a proof that altruists could contribute to the survival of traits shared by kin even as they forfeited direct contribution of genes to the next generation. Kropotkin (1902) offered preliminary explanations. Thereafter, Haldane strongly hinted at a resolution to the issue of biogenically mediated altruism with his intuitive references to what amounted to kin selection (Bowler, 1992).

The formal understanding of altruism in terms of kin selection waited over a century after the publication of *Origin of Species*. Hamilton demonstrated (as discussed in Chapter 4, this volume) with mathematical precision, that individuals can sometimes optimize reproductive success by assisting kin—as a function of relatedness in proportion to altruistic costs (Hamilton, 1963, 1964ab). Soon it was shown that altruism can be directed to others than kin with mutual advantage—reciprocation of assistance can explain altruistic acts exchanged by non-relatives (Trivers, 1971a, b, 1985). More recently, of course, reciprocal altruism is seen as most likely an extension of the basic kin-bonding mammalian affectional mechanisms, not requiring separately evolved circuitries (e.g., see Carter & Keverne, 2002; Cory, 1999, 2002; Buss, 1999).

However, interacting animals must be able to detect and deter those who will not exchange fairly. Such conditions characterize the environment of evolutionary adaptation of most small-group mammals, including primates, hominoids and humans. Also game theoretical mathematics has proven that reciprocal altruism can arise and flourish even within species that initially act selfishly (Maynard Smith, 1974, 1976, 1982; Maynard Smith & Parker, 1976; Maynard Smith & Price, 1973). Thus, sociobiology and evolutionary psychology are emerging as mathematically formal amalgamations of sciences pertaining to natural selection and animal behavior.

Evolution in higher organisms advances primarily via the elaboration of nuances in the neuromental regulation of the social interactions between organisms. Behavior is, therefore, increasingly crucial. The neuromentalities by which reproduction is directly or indirectly maximized is of special relevance to the evolution of sentient creatures to include humans. In the case of humans such neuromentalities become of interest to medicine and psychiatry (Beach, 1974, 1976). Recently, evolutionary neuropsychology has arisen to study the mechanisms subserving these neuromental elaborations of structure and process.

Levels of Evolutionary Selection

There is a complex and ongoing debate about the total concept of evolutionary levels of selection. Much of that debate is largely insubstantial to the present discussion (Dawkins, 1976; Gould & Lewontin; 1979, Wilson & Sober, 1983). However, a few relevant points may be made briefly. Kin selection theory posits that the basic biosocial goal is the survival of genetic material itself with the survival of individuals having less consequence (Barash, 1977; Dawkins, 1976, 1989; Trivers, 1971a, b, 1985; Crook, 1980; Maynard Smith, 1983). While this is likely true, such a supraordinate strategy may be usefully further parsed into a variety of sub-strategies, many of which are evident within the range of individual behavior (Wenegrat, 1984). Still, the principle force driving evolution is, for sociobiologists, the selection of genes into subsequent generations.

Darwin himself noted three types of selection (Darwin, 1859, 1871; Buss, 1988; Stevens & Price, 1996). The first was, of course, natural selection. Natural selection accrues as traits advantageous in the struggle for existence pass on with differential success. Phylogenetically, these characteristics are largely, but not exclusively, physical. The second type of selection was intrasexual competition which accrues as characteristics advantageous in the rivalrous competition for mates, likewise, pass on. These traits are first physical but later become increasingly behavioral. Finally, Darwin identified intersexual competition which accrues in the selection of traits signaling breeding potential. These characteristics are again, both physical and behavioral.

It is the integrated success of these factors which, ultimately, both enables individual reproductive success and influences selection at all levels. Further, these various selective factors can interact to reinforce or interfere with one another; e.g., as when sexually alluring plumage reduces anti-predator status. Strategies of individual reproductive fitness cannot, in themselves, fully explain certain evolved phenomena, notably self-sacrifice or other varieties of altruistic behavior.

Thus arose the notion of inclusive fitness (Hamilton, 1964a, b) discussed in Chapter 4. Inclusive fitness allows social behaviors associated with gene survival to be considered apart from the immediate impact on individuals. With this, phenomena seemingly burdensome to an individual—such as solicitude of young or weak, reciprocal altruism, cooperation and so on—might prove quite adaptive *in toto*. Appropriate affectional circuitry of mammals, therefore, evolved to motivate and sustain such behavior. The CSN Complex, which so elegantly describes this, can properly be seen as capturing the motivational mechanism produced under the terms of inclusive fitness. As was demonstrated in Chapter 4, the social equation expressing the CSN Complex is compatible with the formula for inclusive fitness.

The issue of group versus kinship selection is perhaps less pressing than it may initially appear. After all, hominoids have evolved for some tens of millions of years in small *kinship* groups (Wynne-Edwards, 1962, 1986; Williams, 1966, 1975; Wilson & Sober, 1983). The sense of intimacy converged with consanguinity quite naturally and very strongly in prehistoric hominoid societies and has not had time to have been de-selected. Such was the affectional foundation for the all important characteristic of Empathy. Considerable degrees of genetic relatedness were endemic to the extent that kinship correlated rather directly with familiarity and vice versa. As noted earlier, contrary positions exist as to whether individuals, or genes, drive selection. Thus, debates about group selection are not essential to the present discussion.

The Issue of Cost-Benefit and The Emergence of Global-State Algorithms

With the level of genetic selection left as a side question, an overriding point remains. That is, how inclusive fitness renders organisms capable of analyzing the cost-benefit of behavioral options (Cosmides & Tooby, 1992; Stevens & Price, 1996; Mithen, 1996). This algorithmic capacity places neuromental evaluative competence at the center of social evolution. It appears that unique algorithms have evolved to address specific phylogenetic challenges. There are, perhaps, a number of such algorithms rather than a single neuromental program which solves problems of a general nature. Moreover, each algorithm may have unique, seemingly illogical, considerations which promote stress or conflict.

Such unique algorithms, while substantially innate, are phenotypically modifiable to varying extents. Behavioral algorithms in large measure evolved to afford a plasticity of response to environmental contingencies (Jerison, 1973; Wilson, 1975; Shaw & McEachern, 2001; Kolb, et al. 2003). This essentially neural plasticity, combined with higher cortical development, allows for the merging of independent algorithms into integrated global-state algorithms (Panksepp, 2002; Schulkin, 2002; Deth, 2003, esp. pp. 9–10; Cory, 2004). Such global-state algorithms summate, unify, and focus the unique and perhaps independently evolved algorithms to give the impression of a single evolved mechanism. The cognition, feeling, and speaking (signaling) of "I am being cheated or short-changed" can be experienced generally and applied across a very wide range of interactions. Such interactions and accompanying cognitions or emotions are pervasively social.

Proximal Versus Long-Term Reproductive Tactics and Strategies

Issues of ultimate reproductive success are not the immediate conscious concern of most creatures, humans included. Instead, issues of more direct significance are the focus of affect, emotion, strategic behavior and, where evident, conscious rumination. These proximal issues are essentially biosocial strategies pertaining either to competition or co-operation, the two archetypes of sociality. As we have seen, in most species such biosocial strategies have been phylogenetically interiorized and evolutionarily stabilized.

Several questions emerge. Firstly, what strategies are universal to social mammals as distinct from reptiles? Secondly, by what neurological structures and mental processes are these strategies applied in a tactical sense? Thirdly, how do these strategies influence the social dynamics of proximal resource acquisition which services ultimate reproductive success? Such questions were addressed in part in the development of the CSN Complex in Chapter 2.

Beyond basic material sustenance, social animals have to both compete and cooperate effectively. Competition, in terms of behavioral psychology, generally refers to the struggles evidenced by individuals for material resources; e.g., food, water, nests, as well as for opportunities to reproduce. Struggles can be further parsed as to whether behavior is directed to outshine rivals (intrasexual competition) or to appeal to opposites (intersexual competition) for reproductive purposes.

In each form of competition, an individual vies directly, even personally, with others. The aim may be physical resources or modification of behavior which leads, ultimately, to enhanced survival and reproductive success. In social animals these struggles are channeled through dominance rank hierarchies. Typical competitive tactics may include intimidation, thievery, cheating, attack, humiliation or other efforts to impinge on rivals, increase control of or access to desirables while constraining options for the other(s). Clearly, this is the realm of agonic or egoistic sociality built upon hostile domination.

This individual competition, however, has been phylogenetically succeeded by more hedonic or empathetic interplay. If a species is to socialize even to a limited extent without overt agonisms, evolved social signals are necessary to indicate geniality and the absence of threat (Lorenz, 1965, 1966, 1981; Tinbergen, 1951; Morris, 1977; Chance, 1980; Eibl-Eibesfeldt, 1974, 1975, 1989). In the phylogeny of social evolution, intense individual competition gives way to mechanisms of more symmetric and reciprocal patterns of behavior. Such reciprocity, the tug and pull of Ego and Empathy, is the hallmark of the CSN Complex.

Play as Prototype: Cooperation in the Development of Ego and Empathy

Play constitutes a prototype from which has grown much of the mammalian cooperative capacity (Eisenberg, 1981; MacLean, 1985). It certainly involves the interplay of self-preservation and affection, respect for self and empathy for others. Exercise and development of these fundamental circuits are necessary to achieving the advanced skills required for human social activity—that is, learning the limits of Ego and Empathy, dominance and submission, and how to achieve a dynamic balancing of the two circuits as one interacts with others. Panksepp and Burgdorf (2003) have reported that rats may share with humans an analog of laughing as an expression of joy produced during social play and likely functioning as a motivating factor for such cooperative activity. Social context, a key domain for both challenges and solutions in human life, can drive evolution as robustly as can factors in the physical environment. So it appears cooperation, practiced through play, principally evolved as a means for conspecifics to jointly exploit resources toward a higher level of inclusive fitness than could be achieved by the simpler devices of individual competition alone.

Escalation and De-Escalation

Two types of biosocial strategies or moods—escalating and de-escalating—seemingly emerged in tandem as evolution structured the brain to accommodate the archetypal opposing systems of self-preservation and affection. The two strategic moods evolved with increasing sophistication as a dynamic component of the archetypal opposing systems. As noted in Chapter 2, escalation and de-escalation can be seen to apply across the Ego/Empathy spectrum. That is, one may escalate or de-escalate either empathetic or egoistic behavior. The two strategies of mood, then, should not be seen as applying solely to either of the opposing circuitries, but varying with circumstance. Aaron Beck makes this advisory point when he reminds us to beware of locking horns with a depressed patient or we may be pushed out of the consulting room (quoted in Price, 2002: 110).

The escalation/de-escalation strategies derived from reptilian layers of brain tissue and were widely expressed in reptilian behavior. They were likewise integrated as a dynamic factor with the emergent mammalian circuitry. Thus, these neuropsychological mechanisms combined to resolve a key phylogenetic challenge making social life possible—the dynamic balancing of self-preservation and affection. Accordingly, they have been stabilized in these successive tissue layers and biopatterns over the course of later mammalian evolution.

The phylogenetically older of these two neuromentalities is the midbrain and early limbic system. Here reposes the apparatus of individual competitive advantage via the direct exhibition of power. This is an essentially reptilian self-preservational, form of social governance. Subsequent mammalian evolution extended, modified, and ritualized cooperative strategies in the limbic and neocortical areas, providing at the same time the affectional circuits to motivate and sustain such behavior. Individualistic, competitive neuromentalities thus were overgrown, but not superseded, by the capacity for empathic cooperative sociality. They were stabilized in evolution, for most purposes, as homeostatically regulated social interactions.

The Remarkable Emergence of the Neocortical Capacity for Mediation and Dynamic Balance

Behavior in the reptilian, egoistic range, activates dominance-power, self-preservational repertoires to guard and defend material or gain power over others. Behavior expressed in the mammalian, empathetic range, typically offers support, alliance, and reciprocity (cf., Bateson, 1976). The self-correcting, homeostatic regulation of behavioral tension and stress keeps the opposing circuitries within survival limits when operating correctly.

With the emergent neocortex, however, came cognitive representation and the capacity to consciously modify, direct, and dynamically balance these seemingly dichotomous strategies. This, a truly momentous evolutionary event for higher primates and humans, established the basis for the egalitarian nature of early face to face, kinship-based foraging society. Even today, it provides the foundational driving dynamic for democratic participation and equality in the emerging, virtual face to face global society.

BIOSOCIAL STRATEGIES AND GAME THEORETICAL MATHEMATICS

The immediate aim of the foregoing has been to point out how deeply embedded is the neuromental apparatus which orients behavior to either conflict (egoism) or cooperation (empathy). Nature, generally, is not "red in tooth and claw" since animals do not typically strive to thoroughly destroy intraspecific rivals. Rather as Lorenz (1966, 1981) emphasized, much animal conflict is restrained, formal, stylized—even playful. Ethological bluff and threat replaces deadly earnest aggression so that signals of surrender are recognized by victors and extinguish further attack on the vanquished.

Fundamentally, in most species, the costs and benefits of pugnacity and congeniality are dynamically maintained within survival limits. One cardinal consequence of this dynamic equilibrium has been the evolution of neurobehavioral complexes which subserve decisions, some conscious and others instinctive, to aggress, subordinate, affiliate, or modulate when necessary. Rivals each have alternative channels for such behavioral adjustments.

In higher mammals these channels increasingly rely on affective display or communication. Responding to one display, an individual may joyfully escalate affiliation. Responding to another display, the same individual may angrily escalate aggression. Other displays can evoke both agonic and hedonic, egoistic and empathetic, behaviors and the associated capacities for escalation or de-escalation. In humans, as in some higher primates, the neocortical ability to merge and modulate the two circuits in dynamic balance has greatly enhanced the capacity for ongoing social life.

HAWK-DOVE AND THE TWIN AXES OF EGO/EMPATHY AND ESCALATION/DE-ESCALATION

The Hawk-Dove ESS, although artificial in its dichotomized game theory form, is of direct relevance to an evolutionary conceptualization of human neuromentality, both normal and pathologic. It sets the extreme outer poles of egoistic/empathetic behavior as well as the opposing mood extremes of escalation/de-escalation. It suggests as well the importance of dynamic balance or equilibrium by implying that in continuous social living the more stable range of behavior is equilibrium or a dynamic balance between the extremes.

The essential pathology of the manic-depressive is an extreme dichotomy in both the scale and mood of behavior—along two axes. The swing between opposite extremes of separate underlying circuitries of the Ego/Empathy axis is exacerbated by similar swings between the opposite poles of the escalation/de-escalation axis, i.e., activity level. What the manic-depressive seemingly lacks is a capacity to achieve dynamic balance or equilibrium between the two contesting axes of nature and mood. The cognitive capacity for this harmonization is a later neocortical development, which apparently fails or is overridden in bipolar disorder. The manic-depressive gyrates on sets of circuits, each independently adaptive in historical environments of different evolutionary time periods, but with excessive swings of mood or energy.

In terms of the social equation of the CSN Complex with mania or depression the ratio between Ego and Empathy is diverging from desirable dynamic balance toward the polar extremes. This is accompanied, of course, with mounting stress. The same equation can be applied to express magnitudes of tension and social stress that accompany domination and submission.

Equation 11.1

$$BT = \frac{Ego}{Emp} = \frac{Emp}{Ego} = \frac{Domination}{Submission} = \pm 1$$

As the ratio between Ego and Empathy or domination and submission diverges from the dynamic (or optimal homeostatic) balance of ± 1, it serves as a co-efficient that indexes the increasing level of tension and stress. Additionally, the swing to extremes of escalation/de-escalation of activity level amplifies the stress level. The resulting magnitudes reflect also the potential for psychopathology.

Ego/Empathy (agonic-hedonic) and escalation/de-escalation, then, are not the same thing but are interrelated. The former represents the *nature* of the social response. The latter represents the scale of *mood*. Their interrelationship is best illustrated as two separate, intersecting axes.

The Hawk-Dove response set represents a special case of the intersection of the two axes as illustrated in Figure 11.1. The hawk response begins at escalated ego, traverses the ego-empathy axis diagonally, and de-escalates in the empathy quadrant as the dove response. There is considerable behavioral tension, if not social stress, in this response set as indicated in the deviation from ± 1 along both the Ego/Empathy axis as well as the escalation/de-escalation axis. The deviation from ± 1 along the escalation/de-escalation axis can be factored into the social equation to represent increasing magnitudes of social stress.

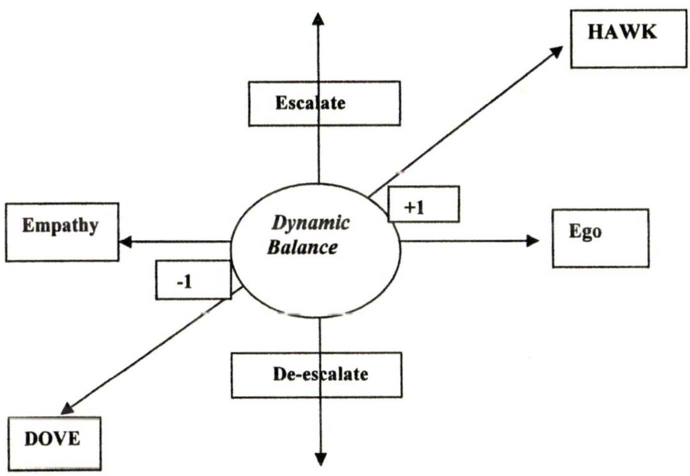

Figure 11. 1. Ego-Empathy axis and escalation/de-escalation axis illustrating Hawk-Dove social interaction (adapted from Cory 1974, 1985, 2003).

12

THE BASIC SOCIOBIOLOGY OF BIPOLAR NEUROMENTALITY

In this chapter we shift to a sociobiological perspective. Although evolutionary sociobiology cannot be clearly separated from evolutionary anthropology (much less evolutionary epidemiology), this slightly alternative vantage entails research that might otherwise be excluded. Anthropology, particularly of the biological type, is largely historical; whereas sociobiology is more functional. Both perspectives are important to a full and proper development of evolutionary epidemiology. From the sociobiological perspective, then, the focus is on functionality—the nature and efficacy of biosocial strategies.

Biosocial strategies are evolutionarily relevant and repay study in and of themselves. Especially revealing, however, is to correlate the evolution of biosocial strategies with major features in the phylogeny of behavioral neurology (Weiss, 1950). Equally important is the consideration of ultimate phylogenetic sources that could have given rise to these alternative strategies. Significant neuromental correlates were thereby evolved. A hierarchical overview of the correlates existing between biosocial strategies and neuromental apparatus follows. Thereafter is a brief consideration of the range of possible evolutionary sources—apart from the stochastic factors delimited in previous chapters.

BACKGROUND OF HIERARCHICAL OVERVIEW

Darwin's theories of natural and sexual selection aver that competition stratifies populations in terms of reproductive fitness in each generation within each species. Evolutionary game theory, as considered earlier in Chapter 11, denotes that the phylogenetical stabilization of social competition requires the neural capacity for two alternative strategies of escalation, *Hawk*, or de-escalation, *Dove* (Maynard Smith, 1982). Furthermore, these two strategies derive from deeply canalized and successive layers of neuromentality (Price, 1996). Tenets of selection taken together with techniques of game theory allow a general dichotomy of behavioral alternatives. That is, elaborations of these alternative strategies can be part of what defines either an entire species genome, or a polymorphism therein (Krebs & Davies, 1981).

Neurobehavioral correlates vary, of course, with the nature of each biosocial strategy, as has been elsewhere exhaustively surveyed (Darwin, 1859, 1871, 1872; Wallace, 1855; Tinbergen, 1951; Price, 1993; Gilbert, 1992). Moreover, conflicts may arise in the priority of different biosocial strategies. Comparative ethologists were at first concerned with the identification of taxon-specific behavior. In other words an ethogram or the "species personality" was sought amid the ubiquities of behavioral typicality. Ethograms were then compared across species clade lines and arranged on a *scala natura*.

Once ethologists turned attention to within-species variations they availed themselves of techniques first developed by human psychologists. This initially focused on a dimensional approach to the analysis of continuous variables such as introversion-extroversion and stability-neuroticism with a concentration on variations from mean values in the population. Not all behavioral traits fell out in Gaussian arrays, such as situations in which an animal could do only one of two alternative actions but not both. So attention then turned to the identification of mutually incompatible strategies which, significantly, were more reconcilable with polymorphic Mendelian rather than non-parametric explanations (Price, 1993; Stevens & Price, 1996).

Price (1993) has classified the evolved factors determinant of alternative strategies. His typology includes epigenesis and a variety of factors mediated by the environment. Of course, epigenetic factors are of particular interest in the present discussion. These are polymorphisms activated when the alternatives of behavior are determined by one or several genes that endow an individual with a single instinctive predisposition. Here the activation and consequences of behavior are largely independent of others in the population.

However, those factors which are more environmentally mediated are not without interest. These phenotypes, in certain circumstances, are complicated by the fact that consequences of behavior can depend on how others in the population behave. That is, they can be expressions of frequency-dependent strategies which are both robustly genetic and environmentally sensitive. Where the payoff is greater as members adopt the same strategy it is positively selected in proportion to its frequency in the population. Where the payoff is greater as an atypical strategy it is selected in inverse proportion to its frequency in the population. Apart from ontogenic implants, Price's typology thus reduces to three genomic possibilities as can be sustained by the three basic mechanisms of selection outlined by Darwin. These are: (1) individual polymorphism, (2) positive frequency-dependent selection and (3) negative frequency-dependent selection.

It has been further suggested that selective priorities ranged from physical needs (for food and warmth) to social and, ultimately, self-actualizing needs (Maslow, 1936, 1937, 1971). But social forces can clearly transcend basic needs, at least temporarily. Gilbert (1989) cites the gathering of social prestige via the voluntary acceptance of pain among schoolboys as a clear and distinct example. The CSN Complex provides that the Ego/Empathy dynamic evolved to adjust for social context in the expression of all human needs (Cory 1999: 45–46; 2004: 181–184).

Here it is useful to yet again recall that evolution retains much from prior adaptive systems as older structures or capacities are reordered or overlain but rarely fully replaced with the new. The evolved redundancy of the neuromental apparatus is extraordinary and complex. Thus, biosocial strategies tend to activate hierarchical networks of general arousal to either bolster or inhibit relevant behavior, cognition, affect and mood.

Recourse to a detailed hierarchical schema is not needed in the present discussion as a reliable taxonomy of phylogenetic goals has been correlated within a matrix of clinically valid epigenetic neuromental strategies for social competition. Table 12.1 is adapted from Stevens and Price (1996) to include the balanced strategy made possible by neocortical evolution as reflected in the CSN Complex (Cory 1974, 1999, 2004).

Table 12.1: Hierarchy of Evolved Biosocial Neuromentality

Alternative strategies for social competition/cooperation at three levels of neuromental phylogeny

BRAIN LEVEL (mentality/mode)	*WIN STRATEGY* (escalating-hawk)	*BALANCED STRATEGY* (modulating-cooperative)	*LOSE STRATEGY* (de-escalating-dove)
NEOMAMMALIAN CORTEX (sociotropic-cognitive/SAHP)	ATTRACT (optimistic, charming)	ALLY	AVOID (pessimistic, ashamed)
PALEOMAMMALIAN LIMBIC (emotive/RHP)	DOMINATE (roused, likely to win)	ALLY	SUBMIT (low, likely to lose)
REPTILIAN MIDBRAIN (territorial-instinctive/RAB)	FIGHT (strident, strong)		YIELD (cowering, weak)

N.B.: Correlate Psychopathologies: Pre-reptilian neuromentalities are essentially vegetative and *asocial*.
Mania = neo-mammalian over-escalation; Depression = reptilian over-submission;
SAHP = Social Attention Holding Power, RHP = Resource Holding Potential, RAB= Ritualised Agonistic Behaviour

Modified from *Evolutionary Psychiatry* *(Stevens and Price, 1996, p.226)*

Normal behavior made possible by neocortical elaboration, of course, allows for the dynamic balancing of the two opposing circuitries in the achievement of alliances and cooperative behavior so characteristic of higher primates and humans. Neocortical evolution also allows for both escalated and de-escalated behavior across all ranges of the CSN Complex spectrum. As the table indicates such balancing capabilities begin to emerge with paleomammalian circuitry but do not exist at the reptilian level.

Indeed, the obvious connection between clinical states of affective disorder and the breakdown of social attachments and alliances (*viz*, losses sustained by death of kin or ally, rejection or other rebuke; the sequelae of manic over assertion) is, in most cases, a secondary development of phenotypic reaction. The dependence of most primates on alliances (or social cooporation) for social competition is a behavioral feature evolved over forty million years in which the loss of kin or ally was the best correlate of alteration in social rank. However, this is a primato-mammalian elaboration upon the phylogenetic foundations of agonistic (self-preservational) escalation and de-escalation which date back some 300 million years. Thus the evolution of the manic-depressive syndrome traces to well before the emergence of the limbic system and cerebral cortex or the parental-affiliative behavior these neuromental innovations allowed (Price, 1993).

The implications of these later mammalian neuromental elaborations are, nevertheless, great for social behavior study in general and psychiatry in particular. The entire human species is ethologically predisposed to competition through prestige and other forms of social approbation leading to alliances and cooperation as the emergence of empathy has come to eclipse the purely agonistic behavior of pre-mammalian vertebrates. This sequence has further been linked to the emergence of human wisdom as distinct from animal intelligence (Csikszentmihalcy & Rathunde, 1990; Kramer, 1990). Like their anthropoid cousins, humans generally reciprocally display attractive qualities promoting balance and cooperation rather than exchange signals of intimidation (Wenegrat, 1984; Glantz & Pearce, 1989; Dunbar, 1988; Goodall, 1986, 1990; Gilbert, 1992; Price, 1996).

These displays involve a new adaptive radiation into roles of group spectator-evaluators and away from dyadic conflicts. The group activity, at least in part relies on cultural determined signals as well as more instinctive criteria (Price, 1996). As sensitivity to approbation was selected, these signals came to re-modulate self-esteem beyond the reptilian substrate. Thus social attention holding power (SAHP), characterized by empathetic reciprocity, co-exists with agonistic self-reference (RHP) (Gilbert, 1992). In the agonic mode of RHP, each individual assesses relative resources (i.e., "am I stronger"?). An affirmative answer ("I am stronger!") promotes self-assertion (threat or attack). A negative assessment ("I have insufficient resources compared to my rival) leads to self-effacement (submission or escape).

In humans, prestige is meted out primarily in hedonic competition via attraction, charm, and charisma (Levi-Strauss, 1944). All of this correlates to proximal SAHP. Much behavior is scaled to these correlations (Price, 1992). The individual assesses his SAHP in a social, not merely dyadic, context (i.e., "will I gather respect?"). A *yes* answer reflects self-esteem and promotes self-assertion as in speaking up or otherwise leading group processes (Price, 1992). A *no* answer reflects low self-esteem and promotes self-effacement as in silence or following group processes. Both yes and no answers and their responses depend crucially on empathic capability.

Whatever qualities form the biocultural basis for prestige, these are not only selected but also tightly linked to both dimensions of leadership and Darwinian fitness (Betzig, 1990). Leadership is covariant in that power is accorded voluntarily by others in an SAHP relationship rather than wrested in agonistic competition. Power is a matter of attracting the favorable attention of others (Gilbert, 1989; 1992). Power is generally an adaptive resource for humans and one directly linked to evolutionary fitness (Chagnon, 1988).

It is, moreover, a matter of common human experience that self-esteem is manipulated by interactions between persons. Praise is welcome, loathing is not. The ethological components of self-esteem reflect this situation as Price (1992) points out. The same signals which raise RHP in the sender are those which lower them in the receiver as an assessment of high relative RHP is displayed. Conversely, signals which display unfavorable relative RHP in the sender raise RHP in the receiver. The main result is an increase in the variance of RHP in a population.

In the agonic mode or egoistic behavior range, then, confidence may be gained with the intimidation of others. However, this agonic system was modified but not replaced with the subsequent evolution of affectional circuitry and SAHP. An important aspect of SAHP is that approbation of attractiveness need not be at the expense of the sender. Confidence is gained with the approval, but not necessarily at the expense, of others. The dynamic balance range of CSN Complex spectrum illustrates more clearly this win-win resolution which is only implicit in prior work on the agonic-hedonic dichotomy.

Signals which boost or put-down are analogous to thymoleptic, or mood-altering, drugs (Price, 1992). The receipt of boosting signals triggers escalation to a high-risk, high-gain hawkish role particularly when engaging primarily egoistic, agonic circuitry. When engaging affectional circuitry, such boosting signals trigger escalation of enhanced cooperation and responsiveness. In humans, such signals maintain self-confidence and brighten mood. The receipt of put-down signals may trigger de-escalation to a low-risk, low-gain dovish role. In humans, such put-down signals have the potential to induce depression or, in special circumstances, rage and even paranoia (Kernberg, 1980).

Therefore, mood regulation is a principle means by which human behavior is scaled in the social sphere (McGuire & Triosi, 1998). Clinical euthymia, or normal mood, requires high constant approbation as in the respect or convivialities accorded by others. As a consequence, almost any person receiving insufficient social signals is prone to reactive depression. Beyond this, certain persons get less approbation than they feel entitled to. They may be victims of societal inequity (their self-assessment is accurate) or they may be prone to psychiatric disorders, either innate or developmental.

In this context, neuroses are typically evident in social comparative assessments which reflect acquired idiosyncrasies wherein self-esteem has become a highly sensitized theme in the course of development. More serious depressive and manic-depressive disorders are typically evident in idiosyncratic neuromental assessments that are the hermeneutical product of variant epigenetic polymorphisms (Crow, 1995). Here thresholds for self-esteem may be set too high or thresholds for submission are set too low. Similarly in this respect, mania is an anomaly of behavioral scaling. Of course, maladaptive clinical phenotypes are most likely when sensitive epigenetic thresholds are traumatized in early development.

The mechanisms of assessment may themselves be grossly aberrant as either an invariant trait or a clinical state. If so, this de-couples internal and external assessments; e.g., the euphoric displays of an overtly manic patient which actually lower SAHP even as the performer perceives it to be prestigious. Likewise, mechanisms of assessment may accurately register as perceived insufficiencies in approbation (or unwarranted disapprobation). If so this can fuel aberrancies in the scaling of behavior in excess of the stimulus, as in hostile or even paranoid drives to remedy perceived social inequity (Kernberg, 1980).

Darwin, in *Origin of Species*, expressed much interest in the evolved basis of such human moral sense and sociability. Yet these interests could not readily mature until the recent era of progress in genetics and neurobiology. With detailed knowledge in these fields having rapidly accumulated in the past several decades it is increasingly feasible to view certain of Darwin's early ideas concerning social competition and cooperation in the fresh light of contemporary science. Moreover, the details of neuromental genetics are increasingly amenable to an integrated viewpoint.

As noted earlier, the social competition hypothesis of mood disorders (Price, et al. 1994) has marshalled a great deal of disparate evidence concerning the evolved basis of social behavior in humans. These lines of evidence start with the social and genetical aspects of mood disorders but can be quite naturally extended to neurotransmitter physiology of a type which evolved to serve social behavior, both competitive and cooperative. A brief extension of the social competition hypothesis into the realm of neurotransmission follows.

EXPANDING THE SOCIAL COMPETITION HYPOTHESIS OF BIPOLAR DISORDER

Darwin, in the good intellectual company of Hume, Mill, and Bentham, had a sound general view of moral theory as it applied to the evolution of standards for conduct within species (Ruse, 1986). All of these thinkers struggled to draw ethics from the muck of Nature and Nature's Law and, thereby, made no appeal to Divine inspiration as the source of or guiding hand for social or ethical interests.

Yet beyond these strong affinities there were important, if subtle, distinctions in the character of their thought. Unlike the others, Darwin was greatly influenced by natural theology, especially that of James Macintosh who emphasized that the human tendency to a moral social purpose is God-given. He further emphasized this tendency caused human sentiments to warm as feelings were first engendered, then possibly reciprocated and, in due course, gratified by the engagement of the socio-moral ethic of doing good unto others (Richards, 1987).

Darwin, replacing Macintosh's God with a more reducible Nature, saw a mechanism of community selection as a link between moral standards and the motive of biological imperative. He thus, far anticipated elements of the definitive work of Fisher (1930), Hamilton (1963, 1964ab), Trivers (1971ab, 1985), and Maynard Smith (1972, 1982). The other enlightenment philosophers meanwhile, and for whatever else their philosophic advances, failed to note this convergence between motives and standards in natural systems.

Despite this robust and most promising start, both Darwinian ethics and related concepts of kinship selection long lay dormant (e.g., see Loye, 2002ab). Neither initiative could prosper in the absence of that cogent grasp of the details of kinship genetics (Hamilton, 1963, 1964ab) and molecular neuroscience (Hyman & Nestler, 1993) which, together, have been attained only in the past generation. Yet with these details manifest, it is now possible to reformulate certain of Darwin's early ideas in the more satisfactory light of contemporary evolutionary and neuroscience. Finally a far more integrated perspective is emerging (Richards, 1987).

In the new field of evolutionary psychology there have been developments which may eventually subsume much of interest in the recent progress in kinship genetics and neurobiology. Among these parallel developments is the specification of the social competition hypothesis of mood disorders in which has been gathered considerable data concerning the evolved basis of moral and social behavior in humans. The lines of evidence buttressing the social competition hypothesis began with the social psychology of mood disorders but is poised to progress briskly with an infusion of research concerning evolutionary neurobiology (e.g., corticolimbic neurotransmitter paths) and social genetics. It is these systems which most proximally subserve not only behavior generally, but social interaction in particular.

RECAPPING SOCIAL COMPETITION

At this juncture it is useful to recap the essential elements of the social competition hypothesis before relating it to kinship genetics or neuroscience. There is an emerging consensus that mood disorders constitute a contemporary phenotypic expression of what was a phylogentically healthy behavioral repertory in the environment of evolutionary adaptation (Lewis, 1934; Hill, 1968; Wenegrat, 1984; Beck, 1987; Glantz & Pearce, 1989; Gilbert, 1992; Wilson, 1993; Stevens & Price, 1996).

Clearly, some variant phenotypes of current mood disorders performed adaptive functions among ancestral hominids. It is thought these variant phenotypes were not identical with contemporary clinical conditions but rather were sub-syndromal and salubrious expressions (Price, 1996). Thus, beyond the constraints of clinical theory and practice, the phylogenetic precursors of what are now diagnosed mood disorders can be better appreciated to have performed some function in connection to social competition (Price, 1967; Sloman, 1976; Gardner, 1982; Gilbert, 1992; Wilson, 1993).

Crucially, the essential result of social competition in any taxon is that winners behave differently than losers (Price, et al. 1994). Yet phylogeny appears to have modulated competitive drives as a key step in the emergence of sociality. Both the Ego-Empathy poles and the agonic-hedonic dichotomy allow the expression of opposing archetypal circuitries in forms that relate closely to the syndrome of depression and mania evident in humans.

In evolutionary psychiatric terms, clinical depression can readily be seen as, in part, a dovish strategy of submission coupled with de-escalation, or loss. Mania, on the other hand, compares to a hawkish strategy of winning via escalated engagement, assertion or charisma (Price, et al. 1994).

A further definition of the hypothesis is that humans retain, among other possibilities, a phylogenetically structured capacity to yield (become depressed) or assert (become expansive) in the face of socially competitive situations. Such social competition can link mood disorders to social competition at several neuromental levels. Thus the hypothesis may also specify hierarchical schemae in a number of domains.

These hierarchies and domains subsume sexual selection, social stratification, ritual antagonistic behavior, resource-holding potential, social attention holding potential and brain phylogeny. In the most fundamental sense, the bases for mood disorders appear to have emerged (1) in taxonomic terms at the reptilian-mammalian interface, (2) in neurobiological terms at the subcortical-limbic interface, and (3) in psychological terms at the Ego and Empathy interface.

Yet such interfaces are not, in themselves, the archetypal aspect of mammalian much less primate sociobiology. Instead, the attainment of social dominance via social appeal and prestige rather than hierarchical threat has come to characterize behavior in the hominoid line (Barkow, 1975, 1980, 1989, 1992; Gilbert, 1992). Put differently, influence rather than raw intimidation is now more commonly successful as means to social dominance and differential reproductive success among primates.

These observations are confirmed in both ethological and socioanthropological field studies. Moreover, this ethological consensus accrues however much the culturogenic-socioconstructivist view has long been favored by academic theorists outside the natural sciences (Paglia, 1994).

STRESS IN SOCIAL COMPETITION

Stress arises in social competition as the need is perceived to either control aversive stimuli or have recourse to safe refuge. This idea of social stress, discussed in Part I, is commonly referred to as the stress-diathesis model (Anisman & Zacharko, 1982; Fowles, 1992). In this model, like the CSN Complex, both submission and dominance behavior can activate psychobiological accompaniments of such stress. Further, evolutionary psychopathology posits both dominance and submission as features normally manifest in evolved behavioral systems. Mechanisms of affect and mood evolved as increasingly crucial signals for social communication as specific behavioral repertoires emerged to mediate socially competitive threats or cooperative safety.

In a review of studies, Gilbert, (1989, 1993, 1995) notes biosocial stimuli are interpreted as to whether they are safe (or rewarding) or threatening (Lazarus, 1966, 1982). If stimuli are perceived as safe they evoke approach behavior or are ignored. Both passive behaviors (e.g., relaxed calmness) and active (e.g., exploration, gregarious assertion) are features of the safety system. Each reduces stress. The safety system has minimal inhibitory components since output varies along a continuum of neuromental arousal. The defensive system relies on attention and appraisal mechanisms to detect aversive inputs, or threats. These increase stress. Algorithms for the evaluation of threats appear to direct a variety of possible outputs, or responses. The output system can induce social escalation (e.g., assertion, dominance display, agonisms) or de-escalation (e.g., fight/flight, inhibition).

Using this approach Gilbert (1989, 1993, 1995) considers states of psychopathology in terms of the simultaneous *assessment* of threat as well as the *response* activated thereby. Thus psychopathologies of extreme escalation and de-escalation derive from sociobiologically normal hawk-dove systems of assessment and behavior. Here submissive behavior is construed as a de-escalating output for defense whereas dominant behavior is an escalating output for assertion. The former is typically a biopsychosocial algorithm of risk avoidance and low gain; the latter is one of risk taking and high gain. Critically, the organism formulates responses in relation to relative risk.

Thus a consensus approach emerges among ethologists, evolutionists, and Darwinian social scientists. These various analyses of dominant and submissive behavior concur that the most common and salient threats to social animals come from competitive conspecifics, as do the many amenities of safety and cooperation. Threats among conspecifics often involve direct down hierarchy attack, loss of status or social approbation, disruption of attachment, individuation and separation, defection of allies, loss of or cuckolding by sexual mates, ostracism, and threats from other groups.

Defense against such threats not only recruits fight-flight mechanisms but also involves displays for the purpose of social communication (Mayr, 1977). That is, defensive responses to threats from conspecifics incorporate complex social behavior and communications which guide the actions of conspecifics, e.g., direct antagonistic attack, ritual agonistic threat displays, submission or appeasement. Deterrent responses include aggression, counter-threat display, or submission. Further nuances include distress calls, gestures of reassurance, and alliance formation.

Upon the neuromental foundation of these nuances were built increasingly elaborate mechanisms promoting social innovations of safety and cooperation. Successful social psychology affects the behavior of conspecifics. These effects can be defensive, as when a successful aggressive display causes the retreat of others, a distress call solicits help from kin or allies and a submissive display inhibits or terminates attack. Similarly, safety signals foster the relaxation of arousal and the redirection of neuromentality to social play, exploration and related phenomena.

Communications based on relative social role thus enable species to organize relationships and, indeed, hierarchies of relationships (e.g., parent-child, dominant-subordinate, supportive ally, sexual partner; pecking orders, etc.). Signals among conspecifics indicate distinct patterns of behavioral interaction (e.g., friendship and alliances, sexual receptivity, and preparedness to assert or yield).

These patterns derive from internal neuromental states that automatically recruit patterns of affect, behavior and appraisal. For example, the submissive response to an unexpected attack is social communication that indicates lower rank which terminates the dominator's attack. This communication is involuntary and associated with increased arousal, fear and specific behavioral displays (Reite & Field, 1985). Other more affiliative states of communication activate equally distinct patterns of emotion, behavior, and social displays that dampen negative affect and enhance social attention. It is the neuromental correlates of affiliation which expand mood to modulate play, leadership and charisma among other exceptional components of advanced sociality. The neural correlates of high and low mood thus repay closer examination.

The brain is in certain respects modular with functional units evolved for certain tasks. In particular, prosocial and affectionate behavior are mediated in part via the orbital frontal cortex, as is evident in meta-analysis of studies in early brain development and social attachment (cf., the discussion in Chapter 3). Thus patterns of social relating are mediated by innate modular mechanisms which subserve specific, evolved social roles.

More precisely, a social neuromentality generates *patterns* of cognition, affect, mood and behavior which guide the enactment of social roles. By way of example, care seeking is a type of submissive dependence reliant on the attention of and proximity to others. In a complementary sense, care-giving is a type of social dominance wherein aggression toward the dependent is at least suppressed and may further affectional and other substrates in the sociobiology of co-operation (Hamburg, 1971).

The communicative aspects of social behavior introduce important new dimensions to a behavioral analysis of the interactive effects of defense, safety and psychopathology. Indeed, from a reductive standpoint, most social behaviors hinge ultimately on defense and safety (Altman, 1981). Therefore powerful emotions are aroused in aggression, ritual agonistic behavior, social hierarchy dynamics of dominance and submission, or alliance and other aspects in the sociobiology of defense and safety (Kroeber, 1952).

Given such modularity, disruptions in discrete brain areas have many specific effects, including the dysregulation-regulation of social information processing and response formulation (MacLean, 1949, 1985, 1990). By extension, psychopathology is manifest when these specific patterns of social defense and safety are dysregulated (McGuire & Triosi, 1998). In humans, social competition may lead to cooperation and affiliation as well as the polar states of yielding or assertion along the Ego/Empathy axis. Cooperation and affiliation reflect dynamically balanced homeostatic regulation of opposing circuitries. On the other hand, yielding or assertion may cause emotional dysregulation, especially where there is a crescendo of interpersonal affects.

These escalations or de-escalations may have been quite helpful in the environment of evolutionary adaptation, especially prior to evolution of neocortical modulation. This is particularly likely given that the dysregulation-regulation of mood in the environment of evolutionary adaptation was limited by more robust modal tendencies of social behavior as is evident in prototypic hunter-gatherers (Dunbar, 1988; Mithen, 1996). The sociobiological cogency of such groups is superior as a regulator of human mood than are massive, industrial societies. It tended to compel an essentially egalitarian dynamic balance of mood and behavior.

Mania and depression may well have been present in the ancestral environment. But contemporary sociocultural contexts constitute a more active substrate for such phenotypic errors. Here, the increasing prevalence of clinically diagnosable manic-depressive illness may be an expression of the putative human hawk-dove polymorphism which is rendered more pathological in the modern setting. As noted at the end of Chapter 11, in terms of the social equation of the CSN Complex, the ratio between Ego and Empathy is divergent from desirable dynamic balance toward the polar extremes, accompanied, of course, with mounting stress. This divergence is further characterized by hawkish escalation in the egoistic range accompanied by dovish de-escalation in the empathetic range. Increasing social stress that goes with such divergence exacerbated by escalation/de-escalation may index a co-efficient of manic-depressive pathology.

With this in mind, neurotransmitter genetics and physiology become relevant as these exist at the direct interface in the phenotypic adjustment of behavior generally and mood disorders in particular. The next chapter will begin to detail this nexus, particularly as it is of such great and direct interest—both clinically and humanistically.

13

TOWARD AN EVOLUTIONARY EPIDEMIOLOGY OF THE NEUROTRANSMISSION OF SOCIAL RANK AND ALLIANCE

Depression, mania, and the many points of phenotypic advantage that traverse the affective spectrum are neuromentalities built upon neurobiological foundations that are themselves the product of genetic selection. The spectrum constitutes a population polymorphism of the human genome in that two or more heritable variants exist. It is, therefore, useful to attempt a more direct connection between the social competition/cooperation hypotheses and recent trends in biomedical research. The endocrine and neurotransmission systems are of crucial importance in supporting and maintaining the expression of conditions such as depression and mania and the related moods of escalation and de-escalation. This chapter will examine each in turn.

NEUROTRANSMISSION

Neurotransmission is a principle means by which neurobehavioral phenotypes are extended from the genome. Much of what is known about neurotransmitter biology has accrued in the past generation. The genomic roots of such extensions were reviewed earlier in Chapter 3. Here the discussion is more proximal and in the service of clinical utility.

Although a variety of neurotransmitters, to include acetylcholine (Janowsky & Overstreet, 2000), GABA (Massat, et al. 2000), norepinephrine (Anand & Charney 2000) have been associated with mood disorders, two neurotransmitter systems in particular have been most closely linked to the pathophysiology of depression and mania—dopamine and serotonin. We will, accordingly, briefly review these neurotransmitters and their interactions with one another. Serotonin and dopamine appear to have co-evolved in the neural evolution of the social mammals and primates, including humans.

Generally serotonin and dopamine have complementary physiological effects. Moreover, they have second-level correlations and interactions in the functional anatomy of the brain. Dopamine is essentially a stimulating, or attention-amplifying (Deth, 2003), neurotransmitter found in excess during manic states. Serotonin is a neurotransmitter that boosts mood and enhances the sense of well-being. These actions are perhaps most clearly reflected by pharmacological manipulations: serotonergic compounds show antidepressant effects whereas dopamine antagonists are potent antipsychotic (anti-manic) agents. Likewise, psychopharmacological progress in relation to mood disorders has lately hinged on therapeutic modulation of these neurotransmitter systems (Nemeroff & Schatzberg, 1995; Oquendo & Mann, 2000; Anand & Charney, 2000).

Even thirty years ago it was obvious that dopamine and serotonin were of special interest to the explanation of mood disorders. With the advent of molecular analytic techniques in the past few years a deeper appreciation of the physiological and genetic complexity of these neurotransmitter systems has emerged (Owen & Risch, 1995; Deth, 2003). For example, it is now known that when neurotransmitters bind receptors, extra-cellular signals transduce to intracellular ones, causing a functional change inside target neurons (Hyman & Nestler, 1993). Also it has been demonstrated that the nervous system contains at least two basic types of receptors, fast and slow. Fast receptors open or close channels as they bind directly on ligand-gates in the cell membrane (e.g., GABA, nicotinic acid). Slow receptor systems modulate second-messenger systems such as cyclic adenosine monophosphate (cAMP) but have a longer duration of action.

DOPAMINE

More specific knowledge pertaining to dopamine has accumulated. Dopamine normally mediates both motor movement as well as affect, mood and cognition while its dysregulation-regulation is implicated in a wide variety of neuropsychiatric disorders including schizophrenia, mania and depression (Owen & Risch, 1995; Stanwood & Zigmond, 2000; Deth, 2003). The first use of chlorpromazine for the treatment of psychosis was an important, if serendipitous, milestone in psychiatry (Delay, et al. 1952). A decade later the antipsychotic efficacy of neuroleptics was related to blockade of dopamine receptors and that conversely depletion of dopamine in the striatum was a key factor in Parkinson's disease (Ehrinnger & Hornykiewicz, 1960; Carlsson & Lindquest, 1963). By 1979, the presence of two dopamine receptor subtypes, D1 and D2, was confirmed and held sway for another ten years (Kebabian & Calne, 1979).

A more complete analysis of the dopaminergic system is now advancing rapidly with combined progress in molecular genetics, brain imaging, and novel receptor-ligand probes (Wong, et al. 1986; Bunny, et al. 1987; Hales & Yudofsky, 1987; Baron, et al. 1989; Onigini & Longo, 1989; van Tol, et al. 1991; Niznick, et al. 1992; Farde, et al. 1987, 1988, 1992; Gingrich & Caron, 1993; Hyman & Nestler, 1993). The principal dopamine projections arise from midbrain and hypothalamic cell bodies of the brain (O?Dowd, 1993). The dopamine neurons form three anatomical groups: the retro-rubral field (A8), the substantia nigra (A9), and the ventral tegmentum (A10) (Ahlskog, et al. 1992).

Of these anatomic arrays, the phylogenetically older A8 and A9 striatal neurons contribute to the extrapyramidal system which initiates and coordinates non-volitional motor activity. The newer A10 neurons enervate mesolimbic and mesocortical tracts, respectively. As it is these A10 neurons that evolved to modulate affect, mood, emotions and cognition; they are critical in pathophysiology of mood disorders, schizophrenia, substance abuse, and a variety of other neuropsychiatric syndromes.

Moreover dopamine, long thought to exist in only two types, is now known to have at least five variant loci, some with considerable point polymorphism (Owen & Risch, 1995; Deth, 2003). Hence the patent inadequacies of this two-receptor model have become quite clear particularly as advances in molecular genetics recently revealed seven distinct dopamine receptors. While, all dopamine receptors are structurally similar with approximately 400 amino acids and stimulate or inhibit via slow G-proteins, minor amino acid variations code for structures which differentiate neurotransmitter subtypes.

Still, it is important to note that these neurotransmitters can still be grouped into two clusters of similar function and distribution. The five dopamine receptors aggregate in line with either the original D1 or D2 typology. Hence they comprise a phylogenic tree. This aggregation reflects similarities in both anatomy and physiology. The distribution of the mRNA of each protein type can serve as an index of neuronal receptor networks in brain tissue (Chio, et al. 1994). Differences in distribution of the subtypes are borne out in mRNA assays.

The D1 receptor is the most abundant in the brain being distributed in the substantia nigra, pars reticulata, caudate, putamen, nucleus accumbens, olfactory tubercle, frontal and temporal cortex. D1 receptors stimulate adenylate cyclase and are strongly implicated in motoric disorders. Their role in psychopathology, if any, is obscure. The D5 subtype is also stimulatory with a profile similar to the prototypic D1 except it more avidly binds the receptor by a factor of ten. This high affinity indicates D5 maintains dopaminergic tone and arousal in its distribution through the limbic system, hippocampus and cortex.

D2 receptors are inhibitory and localized both presynaptically and postsynaptically, and have been identified in two molecular forms, long and short. These classical neuroleptic antipsychotic agents are pharmacologically quite similar and exhibit high affinity for apomorphine, bromocriptine, dopamine, and other agonistics. Indeed, the potency of standard neuroleptics are defined by degree of blockade of D2 receptors. The striatum, substantia nigra and pituitary receive D2 projections. This ability to block the D2 receptor is not uniform throughout the dopaminergic system. For example, clozapine has a moderate affinity for the D2 receptor in the striatum but a much higher affinity for the D2 receptor in the olfactory tubercle, a structure closely tied to the limbic system.

The D3 receptor is a bit more pharmacologically distinct but yet within the D2 cluster (Sokoloff, et al. 1992). It has a fivefold lower affinity for classical neuroleptics and shows little adenylate cyclase activity. D3 appears to be a presynaptic receptor removed from the main site of neuroleptic action and may serve as a regulatory autoreceptor. D3 projects principally to non-cortical areas (olfactory tubercle, nucleus accumbens, striatum, substantia nigra, and hypothalamus) but has little to do with psychiatric pathophysiology.

D4 receptors are similar but do have highly unique binding patterns as demonstrated in the atypical antipsychotic, clozapine (Seeman, et al. 1993). D4 receptors likely are the site of the selective antipsychotic activity of clozapine: marked improvement of both mood and psychotic symptoms without those motoric side effects (as is classically caused by increased D2 blockade, particularly in the striatum). This is quite in keeping with a distribution which concentrates in the frontal cortex and hypothalamus to the exclusion of the substantia nigra.

DOPAMINE RECEPTORS AND BIPOLAR DISORDER (BPD)

D4 is likely the most significant dopamine subtype at least with respect to the psychiatry of manic-depressive illness. More recently, the D4 receptor has been identified as having a polymorphic 16-amino-acid sequence in the third cytoplasmic loop that is repeated from two to eleven times in various genomic alleles (Lichter, et al. 1993: Coyle, 1997; Ronai, et al. 2000). The four-repeat allele predominates as it is found in more than 70 percent of the population. The exon III seven-repeat (EIII7R) allele is of special interest to the present discussion (Kotler, 1997; Prolo & Licinio, 2002). It is implicated in a variety of behavioral repertoires commonly noted in the BPD syndrome—extraversion, novelty-seeking, hyperactivity and substance abuse (Ebstein & Kotler, 2002; cf., Zuckerman, 2002).

The mRNA densities of the D1 cluster follow more primitive extrapyramidal and midbrain tracts which serve territorial and motoric functions than do the densities for the D2 which localize more robustly to the later hypothalamic, limbic and frontocortical areas. The D1 capacities apparently addressed more primitive selective pressures at the time of major reptilian evolutionary advancement. The D2 circuits seemingly mediate challenges arising with mammalian sociality (and also physiology). Related hypothalamic dopamine neurons project through to the anterior pituitary via the tuberoinfundibular tract to the infundibulum. Thus, dopamine impinges directly on the endocrine system, as in the inhibition of prolactin release. These anatomical and functional differences may also account for the fact that it is within the D2 family that most effective and least toxic antipsychotic drugs are to be found (Owen & Risch, 1995).

The D2 cluster is of special interest to the present discussion as its subtypes (D2, D3, D4) appear to be phylogenetically later and designed to subserve the evolution of higher level cognitive-affective expressions selected in response to mammalian social evolutionary pressures. In particular, D2 is linked to mammalian social dominance rank hierarchy attainment and, thereby, to BPD as a dysregulation-regulated form of social competition (Price, 1996).

Moreover, the distinctive anatomical and functional sub-typology of dopamine receptors supports the notion that these arrays derive from a phylogenetically common stock. Each seems to have branched off in resolution of specific selective pressures extant at different times in the environment of evolutionary adaptation (MacLean, 1990; Wilson, 1998). Such branching may have occurred at times as autologous cloning (Wilson, 1993) and clearly arose in conjunction with modulations of neuromentalities needed to accommodate social rank hierarchy as expressed in mammals and primates via hawk-dove escalation and de-escalation strategies.

SEROTONIN

The overall picture for serotonin resembles that of dopamine (Owen & Risch, 1995). Serotonin is, however, even more ubiquitous in the brain and so has perhaps less specific distribution than does dopamine. Nevertheless, serotonin has now some four family clusters with at least fifteen subtypes. The exact number continues to be adjusted upwardly. Indeed, the rapid pace of research makes a fuller account of the current state of knowledge concerning serotonin receptor subtypes especially difficult to summarize and is truly beyond the scope of the present discussion.

Suffice to say, assays of serotonin receptor subtype mRNA densities vary considerably across brain regions. Still, these can also be parsed out into two general types which are distinctive anatomically, clinically and phylogenetically (Wilson, 1997). These assays together establish a clear basis for the mediation by serotonin of widely diverse cognitive and behavioral activities, including attenuation of mood escalation and dominance drive activity (Price, 1996; Wilson, 1997).

As we have seen, MacLean (1990) specified that the circuit which monitors social competition reposes in different brain areas than that which monitors cooperativeness. Further, he delineated that these two circuits activate schematic complexes of highly divergent emotional phylogenetic heritage—reptilian and mammalian, respectively. These can begin to be referenced by their type of neurotransmission. Of special interest to the present discussion are serotonin projections of the raphe-striatum.

The dorsal raphe-projections are phylogenetically earlier features of reptilian brain evolution (a mode similar to D1) than are serotonin projections which mediate the medial raphe-hippocampal system. The raphe-hippocampal serotonin (protoypically, 5-HT_2) distribution thus links to limbic functions. Again, these receptor systems co-evolved to allocate affective-cognitive neuromentalities of mammalian social evolution (MacLean, 1990; Stevens & Price, 1996). Such neuromentalities serve to either escalate or de-escalate mood and related interpersonal affect and cognition in pursuit of social resources (RHP and SAHP) (Wilson, 1997).

COORDINATED ACTIONS OF DOPAMINE AND SEROTONIN

Taken together, these insights required a revised dopamine hypothesis in which it is posited that normal thought and mood is contingent upon orchestrated actions of dopamine and serotonin (and likely including other agents such as norepinephrine, GABA, neurohormones and endogenous opioids. etc.). The recent discovery in the hypothalamic suprachiasmatic nucleus of a master molecular clock regulating circadian rhythms raises provocative questions about the causative source of such wide-ranging, nonlocalized coordinated actions (e.g., see the review by Berns & Nemeroff, 2003). It is even suggested that emergent data in BPD and depression suggest possible mutations in the genes expressing the associated regulating clocks and pacemakers (Mitterauer, 2000; Bunney & Bunney, 2000). Although further research is necessary to elucidate this possible linkage, the particular focus here is on how the conjoint actions of dopamine and serotonin appear to mediate or modulate normal patterns of mood and thought as well as psychosis, abulia, mood, and extrapyramidal effects (e.g., see Ebstein & Auerbach, 2002).

Psychotic and mood symptoms are thought to arise when both the prefrontal cortex and mesolimbic systems become hyperdopaminergic (Janssen, et al. 1988; Davis, et al. 1991; Meltzer, 1992). Serotonin depletion can aggravate symptoms as well (Keck, 1997). This induces initial escalation of mood and activity often followed by increasing irritability and/or thought disorder which can be further accentuated by any reductions in serotoninergic activity. Moreover, compounds with high ratios of serotonin distributed through the raphe-hippocampal-limbic circuit, S_{rhl}, to dopamine of the limbic type, D2, (S_{rhl}/D2) may be especially efficacious in the treatment of psychotic and mood disorders, including severe manic-depression (Meltzer, 1992).

Serotonin inhibits the dopamine projections to the striatum and serotonin re-uptake inhibitors raise mood but also can induce extrapyramidal side effects. Moreover, lesions of the serotonergic dorsal raphe diminish neuroleptic-induced catalepsy. Serotonin also inhibits dopaminergic limbic structures, notably in the accumbens. These serotonin effects on dopamine are modulated by receptors of the 1A, 2 and 3 serotonin subtypes. The increased affinity of clozapine for limbic D4 receptors and its antagonism of 5-HT_2 likely account for its unique profile, both *in vitro* and *in vivo,* as it modulates only mammalian circuits.

The 5-HT_2 moieties can elicit dopamine release downstream in the striatum and thus limit extrapyramidal symptoms induced by dopamine antagonists operating elsewhere in the brain. Abnormalities of the largely mammalian capacities of thought and mood (positive and negative symptoms) can thereby be ameliorated without significant alteration of lower reptilian circuits (involuntary motor symptoms). Likewise, the novel antipsychotic risperidone also improves mood and psychosis with little motoric or endocrine activity via 5-HT_2 antagonism in tandem with D4 (Janssen, et al. 1988).

The conjoint activity of the dopamine and serotonin systems is also of central interest to the elucidation of substance abuse (Ellenwood & Petrie, 1979; Gawin &

Ellenwood, 1988; Wyatt, et al. 1988; Kosten, 1993). This is, in turn, relevant to the present discussion in that cocaine use is mimetic of both manic and depressive symptoms. Cocaine administration acutely increases synaptic dopamine via the inhibition of presynaptic dopamine re-uptake. This produces escalation of mood and activity. If over-stimulation is chronically sustained, D2 receptors eventually down-regulate leading to physical dependence. Thereafter, abrupt cessation of cocaine depletes dopamine. This induces acute agitated depression with a chronic syndrome of anhedonia, dysphoria, lethargy, somnolence, and apathy persisting for up to a year.

For their parts, serotonergic agents moderate both the pre-dependent appetite for cocaine as well as the post-dependent withdrawal syndrome (Rosecan, 1996). Cocaine physiology in humans, then, constitutes a plausible model of manic-depressive, hawk-dove, escalation/de-escalation states. Such significant interplay between social rank hierarchy and particulars of serotonin-dopamine neurotransmission is further borne out in animal studies of escalation and de-escalation noted below.

OTHER NEUROENDOCRINE EFFECTS OF DOMINANCE AND SUBMISSION

Evolved neural circuits modulate both competition and cooperation. In some respects, social neuromentalities are synonymous with the every day term "state of mind" to the extent each embody algorithms of social roles. Gradations in states of mind are to be expected between individuals across the social hierarchy. That is, the state of mind of a dominant differs from those of hierarchical subordinates. Being neuromentalities, these differences in state of mind have proved to be measurable psychobiological variables with both proximal resource and ultimate reproductive consequences. The consequences can be profound (Lack, 1954).

Most remarkably, in some fishes inhibition can cause change of sex (significantly, the subordinate's genes can thus be maintained by mating with dominants; Keenleyside, 1979). Likewise, dominant naked mole rats and New World monkeys have been observed to directly suppress sexual reproductive physiology in subordinates (Abbott, et al. 1989). Studies have also documented major biological distinctions between dominant and subordinate baboons, notably stress hormones which have been linked to the pathophysiology of affective and anxiety disorders (Sapolsky, 1989, 1990ab; Ray & Sapolsky, 1993). Biological profiles of subordinates differ from those of more dominant animals in other species (Henry, 1982; Henry & Stephens, 1977).

Pharmacological probes are also revealing. Drug effects can vary between dominant and subordinate animals, sometimes quite markedly. For example, the response of rhesus monkeys given amphetamine, a dopamine influencing drug, depends on social rank. Dominants showed increased threat, chase and attack behaviors whereas subordinates showed increased submissive behavior, e.g., fear grimaces and turning away (Harbor, et al. 1981).

A later report (Harbor, et al. 1984) described a case in which amphetamine was given to a female rhesus monkey which moved between two groups. In the first group, she was highly subordinate, isolate and fearful with quite limited affiliation. Administration of dopamine-influencing amphetamine accentuated all of these submissive behavioral patterns quite dramatically. However, when moved to a new group she was conspicuously favored by the alpha male. Her behavior changed markedly with this increased rank. When administered amphetamine as the dominant female in this novel social milieu her threats increased dramatically. She did not evidence more submissive behavior as she had done previously, but instead increased her social approach. In this example amphetamine seemed to act as a modulator rather than a causal factor in the nature of the behavior.

Several empirical studies have confirmed that social competition psychobiologically modulates mood states in primates. A recent review of neuroendocrine correlates of social rank confirms that endocrine states are highly sensitive to social feedback (Sapolsky, 1993). Hormone levels in both dominant and subordinate animals vary as a function of group stability. However, instability of the social rank hierarchy affects individuals differently both with respect to directional trends in rank (going up vs. down) as well as basic temperament (Sapolsky, 1994). In this sense dominance-subordination are largely controlled by sociophysiological feedback in the environment.

Perhaps the most significant evidence to date is the correlation of the primate social rank spectrum with serotonin parameters which mirror human mood swings. Physiological changes can be both the cause and consequence of rank changes (Raleigh, et al. 1984, 1991; Hartmann, 1992). Blood levels of serotonin are significantly higher in dominant male vervet monkeys as compared to their subordinates (Raleigh, et al. 1984, 1991).

Sham deposition from social rank of the alpha male was followed by a sharp fall in his serotonin blood level and a sharp rise in that of the subordinate male raised to dominant status. Restoration some weeks later of the formerly dominant male was marked by a reversal of both of these serotonin findings as well as his renewed tenure as the group alpha. Significantly, this restoration led to the fall in both social rank and serotonin level of the "new" dominant below that of his original beta baseline. Thus, social ascendance with subsequent fall in rank produced socio-physiological changes which rendered the subordinate in a lessened biosocial status than had no change occurred.

Sex hormones similarly reflect such social contexts. Androgen levels in humans, both male and female, may vary at baseline both depending upon the degree of social competition or cooperation and also vary directly with success in competitive games (Ehrmann, 1963; Ehrhardt, et al. 1974; Popp & DeVore, 1979; Kemper, 1978, 1990; Bateup, et al. 2002). These rises can include even non-physical contests such as chess (Mazure, et al. 1992). Thus neurotransmitter and endocrine parameters not only reflect general features of social status, affect and mood but also directly link to reproductive biology itself.

So we come full circle. Major subtypes of neurotransmitter receptors, particularly dopamine and serotonin, have unique anatomical distributions and functional consequences. These consequences are quite germane to, among other concerns, the resolution of stress engendered by social competition. Such resolution is most often accomplished via innovative capacities for cognitive-affective assessment and behavior. These capacities increasingly typified the brains and psychology of first reptiles, then social mammals and, most distinctively, the primates and other eusocial species(Dunbar, 1988; Mithen, 1996).

The functional anatomy of neurotransmitters as sketched here is intended to link, in a plausible if preliminary manner, to pathological sociophysiologies of BPD. These pathophysiologies appear to arise more often as the phylogenetic hierarchy set in the triune brain encounters contemporary ontogenic demands.

PEPTIDES AND THE NATURE OF RESPONSE

Before leaving the topic of neurotransmission of social rank and alliance, it is important to acknowledge the importance of peptides. Increasingly, the role of key peptides is seen as determining the *nature* of social response as opposed to the modulating effects of monoamines and some steroid hormones. As noted in Chapter 3, in addition to dopamine, gamma amino butyric acid (GABA) is wide spread in R-complex as an inhibitory neurotransmitter. Glutamate is an even more wide spread excitatory transmitter and there are a variety of opioid peptides. Although the various roles of these other transmitters are yet to be clearly specified, their ubiquitous presence suggests a fundamental influence in determining the nature of the behavioral circuitry and it responses.

Especially relevant in the affiliation and pairbonding necessary to both the concepts of the hedonic mode and to mammalian affectional circuitry facilitating cognitive empathy are the neuropeptides oxytocin, prolactin, and vasopressin. Other important moities include the opioids, corticotrophin-releasing hormone, and yet others which roles are not well specified at this time. Table 13.1 is an elaboration

Table 13.1: Hierarchy of Evolved Biosocial Neuromentality

Alternative strategies for social competition/cooperation at three levels of neuromental phylogeny

BRAIN LEVEL (mentality/mode)	*WIN STRATEGY* (escalating-hawk)	*BALANCED STRATEGY* (modulating-cooperative)	*LOSE STRATEGY* (de-escalating-dove)
NEOMAMMALIAN CORTEX	ATTRACT	ALLY	AVOID
Epithimetikon			
Sociotropic-cognitive/SAHP	Optimistic, charming	Balanced	Pessimistic, ashamed
Cognitive >> Affect	Manic sociopathy		Depressed & Obsessed
S2 > S1 ~ D3, 4 > D1	S2, D3, 4 & NE	S2, D3, 4 &NE	S2 D3, 4 & NE
	High	*Moderate*	*Low?*
Advanced neurohormones & peptides			
PALEOMAMMALIAN LIMBIC	DOMINATE	ALLY	SUBMIT
Thimoedes			
Acquisitive-emotive/RHP	Roused, likely to win	Moderate	Low, likely to lose)
Affect > Cognition	Aggressive & sadistic	Bonding	Morbid & masochistic
S1 > S2 D2, 3, 4 > D1	S1, D3, 4 & NE	S1, D3, 4 & NE	S1, D3, 4 & NE
	High	*Moderate*	*Low*
Primitive neurohormones & peptides			
REPTILIAN MIDBRAIN	FIGHT		FLEE
Logistikon			
Territorial-instinctive/RAB)	Strident, strong		Cowering, weak
S1 >> S2 << D2, 3, 4	D1 & NE S?		D1 & NE S?
	High		*Low*
Few neurohormomes & peptides			

N.B.: Correlate Psychopathologies: Pre-reptilian neuromentalities are essentially vegetative and *asocial.*

Mania = neo-mammalian over-escalation; Depression = reptilian over-submission;

SAHP = Social Attention Holding Power, RHP = Resource Holding Potential, RAB= Ritualised Agonistic Behaviour

S = Serotonin, subreceptors numbered by type, if applicable. D = Dopamine, subreceptors numbered by type, if applicable

NE = Norepinephrine, subreceptors numbered by type, if applicable

Modified from Evolutionary Psychiatry (Steven & Price, 1996, p.226)

of Table 12.1 from Chapter 12 to posit the role of neurotransmitters, neurohormones, and peptides as discussed here.

AFFECTIONAL ESCALATION DE-ESCALATION AND THE LIMITS OF THE HAWK-DOVE DICHOTOMY

Recent findings as to the mood escalating/de-escalating role of dopamine and serotonin suggest these originated in reptilian structures and carried also into later evolving mammalian affectional circuits.

Engagement of escalation/de-escalation amid mammalian affectional circuitry means that both opposing circuits share this capacity. In the agonic reptilian circuits the only options were mood escalation/de-escalation of agonistic behavior, a prototypical hawk-dove/dominate-submit set of responses. The significant involvement of dopamine in the basal ganglia, striatum, or R-complex may account for its apparent relation to feelings of power-control-elation (McGuire & Troisi, 1998: 114).

Cocaine and amphetamines, both dopamine influencing drugs, are often accompanied with agonistic, competitive social behavior. In the mammalian affectional circuitry both Ego and Empathy can escalate or de-escalate mood. Humans can escalate/de-escalate their level of assertiveness as well as their level of cooperation and empathetic supportiveness. Humans can be enthusiastically competitive and loving, or quietly so, as they move back and forth across the Ego/Empathy spectrum. That escalation/de-escalation seems tied to, but somewhat independent of, the nature of either egoistic-empathetic or agonic-hedonic response, is indicated by current research in neurotransmitters and endocrinology.

Given the emerging evidence, one must exercise care not to simplistically equate escalation/de-escalation with the combat analogies of hawk-dove. The latter might be quite descriptive of reptilian functioning but somewhat misleading and inadequate with respect to mammalian circuitry and engagement of it repertoires (see Figure 11.1).

Concerning social behavior, dopamine and serotonin may be primarily, though not exclusively, mood escalator/de-escalators. Dopamine is essentially a stimulating neurotransmitter found in excess during manic states. Serotonin is a neurotransmitter that boosts mood and enhances the sense of well-being. On the other hand, the nature of circuitry responding along the Ego-Empathy axis may be mediated more by the neuropeptides—and, to an extent, steroid hormones.

Only future research will untangle the contributions of these various monoamines, amines, steroids, and peptides in the complexities of the integrated, yet partially independent, functioning of the opposing archetypal circuitries.

14

THE BIPOLAR SPECTRUM AND NEUROPATHOLOGIES OF TALENT

A presumptive case has been made that manic-depressive bipolarity is an enduring population polymorphism of the human genome. However, if the index of contemporary disease burden is a valid consideration, the range of advantageous phenotypic expression appears increasingly mismatched to demands arising in the contemporary developmental environment. How did abiding phylo-epigenetic advantage more recently come to express a greater degree of maladaptive phenotypy and apparent genetic load?

The clarification of phenotypic reactivity as it might pertain to the prevalence of disease expression—a challenging issue in itself—is rendered even more difficult by the inadequacy of primary data sources from which more definitive insights might be forged. Despite the limitations imposed by the primary data, it is possible to consider clinical aspects of manic-depressive epigenesis as a demographic transition between its established phylogeny and very recent trends in ontogeny (Murray & Lopez, 1996).

PHENOTYPIC REACTIVITY DUE TO MISMATCH

Mismatch analysis, referred to previously in this work, is a recent application of basic biological and ecological concepts that has generated novel insights (Bailey, 1987; 1996). Mismatch theory is, varyingly evident in recent work pertaining to developmental economics (Swedlund, 1978), sociobiology (Wilson, 1975), evolutionary psychotherapy (Glantz & Pearce, 1989), evolutionary psychology (Tooby & Cosmides, 1990ab), and Darwinian medicine (Konner, 1983; Nesse &

Williams; 1991). Clearly, evolutionary fluctuations of sociodevelopmental ecology can induce extensive phenotypic reactivity with significant genomic consequences. More specifically, radical shifts away from the environment of evolutionary adaptation predispose to conditions of mismatch within the new selective milieu; e.g., epochal events such as the agricultural or industrial revolutions bring about increased, or at least uniquely differential, incidence of mortality and morbidity.

Reactivity to such shifts can be a factor important in the etiology of epigenetic disease. Thus, it is quite plausible that deleterious phenotypes now expressed by genes linked to bipolar disorder (BPD), rather than constituting intrinsic inborn errors of metabolism, primarily arise as expressions of mismatch between ancestral and current environments. If so, clinical applications of evolutionary epidemiology may become established as an avenue to better understand both phylogenetic and ontogenetic origins of diverse syndromes. Further, evolutionary epidemiology may foster the better treatment of disease states, particularly those recently emergent from genes that had long been fundamentally adaptive.

The term "mismatch" is properly used in reference to insalubrious reactions of human phenotypy induced by phylogenetically novel developmental environs (Bailey, 1995). In standard usage, *mismatch* is limited to specific frictions between abiding adaptations and their current environments of expression. General incongruencies of human nature and culture are referred to as nature-culture *conflict* (Bailey, 1987, 1996). Mismatch results in frictions as evolutionarily stable strategies rub up against novel conditions and give off various clinical sparks (Wilson, 1992).

In the event of either specific mismatch or more general conflict, the etiogenesis of much psychopathology is largely attributable to incompatibilities between evolutionarily stable human nature and the more recent and dynamic aspects of cultural expression. Such incompatibilities provoke diverse mismatches; for example, dietary indiscretion, alcoholism, smoking, senescence, and all manner of psychosocial alienation. Bailey (1996) colorfully avers "... modern humans exist in a sprawling desert of *adaptation-context mismatch* and the frequency and quality of our few 'cases of matching' may be essential for good physical health and psychological optimality". More precisely, the ratio of adaptation match-to-mismatch for an individual in a given environment is plausibly a reliable and valid coefficient for pathologies of both soma and psyche (Bailey, 1996; Gilbert, 1996; Price, 1996; Gardner, 1995; Sloman, 1976).

While interesting in its own right, a thorough historiography of the deeper roots of mismatch research is beyond the present scope. Still, it is worthwhile to also briefly note that mismatch research traces roots through demographic transition theory (Swedlund, 1978) and on to several major intellectual precursors in social psychology. These deeper tap roots notably include alienists such as Marx (1849) with his socialist humanism, Nietszche (1887), with the concept of *ubermensch*, Freud (1930), with his discontents of civilization, Durkheim (1897), concerning suicide, and Weber (1947), with his concerns of *anomie*.

Moreover, mismatch research appears to transcend the limitations of these earlier inquiries as it derives a series of more rigorous inferences drawn from evolutionary genetics. First among these inferences is that taxon-typical morphologies and behaviors were evolved in an environment of evolutionary adaptation sustained in phylogeny (Beck, 1983). With specific reference to humanoids, the genome of *Homo sapiens* is genetically similar to that of 20,000 years ago (Wilson, 1975), or likely even the 80,000 years since the emergence of the species.

This is the case even as massive environmental change has provoked significant phenotypic reaction, largely in the sociocultural sphere. Secondly, humans are often less well matched to current physical, social, and cultural environments than were our Plio-pleistocene ancestors. The quantity and quality of such mismatch characteristics may correlate with individual cases of clinical pathology (Bailey, 1996). Finally, patterns of mismatch pathology may constitute an especially reliable and valid diagnostic taxonomy which, in turn, may lead to improved therapeutic theory and practice (Wilson, 1993).

Mismatch analysis is highly psychological and cognitive, resting as it does on the deduction that human neuromentalities are wired to develop "naturally" in natural circumstances and, less modally, to develop "artificially" in artificial environs (Bailey, 1996). Significantly, neuromental wiring can be transcended via rational cognition. On the other hand, humans, though pleiomorphic in a sociocultural sense, are nevertheless prone to express phenotypes with evolved "default" values (Wilson, 1975).

While contemporary adaptive challenges can resemble ancestral ones, many critical developmental experiences now unfold in situations quite unlike those typical of past evolutionary adaptation. Clearly, adaptive outcomes are more likely when contemporary and ancestral circumstances match insofar as factors subserving biology (e.g., neuromentalities) align with sociocultural needs.

When current environments resemble the ancestral ones, proximate and ultimate adaptive factors reinforce one another. Similarly, human phenotypy is expressed more naturally when free of the often overweening constraints and artifices of cultural, moral or cognitive ethos (Geertz, 1973). Moreover, the severity of both phenotypic reactions and their corollary alteration of selection indices are lessened to the extent the ontogenic environment is akin to the environment of evolutionary adaptation. That is, evolutionary strategic stability is generally maintained.

But even as the current developmental environment is, in certain respects at least, convergent with the past environment of evolutionary adaptation, phenotypes will not necessarily resemble ancestral forms nor be predictably adapted to the new ontogenic milieu. This is particularly true of sociocultural phenomena. Still, creatures generally best do as was done phylogenetically—at least when in environs similar to that typical of evolutionary adaptation at least in the absence of notable exaptational advantages (Aristotle, ca. 450 BCE and as cited in Burton's *The Anatomy of Melancholy*).

Hence, when developmental and evolutionary environmental convergence is sustained, adaptation will become more deeply canalized by phylogeny than would be the case in circumstances highly dissimilar to original ones. Ontogenically emergent phenotypes are, typically, less favored in the struggle for existence. On the physical level there is especially little flexibility. Vital somatic functions must effect a constant fit between body and environs as morbidity or even mortality quickly accrues in the absence of such homeostasis. Thorough-going adaptations are not certain even when the current environment aligns well with phylogeny—optimal fitness is uncommon even in the most modal of developmental environs. Of course it must be said that human culture—mediated as it is by behavior—can mitigate a wide variety of physical constraints (Konner, 1983).

But culture is itself primarily an expression of social psychology. Indeed, considerable innate latitude operates on the sociocultural and psychological level. Such ontogenic latitude is a crucial aspect of how the current human environment differs so radically from that of evolutionary adaptation. A consequence is that fitness of intrinsic elements in human nature can be reduced by the emergent operations of human culture (Nietzsche, 1887). Thus, natural human phenotypes are now rarely expressed as they were in the environs of evolved adaptation, especially in the sociobehavioral realm (Bailey, 1996).

Tendencies of human nature which accrued in phylogeny are now often unsuited to current demands. One central aim of demographic mismatch research is to specify, in both degree and type, those divergencies between evolved adaptations and extant developmental demands which are necessary and sufficient to grasp the etiopathology disease, whether somatic or behavioral.

BIPOLAR PATHOPHENOTYPY AMID THE EMERGENCE OF DEMOGRAPHIC TRANSITION IN THE HOLOCENE

Human nature and nurture are less complementary (matched) as cultural factors gained phenotypic ascendance in the Holocene. Indeed, mismatch has accelerated with the cumulative effects of neocultural revolutions in agriculture, industry (and now, possibly, information). Cultures of massive scale built on complex economic foundations have only recently induced unprecedented degrees of mismatch between human nature and its cultural demands (Calhoun, 1962). Ensuing generations of humans will likely encounter greater degrees of mismatch as this exponential cultural growth manifests no imminent sign of abatement.

All this can be represented visually as in Figure 14.1 below which depicts zones of nature-culture and nature-neoculture mismatch, proceeding from the earliest known hominids, prehumans, early humans, and on to the age of modem *Homo sapiens.* Here, *culture* is roughly synonymous with *ethos and worldview* or *weltanschanung* as in the shared beliefs, customs, values, and symbols of a human social aggregation (Geertz, 1973; Pospisil, 1972). Of course the delineation of nature and culture is, itself, problematic (Wilson, 1975).

FIGURE 14. 1: Biological-Cultural Mismatch and Affective Disorder

MINIMAL MISMATCH:

High Biological & High Cultural Success = Normal Reactivity . . . . F 'guilt' > M

MILD MISMATCH:

High Biological & Low Cultural Success = Social Pathology . . . . .M 'frustration' > F

MODERATE MISMATCH:

Low Biological & High Cultural Success = Neurosis . . . . .F >> M

MAXIMAL MISMATCH:

Low Biological & Low Cultural Success = Severe Psychopathology . . . M >> F

Adapted from Bailey, 1996

In any event, our lineage has been one of small kinship group foragers whose biological and cultural adaptations were rather well complemented through some 99% of human evolution (as well as a substantial, earlier proportion of hominoid, primatoid, mammalian, and reptilian phylogeny). Initially, culture and neoculture were a natural extension of sociocultural function. Formalities and symbolism in ethos and worldview were limited. Neoculture, such as it appears to have been in early hominid and human evolution, derived from basic refinements in food quest activities, toolcraft, shelter provision and so on. These improvements can be termed epi-naturalistic as they appear not to have been the product of highly formalized or symbolic cognitive processes. Moreover, such improvements were likely instituted by those band members who were more analytical, inquisitive, creative, and lucky (Bailey, 1996).

Still, if true innovators were few, most conspecifics were educable. So, human biological, cultural and neocultural evolution reinforced each other at least until processes of *nature-culture separation* were initiated at the dawn of the Holocene. At least since then, culture and neoculture began to grow geometrically and, often, divergently from phylogenetically interiorized patterns. This stimulated an accumulation of knowledge which accelerated as the neocultural innovations of creative individuals spurred the momentum of divergent societal forces (Bailey, 1996).

The distinction between cultural and neocultural innovations is important if somewhat subtle. In particular, neoculture is constituted mostly of historically noteworthy technical advances—often discovered by more creative, non-conformist individuals. Moreover, these advances tend to eclipse or conflict with basic cultural modes as neoculture becomes increasingly formalistic and "objective" (e.g., scientific research, medical or engineering progress, etc.).

Swedlund (1978) notes that unambiguously positive advances (e.g., public sanitation, modern obstetrics and anti-microbial drugs) have, thus far, more than counterbalanced not only emergent mixed blessings (e.g., automobiles) but clear liabilities do accrue (e.g., pollution, trauma). Indeed, neoculture may become an exclusive province of those with unique (largely cortical) gifts; some argue domination by a neocultural elite is well under way (Nietszche, 1887; Bailey, 1996; Itzkoff, 1990). An emphasis of neoculturalists is the distance technology draws us from human nature—natural humanity—itself.

BRIEF SUMMARY OF MEDICO-PSYCHIATRIC SEQUELAE OF DEMOGRAPHIC TRANSITIONS

For some 20,000 years mismatches between systems of human biological and cultural evolution have been progressively expanding. Moreover, the quality and quantity of mismatch is cumulatively more robust with the successive revolutions in agriculture, industry and information. Likewise, the quantity and quality of mismatch is greater in the realm of neoculture than in basic culture insofar as technology is more massive, hierarchical and powerful than is the ethos and worldview of the smaller, more intimate and more egalitarian social groupings which typified the environment of human evolutionary adaptation (Boehm, 1999, Knauft, 1991, 1994). Such egalitarian living marked the dynamic balance of the opposing mammalian and reptilian circuits as depicted in the CSN Complex and social equation. The mismatch between the essentially egalitarian social groups of the evolutionary past is among the most central conflicts in modern society. The modern inequalities of power and wealth probably contribute enormously to stress and pathology. In fact whole psychotherapies (and indeed political and economic philosophies) are built around dealing with inequalities in interpersonal life and society. Fundamentally, the quantity and quality of individual mismatch correlates with physical and psychological pathology.

Mismatch analysis essentially casts basic concepts of phenotypic reactivity into language relevant to medical and psychiatric phenomenology. Thus, etiological considerations can be more readily placed into an evolutionary context. However, several difficulties also arise. Certainly, mismatch concepts relate more directly to descriptions of etiology and diagnosis than they serve as guides to treatment or prognosis. With respect to treatment, naturalistic fallacies are tempting. For example, if pathology is due to individual's mismatch, then is not re-matching curative? On closer examination, however, such phylogenetic regression is not easily or often therapeutic; despite the appeal of Hume (1739) or Rousseau (1755) and others in their wake (Bailey, 1996; Konner, 1983).

Certainly it is simplistic to rely on Haeckel's (1866) notion that biological evolution is necessarily primary while culture is a secondary venue on which phylogeny is successfully and rather passively recapitulated (matched) or not (mismatched). Manifestly, cultural prosperity is a more driving force than is biological success. After all, education, wealth and social prestige each correlates negatively with reproductive success in societies of a strongly neocultural type (Betzig, 1986).

The reconciliation of nature-culture mismatch is perhaps most logical in matters of somatic disease. More compatible environments need to be sought out and habituated accordingly. All else being equal, asthmatics are better adapted to the desert than to Mexico City. Yet, establishing solutions to behavioral mismatches is often a more complex challenge. If one is wholly alienated from the industrial society in which one's phenotype developed, the adoption of a hunter-gatherer lifestyle will, in all probability, be more an affectation than a curative effort.

Indeed, biological and cultural drives are often healthful, when taken separately, but contradictory when considered together. The careerist gratifications of a woman may sharply conflict with traditional family demands. Here evolutionary psychology promises to eventually revise much in the theory and practice of psychotherapy while also resolving many issues of interest to philosophers (e.g., Nietzsche, 1887).

Again, the nature-culture mismatch is derived from strong inferences that the limits of healthy phenotypic expression are increasingly strained. To recap these constituent inferences, humans innately seek both biological and cultural success. However, human nature evolved via natural selection which formed many innate neuromentalities of the brain. Further, *Homo Sapiens* increasingly adapts within neocultures shaped by cultural selection which "usurps" neuromentalities originally fitted for biological success. Still further, demands for biological and cultural success frequently do not match up in the modern world (Hall, 1968).

At this juncture it is timely to consider beneficial aspects which may both inhere to and help explain the high population frequency of the manic-depressive polymorphism whatever its current adaptive value or genetic load.

PUTATIVE ADVANTAGES OF THE BIPOLAR POLYMORPHISM

BPD is a syndrome replete with extraordinary psychological nuances despite its manifestly biogenetic basis. The thought, language, intellect, behavioral and other aspects of manic-depression constitute a highly distinctive and fluctuant temperament well known to patients, family, friends and the world at large. Indeed, adaptive features of the syndrome can be gleaned from merely perusing the two current benchmark diagnostic instruments (i.e., the *International Classification of Disease* 1992 [ICD-10] of the World Health Organization and the *Diagnostic and Statistical Manual* of the American Psychiatric Association 1982 [DSM-III], 1989 [DSM-III-R], 1994 [DSM-IV], 2000 [DSM-IV=TR]).

Indeed, "increased clarity of thought" was identified in the DSM-III as a specific *disease symptom* (sic) of the bipolar spectrum! Of course, modulation is ever an issue. But advantage may well obtain with appropriately expansive mood; increased energy and productivity; optimism and confidence; enhanced gregariousness, jocosity, speech and hedonism amid decreased need for sleep (Wilson, 1994; Davies, et al. 1992). A thorough review of current research—including methodological constraints—is available elsewhere (Goodwin & Jamison, 1990, 2007). Here it is necessary only to survey such empirical research which makes plausible the very notion that manic-depression conferred advantageous characteristics in phylogeny.

The neuromental details of these temperamental elaborations are intimately related to some of the most remarkable clinical phenomena within the manic-depressive syndrome. Surprisingly, some of the strongest empirical evidence that manic-depressive epigenesis can confer advantageous characteristics derives from the realm of clinical phenomenology itself. That the pendulum of mood states within BPD harbored manifold positive features was noted by even early researchers. Esquirol (1845) thusly described the manic-depressive syndrome, with early and quite possibly unequaled eloquence:

> Some maniacs, full of confidence, speak and write with facility and render themselves remarkable by their brilliancy of expression, profundity of thought and their association of the most ingenious ideas .... During their paroxysms all features become animated; the neck swells; the face becomes flushed; the eyes sparkle and all movements are quick and threatening. So many phenomena which, as appertain to the convulsive energy of the organs of life's relations, are associated symptoms; which prove the functions of nutritive life participate in this violent excitement.

Similarly, Kraepelin (1921) emphasized the activity and ingenuity as well as the peculiar mixture of sense and experience of (hypo)-manics. This, in a Darwinian if not a clinical or interpersonal sense, seems an extraordinarily positive commentary concerning a major psychiatric "disease" as offered by one who is perhaps its unsurpassed authority:

> . . .to satisfy their numerous desires, to deceive their surroundings to procure for themselves all kinds of advantages, to secure the property of others for themselves. They usually domineer completely over their fellow patients, use them for profit, report about them . . . act as guardian to them and hold them in check (p. 61).

Kraepelin further noted the rich verbal skills of manic-depressives and observed expanded powers of rhyme and word association. These findings were replicated and extended (Murphy, 1923; Henry, et al. 1971, 1973; Pons, et al. 1985; Calev, et al. 1989). Moreover, these observations are wholly compatible with the social competition model of mood states discussed earlier in detail.

Similarly, Janowsky and colleagues (1970, 1974) empirically identified five types of social control as exercised by persons in elevated mood. The first interpersonal style is the manipulation of self-esteem in others, specifically the raising or lowering of others' self-assessment as a mechanism of social leverage. The second style is perceptiveness of vulnerability and conflict, specifically the capacity to intuit others' sensitivities and reveal or otherwise exploit these. The third style is the projection of responsibility, specifically the technique of attributing responsibility for one's actions to another. The fourth style is the progressive testing of limits, specifically attempts to extend beyond social constraints imposed. The fifth style is the alienation of kith and kin, specifically the distancing from family members.

Like Kraepelin before, Janowsky and colleagues (1970: 254) noted a peculiar admixture of sociability stating persons in elevated mood states are "simultaneously cementing and distancing". Again, these are both of Darwinian significance and entirely compatible with the idea that mood is a potent regulator of social competition.

Moreover, though far from definitive, there is evidence that manic-depressives have greater general intelligence than the population as a whole. Mason (1956) documented that U.S. Army General Classification Test scores of persons with manic-depression were superior to the general population. More specifically hypomania has been related with heightened intellectual capacity (Donnelly. et al. 1982). Robertson and Taylor (1985) found manic-depressives superior to normal controls on several verbal tasks.

However, contrary reports are also available (Silberman, et al. 1983; Gruzelier, et al. 1988). Similarly, important elements of psychomotor and cognitive speed directly co-vary with mood states in manic-depressives—speech, reaction time and psychomotor activity (Miller, 1975; Greden & Carroll, 1981; Widlocher, 1983; Sackheim & Steif, 1988). The preliminary picture is one of manic-depressives suffering from poor concentration and distractibility when highly symptomatic but also having superior associational ability and fluency, particularly when hypomanic.

Empirical studies of interpersonal styles among persons with mood disorder also reveal phenomena with clear consequences in a general Darwinian context and the specific framework of social competition. Communication fluctuations have been reported (Winokur, et al. 1969; Hinchliffe, et al. 1975; Libet & Lewinshon, 1973; Ekman & Friesen, 1974; Miller, 1975; Andreasen, 1975; Akiskal, et al. 1977). Such typically include subtle changes in speech pace and prosody, gesticulation, facial expression, irritability or introversion.

Pronounced gregariousness has been linked to the manic state (Murphy & Beigel, 1974; Akiskal, 1977; Janowsky, et al. 1974). Converse findings typify studies of depression (Paykel & Weissman, 1973; Youngren & Lewinshon, 1980. Coyne (1976; et al. 1987) has evidence that depressives frustrate not only themselves but those nearby and can induce depression in others with their intensely personal discussions.

Hammen and Peters (1977, 1978) further documented both that male depressives are assessed more severely by others than are female depressives and that depressives of either sex are viewed as undesirable potential allies or mates worthy of patent rejection. Gurtman (1986) concludes that the social rejection of depressives is a highly robust finding which consistently arises in studies of diverse design.

Meanwhile, Plutchick (1970a) has established that the manic profile is far more socially desirable than is the depressive presentation. This empirical work supports earlier psychoanalytic formulations that emphasized the "obsessive need" for social approbation which characterizes manic-depressives (Rado, 1928). Likewise, the self-perceptions of manic-depressives comport with these attitudes of others (Jamison, 1996). Thus bipolar persons experience, among many symptoms in the cycle of mood swings, marked changes in self-esteem and self-perception, expressed energy, sociability, and as noted below, sexual identity and sexual intercourse.

Certainly, fluctuations of sexuality are central features of the manic-depressive syndrome. This has obvious connections to natural selection. Not only was this "lewdness in association with manic delirium" noted in antiquity (as by Aretaeus of Cappadocia, c150 AD), but libidinal aspects of BPD constitute major criteria in contemporary diagnostic schemae. Surprisingly, well-designed empirical studies of sexual behavior in manic-depression are relatively rare.

Still, available evidence suggests some two-thirds of manics present with overt hypersexuality while at least a third are so disinhibited as to publicly disrobe or expose their genitalia (Winokur, et al. 1969; Carlson & Strober, 1979; Casper, et al. 1985). Significantly, these libidinal phenomena are empirically independent from considerations of age, religion, social class and duration or recurrence of illness (Allison & Wilson, 1960). This study also documented at least four in five of patients substantially increased sexual intercourse when manic and women were especially prone to provocative displays of libidinal interest. These findings were largely replicated by Winokur and colleagues (1969). Moreover, depression appears to induce a "neutering effect and hypomania or mania had an . . . enhancing effect on sexual identity" (Goodwin & Jamison, 1990: 292).

Spalt (1975) has found that bipolar patients have a threefold greater lifetime experience of extramarital affairs and are far more likely to have greater than ten sexual partners. Such is strong, if indirect, evidence of possible Darwinian selection of manic-depressive genotypy via heightened social competition in reproductive activity (Edwards, 1973). Clearly, more thorough analysis of the reproductive consequences of manic-depression would be useful.

Still, manic-depression has also been tied to issues of Darwinian consequence in other core aspects. Gibson and colleagues (1959) characterized families with manic-depression as much more concerned with issues of prestige and social approbation. Gibson (1963) later specifies how this concern is skillfully gratified as manic-depressives exhibit extraordinary interpersonal perceptions with which they "manipulate and exploit the feelings of others."

This finding echoes proposals made by William James (who with his brother, Henry, suffered severe bouts of affective disorder). James noted that the ardor and excitable character of manic-depression produces, when paired with talent, ". . . the best possible condition for the kind of effective genius that gets into the biographical dictionaries . . . their ideas possess them . . ." (James, 1902: 23–24). Myerson and Boyle (1941: 20) surveyed manic-depressives at McLean Hospital and found ". . . indefatigability, which one so often sees in the hypomanic state may well be a fortunate combination . . ." which confers exceptional social and historical value.

Also, an exhaustive chronicle of biographical studies further supports the idea that bipolar behavior correlates with diverse types of achievement (Goodwin & Jamison, 1990, 2007). Such correlations have been extended to blood-kin who, though seemingly unaffected, may experience sub-syndromal symptoms (Verdoux, et al. 1995). Sub-syndromal symptoms among kin may not meet formal diagnostic criteria, but do suggest kinship selection as a means by which bipolar polymorphism is maintained.

Andreasen and colleagues (1974) first systematically investigated the relationship between mental illness and creativity via well-designed studies based on modern psychiatric diagnostic criteria. Participants at the renown University of Iowa Writers' Workshop were systematically compared with others matched for age, education and sex. Two thirds of the writers had histories of mood disorders, compared with only one-seventh of the controls. Andreasen (1975; 1987) has subsequently extended the study sample over the past twenty years with even more robust findings.

Eighty percent of the writers had been treated for mood disorders compared to 30 percent of controls. Forty-three percent of the writers had some degree of manic-depressive illness, as compared with ten percent of the others. Risk of psychopathology in general and mood disorders in particular among families of writers was also vastly higher than in controls. These findings led Andreasen to propose a familial, epigenetic association between mood disorders and creativity—bipolars are open to contradiction, take risks and defy order in ways which make them both more vulnerable and more original than the general population. The link between bipolar illness and creativity is genuine and extremely robust, Andreasen (1987) maintains and this is consistent with other reports (Karlsson, 1970; McNeil, 1971; Richards, 1981).

Furthermore, these data have been repeatedly replicated. Jamison (1996) surveyed top British artists and writers as to their mood states and history of psychiatric treatment. Her eventual sample included only the heaviest of cultural heavy hitters: members of the Royal Academy, Queens' Gold Medal or Booker Prize winners, Drama Critics awardees, contributors to the *Oxford Book of Twentieth-Century English Verse* among other marks of distinction or accomplishment. She found psychopathology to be epidemic.

Classical bipolar illness affects one percent of the general population and major depression afflicting some fifteen percent. Yet in Jamison's sample more than a third had sought formal treatment for mood disorders. This represents a thirty-fold increase over the general population. Writers experienced the greatest burden of illness with the majority of poets reporting formal psychiatric diagnosis and treatment for depression or mania. One-third of all writers reported mood swings, with poets and novelists having even higher rates. Biographers were less likely "touched with creative fire" as they reported mood swings or elated states at the same rate as the general population.

Jamison concludes intense creative episodes are, in many instances, indistinguishable from hypomania. Again, it seems hypomania may provide intense energy in tandem with perceptions that, percolating through a discerning intellect and creative mind, can induce artistic marked production. Jamison also notes the many elements mood states share with spiritual enlightenment, mystical states and religious inspiration.

Another ongoing study is the personal interviews of Parisian artists, writers and musicians by Akiskal. Akiskal (1996) has previously reported on 750 of his US patients diagnosed with depression, BPD, and schizophrenia. Ten percent of those with more severe BPD were creative artists and writers. But severe manic-depressives had high rates of antisocial behavior, notably violent crimes. The current study compares information from recognized creative individuals with that from a comparison group matched for age, sex, background and achievement in non-artistic areas.

There is, conversely, only one modern study that addresses whether some mentally ill persons are more creative than the general population (Richards, et al. 1988). Subjects were seventeen manic-depressives, sixteen cyclothymics and eleven of their relatives with no psychiatric history. There were two control groups, one without psychopathology altogether and the other with psychiatric diagnoses other than mood disorder. Creativity was assessed by evaluating vocational and avocational activities. The standardized Lifetime Creativity Scale was also administered. Measured levels of creativity were significantly higher among manic-depressives, cyclothymics and their relatives than controls. Cyclothymes and relatives of manic-depressives manifested the highest levels of creativity. It appears creativity is enhanced, on the average, in subjects with mild or even subclinical expressions of the bipolar trait and in lineages with liability to mood disorder.

A variety of other psychometric features converge among manic-depressives and persons of creative temperament. First of all there is emotional reactivity. Artists and manic-depressives tend to be highly sensitive to stimuli in what Andreasen (1987) calls an "extremely fine-tuned" nervous system, sensitive to a wide range of stimuli, including pain. She has speculated that this results from dysfunctional afferent circuits which predispose to altered cognitive filtration of stimuli.

Disinhibition is another trait shared by creative and manic-depressive persons. Tarter, et al. (1985) propose a fundamental breakdown in inhibitory mechanisms is characteristic of most psychopathological conditions. This breakdown, which can also be induced by alcohol or drugs, leads to consilient ideation which connects far flung concepts perhaps afforded by more direct neuromental access to unconscious material. Certainly, thoughts flow freely in mania with loose and novel associations and are otherwise little constrained.

Hypomania is also associated with superior powers of concentration and cognitive absorption evident even in childhood (DeLong & Alderschof, 1983). Studies of children with prodromal signs of manic-depression have significantly richer imaginations than do groups of controls. Manics and hypomanics, often lost in fantasies for hours, can manifest unusual intensity of focus, engage in creative tasks and demonstrate impressive feats of memory or highly detailed concepts or drawings.

This bolsters the notion that manic-depressive epigenesis may facilitate rich and intense experiences due to genetic traits not characteristic of the human genome as a whole. More systematic investigation into the psychopathology of genius may de-romanticize how some extreme individuals have enlarged perceptions of reality (Ellis, 1904). The largest study done to date further confirms the association found by Andreasen (1987), Jamison (1993) and others including Ludwig (1992; the subtitle of the book is indeed telling—*The Price of Greatness: Resolving the Creativity and Madness Controversy*).

For more than a decade Ludwig and associates have sifted through over 2,200 biographies of 1,004 eminent men and women to delineate what characteristics produce high-order creativity. Availing of secondary sources, Ludwig eludes most of the sampling problems for which earlier studies have been criticized. Moreover, Ludwig proposes that, on the whole, biographers who spend years getting to know their subjects make more objective assessments than can a clinician asking standardized questions in a structured questionnaire based on various theoretical or diagnostic assumptions. What one gets in a clinical interview is merely autobiography which at best can be quite selective.

Ludwig reports creative artists, as a group, carry much higher prevalence of mental illness than do counterparts in more organized occupations. Further, between a third of eminent-artists-to-be exhibited psychiatric symptoms as adolescents. These ranged from hypochondria, moodiness and introspective fits to blatant psychosis, hallucinations and suicide attempts. Less than ten percent of future achievers in the sciences, sports and business were affected by similar psychopathology. These differences sustained into adulthood. Rates of psychopathology were between 77 percent for adult poets, musicians and novelists; 68 percent for painters, composers and writers of non-fiction.

Eminent natural scientists, politicians, architects and executives were disproportionately affected compared to the general population, but at a rate considerably lower than that of artists. Ludwig's explanation is circumscribed: creative people with mood or thought disturbance achieve eminence disproportionate to their number and tend to the arts more than commerce or the sciences. The alternate enthusiasm and inhibitions that so often characterize bipolar illness can be a source for inspiration. Indeed, poets have few reasons to constrain flamboyance and the expression of individualism is frequent. However, Ludwig does not accede to the notion of the inspired artist. Instead he emphasizes that greatness is more often a feat of monomaniacal perseverance. Dedication can lead less to madness than a seemingly inevitable pattern of domestic destruction and personal chaos.

There is other important evidence that traits associated with BPD are transmitted via collateral kin (Egeland, 1983; Richards, et al. 1988; Coryell, et al. 1989). Further empirical evidence has been identified which suggest kin of bipolar patients receive experiential contributions from their proband relatives (Greenberg, et al. 1992). This supports the idea that kinship selection may have countervailed, significantly in the environment of evolutionary adaptation if not contemporarily, the *bona fide* disease status of manic-depression (Wilson, 1994).

While psychopathology of the manic-depressive type is most strongly associated with artistic expression, there is a developing body of evidence by which it is linked to other domains of societal leadership. Walter Bagehot (1867) long ago identified the imperious will and energy of leaders who prevail in times of crisis. Subsequently, Thomas (1995) delineates the exceptional prevalence of manic-depressives in on the foundation of the United States Central Intelligence Agency, particularly at the highest echelons.

Similarly, this concept has been the subject of increasingly systematic study (Belloc, 1934; Bluemel, 1948; Goodwin & Jamison, 1990, 2007). Likewise, religious temperaments were early on noted to have a robust linkage to many constituent traits of the bipolar syndrome (Maudsley, 1886; James, 1902). Further, Goodwin and Jamison (1990, 2007) review how the course of religious history has been influenced, repeatedly and directly, by manifestly bipolar persons. Similar analysis has taken root in the history of science (Sulloway, 1996).

SUMMARY

This chapter has been intended, in large part, to demonstrate how the perspectives of evolution can contribute to a fuller understanding of the nature and origins of psychiatric disorder, especially the spectrum of bipolar phenotypy. Therefore, whatever the limitations of this current effort, the Darwinian approach has broad scope and value. Such an approach begins the integration of data gleaned from medical epidemiology through population genetics and on to clinical phenomenology. Further, this approach raises the imminent promise of some therapeutic utility, as well. The preponderance of available evidence strongly suggests that affective temperament is closely allied to a wide range of advantageous endophenotypes. Taken together, these phenotypes begin to account for how the bipolar spectrum includes many "neuropathologies of talent," most notably to do with eusociality.

15

EVOLUTIONARY EPIDEMIOLOGY: A BRIEF REVIEW OF OTHER NEUROPSYCHIATRIC SYNDROMES AND RELATED ISSUES

This book has focused on bipolar disorder (BPD) in explicating the evolutionary perspective and the methods of evolutionary epidemiology. There are many other neuropsychiatric syndromes, however, that may be clarified by a similar analysis. The main thrust of this work has been to demonstrate through the example of BPD that not all so-called diseases or disorders are pathological per se but reflect genotypes that were advantageous in their environments of evolutionary adaptation. Thus, some have been retained above frequency levels for purely pathological mutations. This chapter will briefly consider sociopathy as another "disorder" that likely arose via adaptation as well as other disorders that may have adaptive features to be elucidated by evolutionary epidemiological assessment. It will also touch on related clinical and ethical issues.

SOCIOPATHY

Every society and all ages have a small group of malefactors, including cold hearted egoists. To discuss such persons here, we use the term "sociopathy" meant in a broad sense. However, the technical terminology is more precise and even confusing with a variety of other distinct descriptors, notably *psychopaths* and *antisocial personality* (Hare, 1970; Feldman, 1977; McCord, 1983; Wolf, 1987; Eysenck, 1987; see Mealey, 1995 for an excellent summary in a distinctly evolutionary context).

Sociopaths typically have constrained social sentiments for love, shame, guilt, empathy or remorse despite the classic "veneer of sociability and personal charm.' Indeed, they are aware of how their behavior is discordant with societal expectations, but seem "... to be neither guided by the possibility of such a discrepancy, nor disturbed by its occurrence" (Widom, 1976a, p 614). The essential diagnostic feature is a pervasive pattern of disregard for, and violation of, the rights of others that begins in childhood or early adolescence and continues into adulthood (DSM IV: APA 1994, p. 645).

Sociopathy seems characterized primarily by a failure in the full development and expression of affectional circuitry and neocortical representation of Empathy. This lack of Empathy may even be turned inward towards oneself. It is noteworthy that impulsive aggressive behavior, to include self-mutilation, is a common occurrence in clinical populations with antisocial personality disorder (as is also the case with other related personality disorders) (e.g., see New, et al. 1999; Pattison & Kahan, 1983; Virkkunen, 1976). Yet as a group, sociopaths are not intellectually impaired. They characteristically are skilled at deceptive manipulation of others often with criminal intent and ingenuity in every society, in any economy and in all eras (Robins, et al. 1991).

Frequency of Sociopathy

Only some 3-4% of men and fewer than 1% of women are sociopaths (Strauss & Lahey, 1984; Davison & Neale, 1994; Robins, et al. 1991). The sex differences diminish with inclusion of Borderline Personality, which may be something of a gender-mediated phenotypic equivalent of antisocial personality. Yet sociopaths make up 20% of the prison population (USA) and perhaps as many as 80% of recidivist criminals (Mednick, et al. 1977; Hare, 1980; Harpending & Sobus, 1987). 50% of all crimes are committed by sociopaths (Loeber, 1982; Mednick, et al. 1987; Hare, 1993). Moreover, even non-criminal sociopaths are often irresponsible, impulsive and unreliable with considerable egocentrism and an inability to form lasting personal commitments.

Genetics of sociopathy

For all these alarming facts to do with the disproportionate adverse effects of sociopathy, evidence substantially suggests a robust heritable component (Schulsinger. 1972, 1977; Cadoret, 1978; Crowe, 1974; Cadoret & Cain, 1980; Cadoret, et al. 1987; Cadoret & Stewart, 1991). Related findings include a gene-environment interaction similar to criminality (Cadoret, et al. 1990) as well as a sex effect as male adoptees were more influenced by environmental risks (Cadoret & Cain, 1980, 1981). Moreover, sociopathy correlates with criminal behavior (Eysenck, 1977; Morrison & Stewart, 1971; Cadoret, et al. 1983; Wolf, 1987; Cloninger & Gottesman, 1987; Patterson, et al. 1989). Likewise, adopted-away children of sociopaths have higher rates of criminality (Moffitt, 1987) and, conversely, adopted-away children of criminals are more likely sociopaths (Cadoret & Stewart, 1991; Cadoret, et al. 1990).

The term "antisocial spectrum" has been suggested to incorporate the variety of phenotypic expressions thought to arise from core genotypic traits (Cadoret, 1978). This spectrum is a range of behaviors from extreme to moderate to approaching normal. As symptoms deviate from normality toward extremes, they become individually and socially significant enough to be classified as pathological.

Adaptive Advantage

This genotype confers unique innate temperamental or personality factors as well as autonomic hypoarousal that alter normal socialization and moral development. Such primary sociopaths usually evidence chronically pathologic antisocial behavior. Nevertheless, from an evolutionary epidemiological perspective, sociopathy appears to have been maintained in the genome due to selective advantages that optimize at a relatively low frequency-dependent rate. That is to say, there is a population polymorphism that predisposes some "to acquire and be reinforced for displaying antisociality" (Rowe 1990a, p 122).

Mealey's integrated game-theoretic models of sociopathy is the most comprehensive among the many early efforts (Kenrick, et al. 1983; MacMillan & Kofoed, 1984; Kofoed & MacMillan, 1986; Harpending & Sobus, 1987; Cohen & Machalek, 1988; Wilson, 1993; Mealey, 1995). Mealey identifies sociopaths as cheater-defectors amid societies comprised mostly of mixed-strategy inter-actionists. Here sociopathy is the adaptive product of frequency-dependent evolutionary selection. Such selection greatly predisposes relatively few to an evolved but contrary life-historical biopattern reliant on social deception in pursuit of manipulative or even predatory interactions. So, what is the adaptive advantage of this lack of empathy or disregard of the rights of others?

The Violation of Reciprocity

The CSN Complex of our neural architecture facilitates interplay of reptilian self-preservational circuitry and mammalian affectional circuitry (Ego in conflict with Empathy). This is mediated by frontal neocortical circuitry which, if optimal, achieves dynamically-balanced reciprocity. Reciprocity, the universal norm of society, primitive and modern, is what sociopathic behavior violates (e.g., see McGuire & Troisi, 1998: 191–193; Mealey, 1995; Dugatkin, 1995; Harpending & Sobus, 1987). Such violation creates significant social stress between the nonreciprocator and others, generally with detrimental or objectionable social effects. That is why it is deemed a pathology or personality disorder.

Nonreciprocators such as sociopaths typically are not entirely lacking in empathy. They use their perhaps truncated empathic capacities to deceive others that they will, in fact, reciprocate. Their success at obtaining resources and mates depends on such deceptive capacities. They are often charming, but calculating and egoistic in their ultimate aims. They particularly flourish in the complexities and anonimities of modern mass society, ever inclined to be "looking out for number one" or winning at any cost. Philosophically, they align with Ayn Rand's philosophy of egoism, if also lacking her level of sophistication and ideological justification (Rand 1964, 1967).

It is estimated some 50% of sociopaths go effectively undetected. This is because of their skill in deception and because they are often readily forgiven—except in egregious instances—through the empathic qualities of others.

Such nonreciprocators or limited reciprocators, sustained adaptive advantages over evolutionary history, since their frequency in the population is likely well above presumed mutation rate prevalence. Sociopathy is probably best not viewed as a disease in the strict sense but rather as a behavior repertoire that arose in the environment of evolutionary adaptation to deviate significantly a countervailing force from the normative range along the Ego-Empathy spectrum. However, it seems all but certain that opportunities for sociopaths increasingly abound as human populations continue to move away from the developmental norms of the environment of evolutionary adaptation and toward ever more massive socioeconomies replete with all manner of newly emergent niches, technologies and relationships to exploit.

Further evolutionary epidemiological research is needed to illuminate both the exact modes of transmission as well as key developmental pathways in the expression of sociopathy. Still, their epigenetic neuromentalities appear to be adaptive phenotypic reactions at higher prevalence in societies less able to reliably detect non-reciprocating cheaters, notably mass cultures with complex political-economic systems.

BPD VS. SOCIOPATHY ON THE AXES OF THE CSN COMPLEX

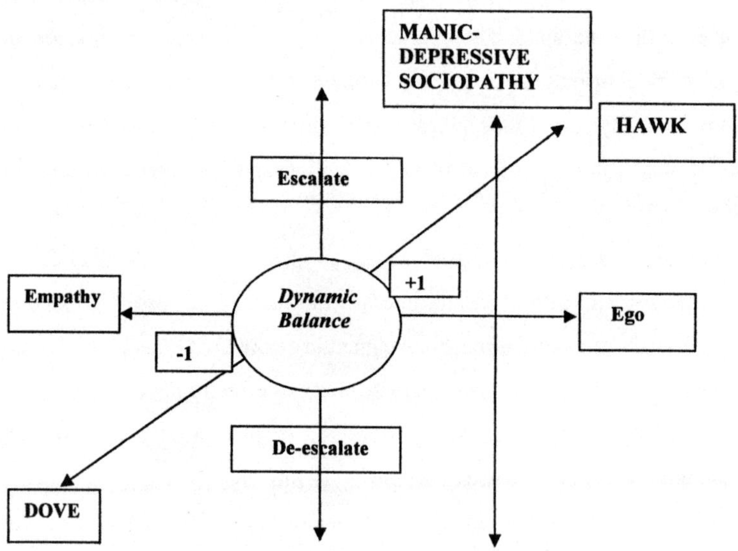

Figure 15.1. BPD and Sociopathy Expressed on Axes of CSN Complex.

In Chapter 11 the bipolar, hawk-dove strategy, was illustrated on the axes of the CSN Complex (Figure 11.1). In Figure 15.1 above, sociopathy is indicated as expressing extreme Ego on the Ego/Empathy axis. Manic-depressive sociopathy is expressed as an escalation/de-escalation intersecting the Ego-Empathy axis at the extreme of Ego, rather than crossing the axis from the egoistic into the empathetic quadrant. Thus the interrelatedness of the two syndromes may be indicated or differentiated by the points of intersection of the two primary axes. Although BPD may achieve dynamic balance of Ego-Empathy at the point of intersection, sociopathy lacks such a capability. In escalated and de-escalated moods, it remains at the extreme of Ego expression.

It is worth noting that the highly simplified, formalized game theory model of hawk-dove, in effect, models only the sociopathic form of BPD. This approximates the foundational reptilian circuitry. Although the game theory model is useful, it does obscure the complicating, counterbalancing effects of mammalian empathic circuitry. Likewise, the momentous event of neocortical capacity to represent and dynamically balance the essentially opposing archetypal circuits is less obvious in this model of dynamic interaction.

APPLICABILITY TO OTHER PSYCHIATRIC SYNDROMES

As previously argued, the clinical and theoretical importance of evolutionary epidemiology extends to both physical and psychological aspects of medicine. Psychiatric genetic epidemiology, as collated from indirect studies of the genome and newer direct studies, points to a robust genetic basis for and significant, species-general expression of several other—but by no means all—important psychiatric syndromes, particularly those that likely surpass thresholds of epigenetic frequency. Some conditions, notably schizophrenia, are likely expressions of transient polymorphisms or stochastic rates of selection and mutation.

Inferences from evolutionary epidemiology will further revise psychopathologic theory and practice as pertains to other syndromes, as well. In addition to manic-depression and sociopathy, the deductive methods of evolutionary epidemiology suggest that natural selection promoted genetic lineages now expressing:

(1) unipolar depression,
(2) anxiety-panic disorder,
(3) obsessive-compulsivity,
(4) neurodevelopmental dyslexias and left-handedness,
(5) the biological underpinnings of personality disorder,
(6) substance abuse syndromes,

(7) psychodynamic conflicts,
(8) homosexuality—both female and male, paraphilias and,
(9) various isolated traits of cognition or personality.

We leave these issues for the future as they are beyond the scope of the present work. The present work has as a focus mania and depression, in part because in all of psychopathology it is the affective spectrum (perhaps with antisociality) that has sufficient epidemiologic and phenomenological data to suggest adaptive polymorphism.

CLINICAL APPLICATIONS AND ETHICAL CONCERNS

BPD, sociopathy, and other disorders each can affect at least one to three percent (and perhaps six to ten percent) of human populations surveyed—hundreds of millions of persons worldwide. The economic impact of such disorders has not been calculated in a comprehensive sense. However, the cost of manic-depression alone was assessed at a total of some $45,000,000,000 in the USA alone in 1990 (Rice, 1997). This estimate included $7,000,000,000 in direct costs, such as hospital or outpatient care as well a legal expenditures. A further $38,000,000,000 was attributed to lost income, reduced productivity as with suicide, and other indirect economic consequences. Cowan and colleagues (2002: 2) estimated that the collective cost of the various psychiatric disorders was in the range of 400 billion, comprising one third of the total health care budget. The World Health Organization reported that by the year 2000, BPD had become the fifth leading cause of disability among the age group 15–44 years (The World Health Report, 2001 as quoted in Ruzickkova, 2003).

In this context and as reviewed more thoroughly earlier, it is striking that a syndrome so clearly epigenetic should also retain such high population prevalence. Even a preliminary re-interpretation of these seemingly contradictory facts concerning BPD must consider evidence derived from the environment of evolutionary adaptation. Such a program of analogical analysis hinges on forms of reason in which the value of conclusions is derived from the value of the premises (Descartes, 1641; Locke, 1690; Newton, 1687). Fortunately, the premises of evolution are themselves the subject of scientific consensus in a general sense.

However, as in application to the evolutionary epidemiology of BPD, there is a further empirical consideration as well. As Bacon (1620) noted, "Nature may only be commanded as She is obeyed." This consideration, in turn, has to do with the questions surrounding the reliability and validity of neuropsychiatric taxonomy (Rose, 1980). Such a situation recalls how Darwinian science could emerge only from a valid Linnaean hierarchy (Wilson, 1993).

Doctors would do well to diagnose disease with the same or similar precision by which biologists classify organisms. Such diagnostic processes should be parallel across environmental range as may include the chance of case identification as well as sociocultural variants. Finally, adjustments must often be made in the conversion of epidemiological prevalence data to evolutionary incidence data. Again as noted earlier, the practice of psychiatric phenomenology only recently has attained a workable, if uneven, level of precision in its epidemiological classification. Thus, the emerging program of psychiatric evolutionary epidemiology is propitiously situated.

However, there remains a dearth of medical-anthropological data relevant to questions of the operations and evolution of epigenetic diseases. This impediment is as remarkable as it is unfortunate. Clearly there is a need for much more research of a systematic and modern type. Information of demonstrable validity and reliability is unavailable in a number of domains absolutely critical to a more thorough understanding of the epigenetic form and function for a variety of syndromes. This is a highly worrisome prospect within the sphere of contemporary human genetics. Technology stimulates scientific understanding as often as it is the product of directed efforts in the application of knowledge (Kuhn, 1970; Price, 1971). That is, in many cases actions precede knowledge of their consequences.

All this is particularly problematic with respect to psychopathology, notably affective disorders (Murray & Lopez, 1996). Residual hunter-gatherer populations have yet to be studied in psychiatric epidemiological surveys of rigorous design and availing of contemporary nosological methods with good reliability and validity. Therefore, the foundation for analogical consideration of the role genes now linked to BPD may have played in the environment of evolutionary adaptation is not solid. Likewise, exceedingly little work has been done in such basic areas as determining fertility rates within psychiatric diagnostic categories (Price, 1996). Similarly, the analysis of positive aspects of psychopathology has come to be formally studied scientifically only recently.

UNIQUE PERSPECTIVES FOR GENETIC RESEARCH AND COUNSELLING

The era of direct molecular genetic investigation is expected not only to identify chromosomal loci, but also to allow transgenic animal models of diseases as well as the important work of specifying modes of transmission, mutation rates, and both cellular and developmental products of specific loci, etc. (Kelsoe, 1996; Rosenberg, 1997). These and other central facts of epigenesis are necessary for any fully informed, much less, ethical consideration of much in the purview of medicine and anthropology.

Yet, in some cases genetic testing for traits may become available before the full nature of phenotypic expression is known. While such information may ultimately be valuable for early therapy, complex ethical issues are engendered by even remote prospects of germ line intervention. Quite clearly, however, genetic manipulation is emerging as a major therapeutic modality. These technologies, now only in the early stages of development, may soon enable treatment or cure of genetic disorders via the administration of genetic material as a direct therapeutic agent. In application to neuromental problems, direct genomic therapy involving germ line surgery is especially complex when there are likely positive endophenotypes.

Psychopathological syndromes present especially formidable hurdles given their unique confluence of germ-line epigenesis and ontogenesis. However, there is every expectation the methods of genomic research will be successfully directed to the elucidation, amelioration or even elimination of at least severe psychiatric disorders (DePaulo, 1996). There are further worries already identified about the potential demand for cosmetic applications in neuromental surgery (Rothstein, et al. 2002).

Current trends in genomic research will provide much more specific knowledge pertaining to debilitating illnesses. For psychiatry, this likely will usher in an era of biological research with manifold and surprising social, legal and ethical consequences. BPD, with its fascinating mix of individual suffering and sociocultural greatness, constitutes a special conundrum. Certainly it is advisable to clarify whether and what are the advantages, individual or social, of the presence of manic-depressive traits in the human genome well before embarkations in the realm of germ-line therapy (Jamison, 1996).

These considerations may differ greatly with diverse perspectives (Trivers, 1971ab, 1985; Badcock, 1990, 1992). Parents, for example, may be less inclined to transmit genes which may burden their offspring with major psychiatric sequelae. On the other hand, society and culture may benefit from such transmission (Jamison, 1996). The experience of manic-depressives themselves is further replete with ambiguities and ambivalences although few convincingly state a preference for a life entirely free of the trait (Gilbert, 1992).

There is, further, the notion of modulating the bipolar trait so as to produce sustained hypomania absent either marked depression or overt mania. While this may be a useful clinical objective, as in titrating pharmacotherapy, it takes on an entirely different aspect when the object becomes one of molecularly re-engineering the character of human life experience. But these are, ultimately, the concerns of philosophers and theologians and, as such, begin to move beyond the scope of the present discussion. Still, there are other, more immediate biomedical and neuropsychiatric, considerations to mention.

FURTHER ISSUES AND POSSIBILITIES

A general deduction obtains that Darwinian advantage inheres to conditions that surpass specific thresholds of epigenesis and prevalence. Initial attention in general medicine has already turned toward the reassessment of, to use Garrod's fine turn of phrase, *common inborn errors of metabolism* (Garrod, 1902, 1909). Again, it may be incidental whether a trait is a Mendelian dominant, recessive, polygenic or mixed. More importantly, the default medical assumption that genomic variants are all purely expressions of disease or otherwise stochastic events may need to be amended.

As has been noted with sickle-cell anemia, highly evolved gene systems cannot be truly termed erroneous. Other traits also spring to mind which are not only common but that can be readily postulated to have conferred advantages in prehistory (Wilson, 1992, 1993). Cystic fibrosis and the anomalies of alpha-1-proteases are obvious candidate genes which may have reduced the burden of infectious disease in the environment of evolutionary adaptation (Wilson, 1992, 1993).

Clearly the Darwinian approach has broad scope and value. Further examples are emerging from a range of medical specialities. Some rheumaetological HLA-typology risk factors appear to co-vary with certain infectious agents (Bodmer & Bodmer, 1978; Thompson, 1981). A variety of ophthalmological conditions such as refractive errors and macular degeneration may be cast in positive evolutionary terms (Lundblad, 1995).

Likewise, the neurology of syndromes such as dyslexiae, attention deficit, Tourette's syndrome and others may harbor adaptive features (Seguin, 1866, on idiot savantry; Maudsley, 1886; Geschwind, 1985, "the neuropathology of talent"; Damasio, 1994). Endocrinology, with its various subclinical humoral anomalies, may reveal certain endophenotypic advantages as well (Tuke, 1878). In fact, a broad array of medical specialties are beginning to study the possibility that some genetic diseases must have definite, if still obscure, advantages (Wilson, 1993, 1994; Nesse, 1995).

SUMMARY

The phylogenetic accumulation of what now constitute etiogenes will be much clarified as the Human Genome Project and related protocols move forward. Likewise, such projects shall clarify much more the epidemiology of diseases (both genotypic and phenotypic). The time is nearly at hand when it will be feasible to routinely assess whether a wide range of genes linked to contemporary pathophenotypes were, by the criteria of evolutionary epidemiology, selected into the human genome. Already, it is clear at least some contemporary pathologic syndromes are expressions not of genetic error but of past advantageous selection stretched by the current environment beyond the plasticity of their healthy range of phenotypic reaction.

Such knowledge may well constitute an unavoidable temptation to therapeutic adventurism and misadventurism. It is important that any genetic therapies not assume disease is simply disease. Certain polymorphisms of at least possible utility are at risk of misguided therapy. Surely other gene systems now notable only as causes of individual disease will come to be seen, in the full light of evolutionary epidemiological analysis, as fundamentally salubrious characteristics.

PART IV

CONCLUSION

16

CONCLUSION: BETWEEN OPTIMISM AND CAUTION

The present effort to fully develop evolutionary epidemiological analysis is certainly limited by constraints of available data. Still, this explication of a Darwinian approach has begun the further extraction and assembly of data from medical epidemiology and facilitated its integration with principles of population genetics. These efforts lead toward a reconceptualization of clinical phenomenology. Therapeutic utility seems an imminent promise, as well. Nevertheless, caution is required in moving forward.

In doing this evolutionary epidemiological analysis, we have focused on phenotypically expressed global state variables, that is, the core motivational states produced by interactions of archetypal self-preservational and affectional circuitry of our reptilian and mammalian ancestry. These interactive, global motive states are biological functions that have long served our lineage in the organization of social behavior and permitted adaptation to changing demands of the physical and social environments. Their homeostatic functioning within normal limits—made possible by socially supported and internalized neocortical mediation—makes our advanced social life possible. Normal limits have, of course, long been defined in terms of the existing social context.

When behavior exceeds normal limits as defined in the current social environment, it is pathological because of the significant individual and social stress that it inflicts. The illnesses we define as bipolar disorder, sociopathy, and probably many other syndromes, fall outside of current normal limits—although they were likely adaptive in earlier environments. A full explanation of such motivated states cannot be achieved simplistically at a single level of integration or reduction. Full understanding will require the integration of genetic, molecular, physiological, neural, as well as behavioral-psychological levels of explanation perhaps entailing multiple mechanisms. Such integration is possible only in an evolutionary context (cf., Panksepp, 2002; Schulkin, 2002). In this work we have attempted to advance this inclusive process.

Care must be taken not to oversimplify the task. Despite the fundamental importance of genes, we cannot simply make the direct linear leap from genes to social behavior and psychology. From genes, we must move to biochemical pathways and from these to cellular physiology and function. From cells we must move to anatomical organization and interaction among brain structures themselves. Only at the level of the whole brain, the whole organism, or groups of socially interacting organisms, can we fully define, explain, and appreciate the function and adaptive value of such global motivation states (e.g., see Balaban, 2002; Bakker, et al. 2002).

Interestingly, Müller and colleagues, in their discussion of the genetics of endocrine-behavior interactions, finalize their review with the caution that in looking back we may come to recognize that "genes were easy." The analysis of the proteome, which the genes create through incorporation of environmental nutrients, may constitute the greater challenge. The proteome comprises a much greater level of complexity than the genome itself and contains a much larger number of molecules than could be predicted from the 20,000 to 40,000 genes comprising the genome. Protein analysis will be essential to an explanation of that most complex form of biological organization—the endocrine and neuro-endocrinetransmitter activity underlying behavioral expression (Müller, et al. 2002: 289–290; see also the review by Patterson & Aebersold, 2003).

In conjunction with current efforts to understand the complex transition from genotype to phenotype, the time is at hand for a thorough-going archaeology of brain and mind which might better exhume and clarify the origins of our social brain and the triumphs and tragedies with which both phylogeny and ontogeny imbue human life. Such explorations can only enliven discourse pertaining to the evolutionary origins and ethics of human behavior. It was this very discourse which so particularly fascinated Darwin late in his life. Such exploratory efforts and discussion are otherwise crucial to any complete understanding of a number of major medical syndromes as well as our successful adaptation to the evolving complexities of the emerging global social environment.

The challenges that confront society, including psychiatry and its related disciplines, based upon the new findings and insights emerging from an evolutionary perspective, are formidable. In the modern social context, we face an environment significantly different from the hunter-gatherer or foraging societies in which our archetypal motivational systems evolved. The most significant social mismatch with the environment of evolutionary adaptation is the shift from an essentially egalitarian to a hierarchical, large scale society with manifold tableaux of interaction. Much of our recent history, involving the struggles with totalitarianism and demands for greater equality, has been driven by the stress of this mismatch. Much of our current conflict in the virtual global society may, likewise, be seen as proceeding from this mismatch. The social stress emanating from mismatch gives rise at the societal level to diverse conflicts, both economically and politically.

At the individual level, driven by tension in our archetypal motivations, we have become increasingly ambitious. We seek not just the minimal adaptiveness for survival but desire much more. We seek to manage our conflicting archetypal motivations within desirable normative limits. We further seek to achieve even more—the optimal health and happiness of all. Such high—even megalomanic—expectations could never be guaranteed by the process of natural selection or inclusive fitness.

The achievement of worldwide self-governance consonant with our evolutionary origins, coupled with the unprecedented stress-driven social objectives of optimality, will require a full appreciation of the evolved roots of human sociality—both the nature of social competition as well as its dynamically balanced expression in social cooperation. The development of appropriate methodologies as well as social strategies must be on a realistic assessment of our human limitations and potentialities or they may prove tragically abortive. The challenge is worthy.

BIBLIOGRAPHY

Aboitiz, A.; Montiel, J.; Morales, D.; & Concha, M. 2002. "Evolutionary Divergence of the Reptilian and the Mammalian Brains: Considerations on Connectivity and Development." *Brain Research Reviews*. 39: 141–152.

Abbott, D; Barrette, J; Faulkes, C; George, A. 1989. "Social contraception in naked moles rats and marmoset monkeys." *Journal of Zoology*. London, 219: 703–710.

Abrams, D; Cochrane, S; Hogg, M; Turner, J. 1990. "Knowing what to think by knowing who you are: Self categorisation and the nature of norm formation, conformity and group polarization." *British Journal of Social Psychology*. 29: 97–119.

Ahlskog, J; Muenter, M; Bailey, P; Stevens, P. 1992. "Dopamine agonist treatment of fluctuating Parkinsonism." *Arch Neurol*, 49: 560–568.

Aita, V., Liu, J., Knowles, J., Terwilliger, J., Baltazar, R., et al. 1999. "A Comprehensive Linkage Analysis of Chromosome 21q22 Supports Prior Evidence for a Putative Bipolar Affective Disorder Locus." *American Journal of Human Genetics*. 64: 210–217.

Akiskal, H; McKinney, W. 1973. "Psychiatry and pseudopsychiatry." *Arch Gen Psychiatry*. 28: 367–373.

Akiskal, H; Djenderedijian, A; Rosenthal, R; Khani, M. 1977. "Cyclothymic disorder: Validating criteria for inclusion in the bipolar affective group." *Am J Psychiatry*. 134: 1227–1233.

Akiskal, H. 1989. "Psychopathology, temperament and past course in primary major depressions. 1. Review of evidence for a bipolar spectrum." *Psychopathology*. 22: 268–277.

Akiskal, H. 1995. "The bipolar spectrum: Research and clinical perspectives." *Encephale*. 21 Spec 6: 3–11.

Akiskal, H. 1996. "The prevalent clinical spectrum of bipolar disorders: Beyond DSM-IV." *J Clin Psychopharmacology*. 16(2 Suppl 1): 4–14.

Akiskal, H. 1997. "Need for recognising bipolar disorders in all its forms (Editorial)." *Encephale*. 23 Spec 1: 1.

Albers, H.; Huhman, K.; & Meisel, R. 2002. "Hormonal Basis of Social Conflict and Communication." Pp. 393–434 in *Hormones, Brain, and Behavior*. Ed. by Pfaff, D. V. I. NY: Academic Press.

Alexander, J; Lerer, B; Baron, M. 1992. "Ethical issues in genetic linkage studies of psychiatric disorders." *Br J Psychiatry*. 160: 98.

Alexander, R. 1987. *The Biology of Moral Systems*. Hawthorne, NY: Aldine de Gruyter.

Allen, M; Cohen, S; Pollin, W; Greenspan, S. 1974. "Affective illness in veteran twins: A diagnostic review." *Am J Psychiatry*, 131: 1234–1239.

Allison, A. 1954a. "The distribution of the sickle-cell trait in East Africa and elsewhere, and its apparent relationship to the incidence of subtertian malaria." *Trans R Soc Trop Med Hyg.* 48: 312.

Allison, A. 1954b.. "Protection by the sickle-cell trait against subtertian malarial infection." *Br Med J.* 290.

Allison, A. 1954c. "Notes on sickle-cell polymorphism." *Ann Hum Gen.* 13: 39.

Allison, A. 1964. "Polymorphism and natural selection in human populations." *Cold Spring Harbor Symp Quant Biol.* 29: 137.

Allison, J; Wilson, W. 1960. "Sexual behaviors of manic patients: A preliminary report." *South Med J* 53: 870–874.

Allman, J. 1999. *Evolving Brains.* NY: Scientific American Library.

Altman, S: "Dominance relationships: The Cheshire cat's grin? *Brain and Behavioral Sciences.* 4: 430–431, 1981.

American Psychiatric Association: *Diagnostic and Statistical Manual of Mental Disorders* Washington, DC: American Psychiatric Association, (3rd ed) 1982, (3rd rev ed) 1989, (4th ed) 1994.

Anand, A. & Charney, D. 2000. "Abnormalities in Catecholamines and Pathophysiology of Bipolar Disorder." Pp. 59–94 in *Bipolar Disorder: Basic Mechanisms and Therapeutic Implications.* Ed. by Soares, J. & Gershon, S. NY: Marcel Dekker.

Andersen, S. 2003. "Trajectories of brain development: point of vulnerability or window of opportunity." *Neuroscience and Biobehavioral Reviews.* 27: 3–18.

Andreasen, N; Canter, A. 1974. "The creative writer: Psychiatric symptoms and family history." *Compr Psychiatry* 15: 123–131.

Andreasen, N; Powers, P. 1975. "Creativity and psychosis: An examination of conceptual style." *Arch Gen Psychiatry.* 32: 70–73.

Andreasen, N. 1987. "Creativity and mental illness: Prevalence rates in writers and their first-degree relatives." *Am J Psychiatry.* 144: 1288–1292.

Andreasen, N; Rice, J; Endicott, J; Coryell, W; Grove, W; Reich, T. 1987. "Familial rates of affective disorder: A report from the National Institute of Mental Health Collaborative Study." *Arch Gen Psychiatry.* 44: 461–469, {with published erratum in Arch Gen Psychiatry 45: 776},1988.

Andreoli, V. 1992. "The complexity of psychiatric nosography and the'"simplicity' of molecular genetics." *J Psychiatr Res*, 26: 279–284.

Anisman, H; Zacharko, R. 1982. "Depression: The predisposing influence of stress." *Beh Brain Sci.* 5: 89–137.

Angst, J. 1986. "The course of major depression, atypical bipolar disorder and bipolar disorder." In H Hippius; G Klerman; N Matussek, (eds). *New Results in Depression Research.* Berlin: Springer-Verlag,

Angyal, A. 1958. *Foundations for a Science of Personality.* Cambridge, MA: Harvard University Press.

Aquinas, St Thomas, 1960 (1265). *Summa Theologiae*. In P Mandonnet; J Destrez (Eds) *Bibliographie Thomiste*, 2nd rev ed. Paris: Ecole Superiore.

Archer, J. 1988. *The Behavioural Biology of Aggression*. Cambridge: Cambridge UniversityPress.

Aretaeus of Cappadocia. 1856. In F. Adams, (Ed and transl) *The Cappadocian*. London: Sydenham Society.

Aristotle. 1908-1952 (*circa* 450 BCE). In J Smith; W Ross (eds) *The Works of Aristotle*, 12 Vols. Oxford: Clarendon Press.

Asherson, P., Mant, R., Williams, N., et al. 1998. "A Study of Chromosome 4p Markers and Dopamine D5 Receptor Gene in Schizophrenia and Bipolar Disorder." *Molecular Psychiatry*. 3: 310–320.

Asquith, P. 1984. "The inevitability and utility of anthropomorphism in description of primate behavior." R Harre; V Reynolds, (eds) *The Meaning of Primate Signals*, pp 138–176. Cambridge: Cambridge University Press,

Astington, J., Harris, P., and Olson, D., (eds). 1988. *Developing Theories of Mind*. Cambridge University Press.

Avramopoulos D, Willour V, Zandi P, Huo Y, MacKinnon DF, Potash J, DePaulo J Jr, McInnis M. 2004. "Linkage of bipolar affective disorder on chromosome 8q24: follow-up and parametric analysis." *Mol Psychiatry*. Feb;9(2):191–196.

Axelrod, R. & Hamilton, W. 1981. "The Evolution of Cooperation." Pp.1290–1396 in *Science*. V. 211.

Azmitia E. 2001. "Modern views on an ancient chemical: serotonin effects on cell proliferation, maturation, and apoptosis." *Brain Research Bulletin*. Nov 15; 56(5): 413–424.

Baars, B. 1997. *In the Theatre of Consciousness*. Oxford University Press.

Baars, B. 1988. *A Cognitive Theory of Consciousness*. Cambridge University Press.

Bachevalier, J. 2000. "The Amygdala, Social Cognition, and Autism." Pp. 509–543 in *The Amygdala: A Functional Analysis*. ed. by J. P. Aggleton. Oxford University Press.

Bacon, F. 1965 (1620). *Novum Organon: Indications Respecting the Interpretation of Nature*. London, Penguin.

Badcock, C. 1986. *The Problem of Altruism: Freudian-Darwinian Solutions*. Oxford: Blackwell.

Badcock, C. 1990. *Oedipus in Evolution*. Oxford: Blackwell.

Badcock, C. 1992. *The Essential Freud*. Oxford, Blackwell.

Badenhop R., Moses M., Scimone A., Mitchell P., Ewen K., Rosso A., Donald J., Adams L., Schofield P. 2001. "A genome screen of a large bipolar affective disorder pedigree supports evidence for a susceptibility locus on chromosome 13q." *Molecular Psychiatry*. Jul;6(4):396–403.

Badner, J. & Gershon, E. 2002. "Meta-Analysis of Whole-Genome Linkage Scans of Bipolar Disorder and Schizophrenia." *Molecular Psychiatry*. 7: 405–411.

Bagehot, W. 1867. *The English Constitution*. London: Chapman and Hall.

Bailey, K: Personal Communication. 1996

Bailey, K. 2002. "Upshifting and Downshifting the Triune Brain: Roles in Individual and Social Pathology." Pp. 317–343 in *The Evolutionary Neuroethology of Paul MacLean: Convergences and Frontiers.* Ed. by G. Cory & R. Gardner. Westport, CT: Greenwood/ Praeger.

Bailey, K. 1987. *Human Paleopsychology: Applications to Aggression and Pathological Processes*. Hillsdale, NJ: Erlbaum.

Bakker, C., Gardner, R., Koliatsos, V. Kerbeshian, J., Looney, J., Sutton, B., Swann, A., Verhulst, J., Wagner, K., Wamboldt, F., Wilson, D. R. 2002. "The Social Brain: A Unifying Foundation for Psychiatry." *Academic Psychiatry*. 26: 219, Sep.

Balaban, E. 2002. "Human Correlative Behavioral Genetics." Pp. 293–314 in J. Benjamin, R. Ebstein, and R. Belmaker (eds). *Molecular Genetics and the Human Personality*. Washington, D.C.: American Psychiatric Publishing, Inc.

Baldessarini, R. 1983. *Biomedical Aspects of Depression and its Treatment.* Washington, DC: American Psychiatric Association Press.

Barash, D. 1977. *Sociobiology and Behavior*. London: Heineman.

Bard, P; Mountcastle, V. 1948. "Some forebrain mechanisms involved in expression of rage with special reference to suppression of angry behavior." *Res Publ Ass Nerv Ment Dis*. 27: 362–404.

Barden, N., Plante, M., Rochette, D., et al. 1996. "Genome-Wide Microsatellite Marker Linkage Study of Bipolar Affective Disorders in a Very Large Pedigree Derived from a Homogeneous Population in Quebec Points to Susceptibility Locus on Chromosome 12." *Psychiatric Genetics*. 6: 145–146.

Barkow, J. Cosmides, L. & Tooby, J. (eds). 1992. *The Adapted Mind: Evolutionary Psychology and the Generation of Culture.* Oxford University Press.

Barkow, J. 1975. "Prestige and culture: A biosocial interpretation." *Current Anthropology*. 16: 533–572, 1975

Barkow, J. 1980. "Prestige and self-esteem: A biosocial interpretation." In D. Omark, D. Strayer, J Freedman (eds), *Dominance relations: An ethological view of social conflict and social interaction.* NY: Garland STMP Press.

Barkow, J. 1989. *Darwin, Sex and Status*. Toronto: University of Toronto Press.

Barlow, N. (ed). 1959. *The Autobiography of Charles Darwin.* NY: Harcourt Brace.

Baron-Cohen, S. 1995. *Mindblindness*. The MIT Press.

Baron, J.; Martinot, J.; Cambon, H.; 1989. "Striatal dopamine receptor occupancy during and following neuroleptic treatment: Correlative evaluation by PET and plasma prolactin levels." *Psychopharmacology.* 99: 463–472.

Baron, M. 1977. "Linkage between an X-chromosome marker (deutan color blindness) and bipolar affective illness: Occurence in the family of a lithium carbonate-responsive schizo-affective proband." *Arch Gen Psychiatry*. 34:721–725.

Baron, M; Klotz, J; Mendlewicz, J; Rainer, J. 1981. "Multiple-threshold transmission of affective disorders." *Arch Gen Psychiatry*. 38:79–84.

Baron, M; Gruen, R; Asnis, L; Kane, J. 1982. "Schizoaffective illness, schizophrenia and affective disorders: Morbidity risk and genetic transmission." *Acta Psychiatr Scand*. 65: 253–262.

Baron, M; Risch, N; Mendelwicz, J. 1983. "Age at onset in bipolar-related major affective illness: Clinical and genetic implications." *J Psychiatric Res*, 17: 5–18.

Baron, M; Risch, N; Hamburger, R; Mandel, B: Kushner, S; Newman, M; Drumer, D; Belmaker, RK. 1987. "Genetic linkage between X-chromosome markers and bipolar affective illness." *Nature*. 326: 289–292.

Baron, M. 1991. "Schizophrenia and affective disorder: Are they genetically linked?" *Br J Psychiatry*. 159: 267–270.

Baron, M; Friemer, NF; Risch, N; Alexander, JR; Straub, RE; Asokan, S; Das, K; Peterson, A; Amos, J; Endicott, J; Ott, J; Gilliam, TC. 1993. "Diminished support for linkage between manic depressive illness and X-chromosome markers in three Israeli pedigrees." *Nature Genetics*. 3: 49–55.

Baron, M; Straub, RE; Lehner, T; Endicott, J; Ott, J; Gilliam, TC; Lere, B. 1994. "Bipolar disorder and linkage to Xq28." *Nature Genetics*. 7: 461.

Baron, M. 1977. "Linkage Between and X-Chromosome Marker (deutan color blindness) and Bipolar Affective Illness. Occurrence in the Family of a Lithium Carbonate-responsive Schizo-affective Proband." *Archives of General Psychiatry*. 34: 721–725.

Baron, M., Risch, N., Hamburger, R., Mandel, B., Kushner, S., et al. 1987. "Genetic Linkage Between X-Chromosome Markers and Bipolar Affective Illness." *Nature*. 326: 289–292.

Baron, M., Freimer, N., Risch, N., Lerer, B., Alexander, J., et al. 1993. "Diminished Support for Linkage Between Manic Depressive Illness and X-chromosome Markers in Three Israeli Pedigrees." *Nature Genetics*. 3: 49–55.

Barondes, S: Personal Communication, 1996

Bassett, AS. 1992. "Chromosomal aberration and schizophrenia. Autosomes." *British Journal of Psychiatr.*, 23: 305–309.

Bateson, W. 1902. *Mendel's Principles of Heredity*. London: Cambridge University Press.

Bateson, PPG. 1976. "Rules and reciprocity in behavioural development." In PPG Bateson; RA Hinde (eds) *Growing Points in Ethology*. Cambridge University Press.

Bateup, H., Booth, A., Shirtcliff, E., Granger, D. 2002. "Testosterone, Cortisol, and Women's Competition." *Evolution and Human Behavior*. 23: 181–192.

Batson, C. 1991. *The Altruism Question: Toward a Social-Psychological Answer*. Hillsdale, NJ: Erlbaum.

Beach, F. 1974. "Human sexuality and evolution." In W. Montagna; WA Sadler (eds) *Reproductive Behavior*. New York: Plenum.

Beach, F. 1976. "Cross-species comparisons and the human heritage." *Archives of Sexual Behavior*. 5: 469-485.

Beardslee, W; Bemporad, J; Keller, M; Klerman, G. 1983. "Children of parents with major affective disorder: A review." *Am J Psychiatry*. 140:825–832.

Beck, A. 1974a. "The development of depression." In RJ Fiedman, MM Katz (eds), *The Psychology of Depression: Contemporary Theory and Research*. New York: Winston-Wiley.

Beck, A. 1974b. "Cognition, affect and psychopathology." In H London, R E Nibett (Eds), *Thought and Feeling*. Chicago: Aldine.

Beck, A. 1983 Cognitive therapy of depression: New perspectives. In PJ Clayton, JE Barrett (eds), *Treatment of Depression: Old Controversies and New Approaches*. NY: Raven Press. p.274

Beck, A. 1987. "Cognitive models of depression." *J Cogn Psychotherapy*. 1: 5–38.

Beck, A; Emery, G; Greenberg, R. 1985. *Anxiety Disorders and Phobias: A Cognitive Approach.* NY: Basic Books.

Becker, J.; Breedlove, S.; & Crews, D. (eds) 1992. *Behavioral Endocrinology*. MIT Press.

Beckman, L; Cedergreen, B; Perris, C Strandman, E. 1978. "Blood groups and affective disorders." *Hum Hered*, 28: 48–55.

Beckwith, J. and Alper, J. 2002. "Genetics of Human Personality: Social and Ethical Implications." Pp. 315–331 in *Molecular Genetics and the Human Personality*. Ed. by Benjamin, J., Ebstein, R., & Belmaker, R. Washington, D. C.: American Psychiatric Publishing, Inc.

Beer. C. 1973. "Species-Typical Behavior and Ethology." Pp. 21–77 in *Comparative Psychology: A Modern Survey.* Ed. by D. Dewsbury & D. Rethlingshafer. NY: McGraw-Hill.

Belloc, H. 1934. *Cromwell.* London: Cassell and Company.

Bellodi, L; Sciuto, G; Diaferia, G; Ronchi, P; Smeraldi, E, 1992. "Psychiatric disorders in the families of patients with obsessive-compulsive disorder." *Psychiatry Res.* 42: 111–120.

Benedetti F, Serretti A, Colombo C, Barbini B, Lorenzi C, Campori E, Smeraldi, E. 2003. "Influence of CLOCK gene polymorphism on circadian mood fluctuation and illness recurrence in bipolar depression." *Am J Med Genet.* Nov 15;123B(1): 23–26.).

Benjamin, J., Ebstein, R., & Belmaker, R. (eds) 2002. *Molecular Genetics and the Human Personality*. Washington, D. C.: American Psychiatric Publishing, Inc.

Bentham, J. 1968. "An Introduction to the Principles of Morals and Legislation." In J Burns (ed) *The Collected Works of Jeremy Bentham*. Oxford University Press.

Benton, A. 1979. "Visuoperceptive, visuospatial and visuoconstructive disorders." In K Heilman; E Valenstien (eds) *Clinical Neuropsychology*. New York: Oxford University Press.

Bergson, H. 1907. *Evolution Creative*. Paris: Alcan.

Berns, G. and Nemeroff, C. 2003. "The Neurobiology of Bipolar Disorder." *American Journal of Medical Genetics*. Part C. 123C: 76–84.

Berretini, W; Goldin, L; Gelernter, J; Gejman, PV; Gershon, E; Detera-Wadleigh, S. 1990. "X-chromosome markers and manic-depressive illness: Rejection of linkage to Xq28 in nine bipolar pedigrees." *Arch Gen Psychiatry*. 47:366–373.

Berretini, W; Detra-Wadleigh, S; Goldin, L; Martinez, M; Hseih, W, Hoehe, M; Choi, H; Muniec, D; Ferraro, T; Guroff, JG; Kazuba, D; Harris, N; Kron, E; Nurnberger, J; Alexander, R; Gershon, E. 1992. Genomic screening for genes predisposing to bipolar disease. *Psychiatric Genetics.*:191–208, 1992.

Berretini, W; Ferraro, T; Goldin, L; Weeks, D; Detra-Wadleigh, S; Nurnberger, J; Gershon, E. 1994. "Chromosome 18 DNA markers and manic-depressive illness: Evidence for a susceptibility gene." *Proceedings of the National Academy of Science* USA, 91:5918–5921.

Berretini, W; Pekkarinen, P. 1996. "Molecular genetics of bipolar disorder." *Ann Med.* 28: 191–194.

Berrettini, W. 1997. "A linkage study of bipolar illness." *Arch Gen Psychiatry*. 54: 27–35.

Berrettini, W. 2001. "Molecular Linkage Studies of Bipolar Disorders." *Bipolar Disorders*. 3: 276–283.

Berrettini, W. 2003. "Evidence for Shared Susceptibility in Bipolar Disorder and Schizophrenia." *American Journal of Medical Genetics*. Part C. 123C: 59–64.

Berrettini, W., Ferraro, T., Goldin, L. Weeks, D., Detera-Wadleigh, S., et al. 1994. "Chromosome 18 DNA Markers and Manic-Depressive Illness: Evidence for a Susceptibility Gene." *Proceedings of the National Academy of Science USA*. 91: 5918–5921.

Berrettini, W., Goldin, L., Gelernter, J., Gejman, P., et al. 1990. "X-Chromosome Markers and Manic-Depressive Illness. Rejection of Linkage to Xq28 in Nine Bipolar Pedigrees." *Archives of General Psychiatry*. 47: 366–373.

Berridge, K. 2003. "Comparing the Emotional Brains of Humans and other Animals." Pp. 25–51 in *Handbook of Affective Sciences*. Ed. by R. Davidson, K. Scherer, and H. Goldsmith. Oxford University Press.

Bertelson, A., Harvald, B., Hauge, M. 1977. "A Danish Twin Study of Manic-Depressive Disorders." *British Journal of Psychiatry*. 130: 330–351.

Bertelsen, A. 1979. "A Danish twin study of manic-depressive disorders." In M Schou, E Stromgren (eds) *Origin, Prevention and Treatment of Affective Disorders*. London: Academic Press.

Betzig, L. 1986. *Despotism and Differential Reproduction: A Darwinian View of History*. NY: Aldine.

Betzig, L; Borgerhohoff Mulder, M; Turke, P. 1988. *Human Reproductive Behaviour: A Darwinian Perspective*. Cambridge: Cambridge University Press.

Birtchnell, J. 1999. *Relating in Psychotherapy*. Westport, CT: Praeger.

Birtchnell, J. 1993. *How Humans Relate*. Westport, CT: Praeger.

Blackwood, D; Lin, H; Morris, S; McLean, A; Whitton, C; Thomson, M; Walker, M; Woodburn, K; Sharp, C; Wright, A; Shibasaki, Y; St Clair, D; Porteous, D; Muir, W. 1996. "A locus for bipolar affective disorder on chromosome 4p." *Nature Genetics*. 12: 427–430.

Blanchard, D.; McKittrick, C.; Hardy, M.; & Blanchard, R. 2002. "Effects of Social Stress on Hormones, Brain, and Behavior." Pp. 735–772 in *Hormones, Brain, and Behavior*. Ed. by Pfaff, D. V. I. NY: Academic Press.

Bleuler, E. 1950 (1911). *Dementia Praecox or the Group of Schizophrenias*. New York: International Universities Press.

Bluemel, C. 1948. *War, Politics and Insanity*. Denver: World Press.

Bodmer, WF; Bodmer, J. 1978. "Evolution and function of the HLA system." *Br Med Bull*. 34: 309.

Boehm, C. 1999. *Hierarchy in the Forest: The Evolution of Egalitarian Behavior*. Harvard University Press.

Bomze, IM; Schuster, P; Sigmund, K. 1983. "The role of Mendelian genetics in strategic models of animal behaviour." *J Theor Biol*. 101:19–38.

Borel, E. 1953a. "Theory of play and integral equations with skew symmetric kernals." *Econometrica*. 21: 91-100, 1953a (1921–1927).

Borel, E. 1953b. "On games that involve chance and the skill of the players." *Econometric.*, 21: 101–115, 1953b (1921–1927).

Borel, E. 1953c. "On systems of linear forms of skew symmetric determinants and the general theory of play." *Econometrica*. 21: 116–117, 1953c (1921–1927).

Botstein, D; White, RL; Skolnick, M; Davis, R. 1980. "Construction of a genetic linkage map in man using restriction fragment length polymorphism." *Am J Hum Genetics*. 32: 314–331.

Botstein, D. and Risch, N. 2003. "Discovering Genotypes Underlying Human Phenotypes: Past Successes for Mendelian Disease, Future Approaches for Complex Disease." *Nature Genetics Supplement*. 33: 228–237.

Bouchard, TJ; Lykken, DT; McGue, M; Segal, NL; Tellegen, A. 1990. "Sources of human psychological differences: The Minnesota study of twins reared apart." *Science*. 250: 223–228.

Bovari, T. 1902. "Chromosomes." *Verh Phys Med Ges*, 35: 67–90.

Bowers, MB; Freedman, DX. 1975. "Psychoses associated with drug use." In S Arieti (ed) *American Handbook of Psychiatry,* Vol 4. New York: Basic Books.

Bowers, W.; Zacharko, R.; & Anisman, H. 1987. "Evaluation of stressor effects on intracranial self-stimulation from the nucleus accumbens and the substantia nigra in a current intensity paradigm." *Behavioral Brain Research.* 23: 85–93.

Bowlby, J. 1988. *A Secure Base: Parent-Child Attachment and Healthy Human Development.* NY: Basic Books.

Bowlby, J. 1969. *Attachment.* Vol.1. NY: Basic Books.

Bowlby, J. 1973. *Attachment and Loss,* Vol 2. NY: Basic Books.

Bowlby, J. 1980. *Attachment and Loss.* Vol 3. NY: Basic Books.

Bowler, P. 1992. *The Environmental Sciences.* London: Fontana.

Bowles, S. & Gintis, H. 1998. *Recasting Egalitarianism.* London: Verso.

Boyd, C. 1979 "Chemical neurotransmission: an hypothesis concerning the evolution of neurotransmitters substances." *J. Theor. Biol.,* 76, 415–417.

Brady, JV. 1958. "The paleocortex and behavioral motivation." In H Harlow; C Woolsey (eds) *Biological and Biochemical Bases of Behavior.* Madison: University of Wisconsin Press.

Brady, JV. 1975. "Toward a behavioral biology of emotion." In L Levi (ed) *Emotions: Their Parameters and Measurement.* New York: Raven Press.

Brain, Lord. 1960. *Brain's Clinical Neurology.* Oxford: Oxford University Press.

Brann, D. and Mahesh, V. 1994. "Excitatory Amino Acids: Function and Significance in Reproduction and Neuroendocrine Regulation." *Front Neuroendocrinol.* Mar; 15 (1): 3–49.

Bray, D. 1995. "Protein Molecules as Computational Elements in Living Cells." *Nature.* V. 376 (27 Jul): 307–312.

Bridges, B; Drake, JW; Fishel, R; Foster, P; Monnat, R; Ripley, LS; Taddei, F. 1997. "Optimum mutation rates in evolution and disease: A debate. In HMS Beagle." *A BioMedNet Publication* (http://hmsbeagle.com) 13, 1997

Broca, P. 1878. "Le grande lobe limbique et le scissure limbique dans le serie des mammiferes." *Rev d'Anthropologie 1*, Ser 2: 385–498.

Brothers, L. 1995. "Neurophysiology of the Perception of Intentions by Primates." Pp. 1107–1115 in *The Cognitive Neurosciences.* Ed. by M. Gazzaniga. MIT Press.

Brothers, L. 1989. "A Biological Perspective on Empathy." *American Journal of Psychiatry,* 146, 10–19.

Brown, J. 1977. *Mind, Brain, and Consciousness.* NY: Academic Press.

Brown, S; Schafer, E. 1888. "An investigation into the function of the occipital and temporal lobes of the monkey's brain." *Phil Trans R Soc.* Ser B, 179: 303–327.

Brownell, H. & Martino, G. 1998. "Deficits in Inference and Social Cognition: The Effects of Right Hemisphere Brain Damage on Discourse." Pp. 309–328 in *Right Hemisphere Language Comprehension: Perspectives from Cognitive Neuroscience*. Ed. by M. Beeman & C. Chiarello. Mahwah, NJ: Lawrence Erlbaum Associates.

Buber, M. 1958. *I and Thou*. 2nd ed., Trans. by R. Smith. NY: Scribners.

Buckingham, J. 2000. "Glucocorticoids, Role in Stress." Pp. 261–269 in *Encyclopedia of Stress*. V. 2. Ed. by G. Fink. San Diego, CA: Academic Press.

Bucy, PC; Kluver, H. 1955. "An anatomical investigation of the temporal lobe in the monkey (Macaca mulatta)." *J Comp Neurol*. 103: 151–251.

Bunney, WE; Murphy, D; Goodwin, FK; Borge, G. 1972a. "The 'switch process' in manic-depressive illness, I: A systematic study of sequential behavior change." *Arch Gen Psychiatry*.

Bunney, WE; Goodwin, FK; Murphy, DL; House, KM; Gordon, E. 1972b. "The 'switch process' in manic-depressive illness, II: Relationship to catecholamines, REM sleep and drugs." *Arch Gen Psychiatr.*, 27: 304–309.

Bunney, WE; Goodwin, FK; Murphy, D. 1972c. "The 'switch process' in manic-depressive illness, III: Theoretical implications." *Arch Gen Psychiatry*. 27:312–317.

Bunney, W. and Bunney, G. 2000. "Molecular Clock Genes in Man and Lower Animals: Possible Implications for Circadian Abnormalities in Depression. *Neuropsychopharmacology* 22: 335–345.

Bunny, B; Sesack, SR; Silva, N. 1987. "Midbrain dopaminergic systems: neurophysiology and electrophysiological pharmacology." In H Meltzer (ed) *Psychopharmacology: The Third Generation of Progress*. NY: Raven Press.

Burton, R. 1621. *Anatomy of Melancholy*. London.

Buss, A.; Plomin, R. 1984. *Temperment: Early Developing Personality Traits*. London: Lawrence Erlbaum.

Buss, A. 1988 *Personality: Evolutionary Heritage and Human Distinctiveness*. Hillsdale, NJ: Lawrence Erlbaum.

Buss, D. 1988. "The evolution of human intrasexual competition: Tactics of mate attraction." *J of Personality and Soc Psychology*.

Buss, D. 1991. "Evolutionary personality psychology." *Ann Rev of Psychology*. 42: 459-491.

Buss, D. 1992. "Sex differences in jealousy: Evolution, physiology and psychology." *Psychological Science*, 3: 251–255.

Buss, D. 1994. *The Evolution of Desire: Strategies of Human Mating*. New York: Basic Books.

Buss, D. 1999. *Evolutionary Psychology*. Boston: Allyn and Bacon.

Byrne, R; Whiten, A. 1988. *Machiavellian Intelligence: Social Expertise and the Evolution of Intellect in Monkeys, Apes and Humans*. Oxford: Science Publishers.

Cacioppo, J; Berntson, G. 1992. "Social psychological contributions to the decade of the brain." *American Psychologist.* 47: 1019–1028.

Cacioppo, J.; G. Berntson; R. Adolphs; C. Carter; R. Davidson, M. McClintock; B. McEwen; M. Meaney; D. Schacter; E. Sternberg; S. Suomi; & S. Taylor, (eds). 2002. *Foundations in Social Neuroscience*. MIT Press.

Cadoret, R; Winokur, G. 1975. "X-linkage in manic-depressive illness." *Ann Rev Med.,* 26:21–25.

Cadoret, R. 1978. "Psychopathology of adopted-away offspring of biologic parents with antisocial personality." *Arch Gen Psychiatry*, 35: 176–184.

Cadoret, R.; Cain, C; Grove, W. 1979. "Development of alcoholism in children raised apart from alcoholic biologic relatives." *Arch Gen Psychiatry*. 37: 561–563.

Cadoret R. 1986. "Adoption studies: Historical and methodological critique." *Psychiatr Dev*: 45–64.

Cadoret, R. 1978. "Psychopathology of adopted-away offspring of biologic parents with antisocial personality." *Archives of General Psychiatry* 35: 176–184.

Cadoret, R. 1982. "Genotype-environment interaction in antisocial behavior." *Psychosomatic Medicine* 12: 235–239.

Cadoret, R. 1986. "Epidemiology of antisocial personality." In: *Unmasking the psychopath: Antisocial personality and related syndromes*. (eds). W. Reid, D. Dorr, J. Walker & J. Bonner, III. W.W. Norton.

Cadoret, R. & Cain, C. 1980. "Sex differences in predictors of antisocial behavior in adoptees." *Archives of General Psychiatry* 37: 1171–1175.

Cadoret, R.. & Cain, C. 1981. "Environmental and genetic factors in predicting adolescent antisocial behavior." *The Psychiatric Journal of the University of Ottawa* 6(4): 220–225.

Cadoret, R., Cain, C. & Crowe, R. 1983. "Evidence for gene- environment interaction in the development of adolescent antisocial behavior." *Behavior Genetics* 13(3): 301–310.

Cadoret, R.. & Stewart, M. 1991. "An adoption study of attention deficit/hyperactivity/aggression and their relationship to adult antisocial personality." *Comprehensive Psychiatry* 32(1): 73–82.

Cadoret, R., Troughton, E., Bagford, J. & Woodworth, G. 1990. "Genetic and environmental factors in adoptee antisocial personality." *European Archives of Psychiatry and Neurological Sciences* 239:231–240.

Cadoret, R., Troughton, E. & O'Gorman, T. 1987. "Genetic and environmental factors in alcohol abuse and antisocial personality." *Journal of Studies on Alcohol.* 48(1): 1–8.

Cairns, J; Foster, P. 1991. "Adaptive reversion of a frameshift mutation in Escherichia coli." *Genetics*. 128: 695–701.

Calev, A; Nigal, D; Chazan, S. 1989. "Retrieval from semantic memory using meaningful and meaningless constructs by depressed, stable bipolar and manic patients." *Br J Clinical Psychology*. 28: 67–73.

Calhoun, J. 1962. "Population density and social pathology." *Scientific American*. 206: 139–148.

Campbell, D. 1975. "On the conflicts between biological and social evaluation and between psychology and moral tradition." *American Psychologist*. 117: 1103–1126.

Campbell, J. 1953. *Manic Depressive Disease: Clinical and Psychiatric Significance*. Philadelphia: JB Lippincott.

Campbell, J; Crowe, RR; Goeken, N; Pfohl, B; Pauls, D; Palmer, D. 1984. "Affective disorder linked to HLA in a large bipolar kindred." *J Affective Dis*. 7: 45–51.

Campbell, N. 1990. *Biology*. Benjamin/Cummings; p.443

Cannon, W. B. 1932. *The Wisdom of the Body*. NY: Norton.

Cannon, W. B. 1929. *Bodily Changes in Pain, Hunger, Fear, and Rage*. NY: D. Appleton & Co.

Cardno, A. Marshall, E., Coid, B., Macdonald, A., et al. 1999. "Heritability Estimates for Psychotic Disorders: the Maudsley Twin Psychosis Series." *Archives of General Psychiatry*. 56: 162–168.

Cardno, A., Rijsdijk, F., Sham, P., Murray, R., McGuffin, P. 2002. "A Twin Study of Genetic Relationships Between Psychotic Symptoms." *American Journal of Psychiatry*. 159: 539–545.

Carlson, G; Strober, M. 1979. "Affective disorders in adolescence: Issues in misdiagnosis." *J Clin Psychiatry*. 39: 59–66.

Carlsson A, Lindquest M. 1963. "Effects of chlorpromazine or haloperidol on formation of 3-methoxytyramine and normetanephrine in mouse brain." *Acta Pharmacol Toxicol*. 20: 140–144.

Carpenter, C. 1978. "Ritualistic social behaviors in lizards." In N Greenberg; P MacLean (eds) *Behavior and Neurology of Lizards*. Washington, DC: National Institute of Mental Health USA.

Carter, C. and Keverne, E. 2002. "The Neurobiology of Social Affiliation and Pair Bonding." Pp. 299–337 in *Hormones, Brain, and Behavior*. Ed. by Pfaff, D. V. I. NY: Academic Press

Carter, C., Lederhendler, I., & Kirkpatrick, B. eds. 1997. *The Integrative Neurobiology of Affiliation*. NY: New York Academy of Sciences, Vol. 807.

Caspari, E. 1972. "Sexual selection in human evolution." In B Campbell (ed). *Sexual Selection and the Descent of Man 1871-1971*. Chicago: Aldine.

Casper, R; Redmond, E Jr; Katz, M; Schaffer, C; Davis, J; Koslow, S. 1985. "Somatic symptoms in primary affective disorder: Presence and relationship to the classification of depression." *Arch Gen Psychiatry*. 42: 1098–1104.

Catalano, R. 2000. "Economic Factors and Stress." Pp. 9–15 in *Encyclopedia of Stress*. V. 2. Ed. by G. Fink. San Diego, CA: Academic Press.

Cavalli-Sforza, L; Bodmer, W. 1976. *The Genetics of Human Populations*. San Francisco: Freeman.

Cavalli-Sforza, L; Feldman, M. 1973. "Models for cultural inheritance." *Theoretical Population Biol*. 4: 42–55.

Chagnon, N. 1968. *Yanomamo: The Fierce People*. New York: Holt, Rinehart and Winston.

Chagnon, N. 1988. "Life histories, blood revenge and warfare in a tribal population." *Science*. 239: 985–992.

Chance, M. 1967. "Attention structure as the basis of primate rank orders." *Man*. 2: 503–518.

Chance, M. 1980. "An ethological assessment of emotion." R Plutchick; H Kellerman (eds). *Emotions: Theory, Research and Experience*. Vol 1. New York: Academic Press.

Chance, M. 1988. "A systems synthesis of sociality." In M Chance (ed). *Social Fabrics of the Mind*. London: Lawrence Erlbaum.

Chang, F.; Kidd, J.; Livak, K.; Pakstis, A. & Kidd, K. 1996. "The World-Wide Distribution of Allele Frequencies at the Human Dopamine D4 Receptor." *Human Genetics*. 98: 91–101.

Chaudhari, N; Hahn, W. 1983. "Genetic expression in the developing brain." *Science*. 220: 924–928.

Chen H, Huo Y, Patel S, Zhu X, Swift-Scanlan T, Reeves RH, DePaulo R Jr, Ross CA, McInnis MG. 2000. "Gene identification using exon amplification on human chromosome 18q21: implications for bipolar disorder." *Mol Psychiatry*. 2000 Sep;5(5): 502–9.

Chen, YW; Dilsaver, S. 1995. "Comorbidity of panic disorder in bipolar illness: Evidence from the Epidemiological Catchment Area Survey." *American Journal of Psychiatry*. 152: 280–282.

Chetverikov, S. 1926. "On certain features of the evolutionary process from the viewpoint of modern genetics." *J Exp Biol*. 2: 3-54 (as translated in Proc Amer Phil Soc, 105:167–195, 1961).

Chio, C, Lajiness, M. Huff, R. 1994. "Activation of Heterologously Expressed D_3 Dopamine Receptors: Comparison with D_2 Dopamine Receptors." *Molecular Pharmacology*. 45: 51–60.

Cichon S, Schumacher J, Muller D, Hurter M, Windemuth C, Strauch K, Hemmer S, Schulze T, Schmidt-Wolf G, Albus M, Borrmann-Hassenbach M, Franzek E, Lanczik M, Fritze J, Kreiner R, Reuner U, Weigelt B, Minges J, Lichtermann D, Lerer B, Clark, DC; Clayton, PJ; Andreasen, NC; Lewis, C;

Fawcett, J; Scheffner, W. 1985. "Intellectual functioning and abstraction ability in major affective disorders." *Compr Psychiatry*. 26: 313–325.

Clerget-Darpoux, F; Goldin, LR; Gershon, E. 1986a. "Clinical methods in psychiatric genetics, III: Environmental stratification may simulate a genetic effect in adoption studies." *Acta Psychiatr Scand.* 74: 395–311.

Clerget-Darpoux, F; Bonaiti-Pellie, C; Hochez, J. 1986b. "Effects of misspecifying genetic parameters in lod score analysis." *Biometrics*. 42: 393–399.

Cloninger, C. 2002. "Relevance of Normal Personality for Psychiatrists." Pp. 33–42 in *Molecular Genetics and the Human Personality*. Ed. by Benjamin, J., Ebstein, R., & Belmaker, R. Washington, D. C.: American Psychiatric Publishing, Inc.

Cloninger, C. 1995. The psychobiological regulation of social cooperation." *Nature Medicine*. 1: 7: 623–625.

Cloninger, C; Christiansen, KD; Reich, T; Gottesman, II. 1978. "Implication of sex differences in the prevalence for antisocial personality, alcoholism and criminality for familial transmission." *Arch Gen Psychiatry*. 35: 941–951.

Cloninger, C; Rice, J; Reich, T. 1979. "Multifactorial inheritance with cultural transmission and assortative mating." *Am J Hum Gen*. 31: 176.

Cloninger, C.; Reich, T; Yokoyama, S. 1983. "Genetic diversity, genome organization and investigation into the etiology of psychiatric diseases." *Psychiatric Developments*. 3: 775–796.

Cloninger, C. 1988. "Schizophrenic disorders: Genetic etiologic factors." In H Kaplan; B Sadock (eds) *Comprehensive Textbook of Psychiatry*, 5th Edition. Baltimore: William and Wilkins.

Cloninger, C; Reich, T; Wetzel, R. 1976b. "Alcoholism and affective disorders: Familial associations and genetic models." In D. Goodwin; C. Erickson (eds). *Alcoholism and Affective Disorders: Clinical, Genetic and Biochemical Studies.* NY: SP Medical and Scientific Books, 1979b

Cloninger, C. & Gottesman, I.I. 1987. "Genetic and environmental factors in anti-social behavior disorders." In: *The causes of crime: New biological approaches*. (eds). S.A. Mednick, Terrie E. Moffitt, & S.A. Stack. Cambridge University Press.

Clutton-Brock, T. 1988. (ed) *Reproductive Success: Studies of Individual Variation in Contrasting Breeding Systems*. Chicago: University of Chicago Press.

Clutton-Brock, T. 1991. *The Evolution of Parental Care*. Princeton University Press.

Cobb. S.; Buhl, E.; Halasy, K.; Paulsen, O., Somogyi, P. 1995. "Synchronization of Neuronal Activity in Hippocampus by Individual GABAergic Interneurons." *Nature*. 378: 75–78.

Coe, C. 1990. Psychobiology of Maternal Behavior in Nonhuman Primates." Pp. 157–183 in *Mammalian Parenting: Biochemical, Neurobiological, and Behavioral Determinants.* Ed. by N. Krasnegor & R. Bridges. Oxford University Press.

Cohen, L.E. & Machalek, R. 1988. A general theory of expropriative crime: An evolutionary ecological approach. *American Journal of Sociology.* 94(3): 465–501.

Comings, D., Saucier, G., MacMurray, J. 2002. "Role of DRD2 and Other Dopamine Genes in Personality Traits." Pp. 165–191 in *Molecular Genetics and the Human Personality.* Ed. by Benjamin, J., Ebstein, R., & Belmaker, R. Washington, D. C.: American Psychiatric Publishing, Inc.

Comte, Auguste. 1853 (1830–1842) *Course of Positive Philosophy.* Paris.

Correns, C. 1900. "G. Mendel's law concerning the behavior of progeny of varietal hybrids." *Berichte der Deutchen Botanischen Gesellschaft* 18: 158–168.

Cory, G. 1999. *The Reciprocal Modular Brain in Economics and Politics: Shaping the Rational and Moral Basis of Organization, Exchange, and Choice.* NY: Plenum Press.

Cory, G. 2000a. "From MacLean's Triune Brain Concept to the Conflict Systems Neurobehavioral Model: "The Subjective Basis of Moral and Spiritual Consciousness." *Zygon: Journal of Religion & Science,* Vol 35. No.2, pp. 385–414.

Cory, G. 2000b. *Toward Consilience: The Bioneurological Basis of Thought, Behavior, Language, and Experience.* NY: Kluwer Academic/Plenum.

Cory, G. 2001a. "Neural Network Theory and Neuroscience: Applications to Socio-Economic Theory." *Paper presented at the 13th Annual Meeting on Socio-Economics,* University of Amsterdam, June 28–July 1, 2001.

Cory, G. 2001b. "Transaction Costs, the Firm and Evolved Neural Architecture." *Paper presented at the 76th Annual Conference of the Western Economic Association International,* San Francisco, July 4–8.

Cory, G. 2002a. "Reappraising MacLean's Triune Brain Concept." Pp. 9–27 in *The Evolutionary Neuroethology of Paul MacLean: Convergences and Frontiers.* Ed. by G. Cory & R. Gardner. Westport, CT: Greenwood/Praeger.

Cory, G. 2002b. "Algorithms of Neural Architecture, Hamilton's Rule, and the Invisible Hand of Economics." In *The Evolutionary Neuroethology of Paul MacLean: Convergences and Frontiers.* Ed. by G. Cory and R. Gardner. Westport CT: Greenwood/Praeger

Cory, G. 2003. "MacLean's Evolutionary Neuroscience and the CSN Model: Some Clinical and Social Policy Implications." Pp. 161–180 in *Human Nature and Public Policy.* Ed. by Somit, A. and Peterson, S. NY: Palgrave MacMillan.

Cory, G. 2004. *The Consilient Brain: The Bioneurological Basis of Economics, Society, and Politics.* Kluwer Academic/Plenum.

Cory, G. 2006a. "Physiology and Behavioral Economics: The New Findings from Evolutionary Neuroscience." Pp. 24–49 in *Handbook of Contemporary Behavioral Economics: Foundations and Developments*. Ed. By Altman, M. Armonk, NY: MESharpe.

Cory, G. 2006b. "A behavioral model of the dual motive approach to behavioral economics and social exchange." *The Journal of Socio-Economics*. V. 35. N. 4 (August): 592–612.

Cory, G. and Gardner, R., eds. 2002. *The Evolutionary Neuroethology of Paul MacLean: Convergences and Frontiers*. Westport CN: Greenwood/Praeger.

Coryell, W; Endicott, J; Reich, T; Andreasen, NC; Keller, M. 1984. "A family study of bipolar II disorder." *Br J Psychiatry*.145: 49–54.

Coryell W, Endicott J, Keller M, Andreasen N, Grove W, Hirschfeld RM, Scheftner W. 1989. "Bipolar affective disorder and high achievement: A familial association (National Institute of Mental Health Collaborative Program on the Psychobiology of Depression)." *Am J Psychiatry*. 146: 983–988.

Cosmides, L. & Tooby, J. 1989. "Evolutionary Psychology and the Generation of Culture, Part II." Pp. 51–97 in *Ethology and Sociobiology*. V. 10.

Cosmides, L; Tooby, J. 1992. "Cognitive adaptations for social exchange." In J Barkow, L Cosmides, J Tooby (eds) *The Adapted Mind*. Oxford: Oxford University Press.

Cournot, A. 1929 [1838] *Researches into the Mathematical Principles of the Theory of Wealth*. Trans. N. Bacon. Homewood, Ill: Richard D. Irwin.

Cowan, W., Jessell, T., & Lipursky, S., eds. 1997. *Molecular and Cellular Approaches to Neural Development*. Oxford University Press.

Cowan, W., Kopinski, K., and Hyman, S. 2002. "The Human Genome Project and Its Impact on Psychiatry." Pp. 1–50 in *Annual Review of Neuroscience*. 25.

Coyle, JT: Personal Communication. 1997

Coyne, JC. 1976. "Depression and the response of others." *J Abnormal Psychology*. 85:186–193.

Coyne, JC; Kessler, TC; Tal, M; Turnbull, J; Wortman, CB; Greden, JF. 1987. "Living with a depressed person." *J Consult Clin Psychology*. 55: 347–352.

Craddock, N., Owen, M., Burge, S., Kurian, B., Thomsas, P., McGuffin, P. 1994. "Familial Cosegregation of Major Affective Disorder and Drier's Disease." *British Journal of Psychiatry*. 164: 355–358.

Craddock, N. 1995. "Mathematical limits of multilocus models: the genetic transmission of bipolar disorder." *Am J Hum Genet*. 57: 690–702.

Craddock, N; McGuffin, P. 1993. "Approaches to the genetics of affective disorder." *Annals Med*. 25:317–322.

Crawford, C. and Krebs, D., (eds). 1998. *The Handbook of Evolutionary Psychology*. Mahwah, NJ: Lawrence Erlbaum.

Crawford, C. and Krebs, D., (eds). 1998. *The Handbook of Evolutionary Psychology*. Mahwah, NJ: Lawrence Erlbaum.

Crews, D. 1997. "Species Diversity with the Evolution of Behavioral Controlling Mechanisms. Carter, C., Lederhendler, I., & Kirkpatrick, B. (eds.). *The Integrative Neurobiology of Affiliation.* NY: New York Academy of Sciences, Vol. 807: 1–21.

Crick, F. 1994. *The Astonishing Hypothesis: The Scientific Search for the Soul.* NY: Charles Scribner's Sons.

Craddock, N; Owen, M. 1994. "Chromosomal aberrations and bipolar affective disorder." *Br J Psychiatry.* 164:507–512.

Cronin, H. 1991. *The Ant and the Peacock: Altruism and Sexual Selection from Darwin to Today.* New York: Cambridge University Press.

Crook, J. 1980. *The Evolution of Consciousness.* Oxford: Clarendon Press.

Crow, T. 1980. "Molecular pathology of schizophrenia: More than one disease process?" *Br Med J.* 280: 66–68.

Crow, T. 1986. "The continuum of psychosis and its implication for the structure of the gene." *Br J Psychiatry*, 145: 243–253.

Crow, T. 1988. "Sex chromasomes and psychosis: The case for a pseudoautosomal locus." *Br J Psychiatry.*153: 675–683.

Crow, T. 1993. "Sexual selection, Machiavellian intelligence and the origins of psychosis." *Lancet.* 342: 8871: 594-598.

Crow, T: 1995. A Darwinian approach to the origins of psychosis. Br J Psychiatry, 167: 1: 12–25.

Crowe, R.. 1974. An adoption study of anti-social personality. *Archives of General Psychiatry* 31: 785–791.

Crowe, T. 1998. "Commentary: From Kraepelin to Kretschmer Leavened by Schneider: The Transition From Categories of Psychosis to Dimensions of Variation Intrinsic to *Homo sapiens.*" *Arch Gen Psychiatr.*, Jun 1998; 55: 502–504.

Csaba, G. 1981. *Ontogeny and phylogeny of hormone receptors.* Basel, Karger.

Csaba G. 1984. "The present state in the phylogeny and ontogeny of hormone receptors." *Hormone Metabol. Res.*, 16, 329–335.

Csikszentmihaly, M; Rathunde, K. 1990. "The psychology of wisdom: An evolutionary interpretation." In R Steinberg (ed) *Wisdom: It Nature, Origins and Development.* Cambridge: Cambridge University Press.

Cummings, J. 1993. "Frontal-Subcortical Circuits and Human Behavior." in *Archives of Neurology* V. 50 (Aug): 873–880.

Cummins, D. 1998. "Social Norms and Other Minds: The Evolutionary Roots of Higher Cognition." Pp. 30–50 in *The Evolution of Mind.* Ed. by Cummins, D. & Allen, C. Oxford University Press.

Curtis D, Kalsi G, Brynjolfsson J, McInnis M, O'Neill J, Smyth C, Moloney E, Murphy P, McQuillin A, Petursson H, Gurling H. 2003. “Genome scan of pedigrees multiply affected with bipolar disorder provides further support for the presence of a susceptibility locus on chromosome 12q23-q24, and suggests the presence of additional loci on 1p and 1q.” *Psychiatr Genet.* 2003 Jun:13(2):77–84.

Daly, M; Wilson, M. 1988. *Homicide.* Hawthorne, NewYork: Aldine de Gruyter.

Daly, M; Wilson, M. 1983. *Sex, Evolution and Behavior*. Boston: Willard Grant.

Damasio, A. 1979. “Frontal lobes.” In K Heilman; E Valenstien (eds) *Clinical Neuropsychology*. New York: Oxford University Press.

Damasio, A. and Benton, A. 1979. “Impairment of Hand Movements under Visual Guidance.” *Neurology*. 29: 170–174.

Damasio, A; Van Hoesen, G. 1983. “Emotional disturbances associated with focal lesions of the limbic frontal lobe.” In K Heilman; P Satz (eds). *Neuropsychology of Human Emotion*. New York: Guilford Press.

Damasio, A. 1999. *The Feeling of What Happens*. NY: Harcourt.

Damasio, A. 1994. *Descartes Error: Emotion, Reason, and the Human Brain.* NY: Grosset/Putnam.

Darwin, C. 1868. *The Variation of Plants and Animals under Domestication*, 2 Vol. London: John Murray.

Darwin, C. 1871. *The Descent of Man and Selection in Relation to Sex*. London: John Murray.

Darwin, C. 1965 (1872). *The Expression of Emotions in Man and Animals.* Chicago: University of Chicago Press.

Darwin, C. 1964 [1859]. *The Origin of Species.* Facsimile of the First Edition with introduction by E. Mayr. Harvard University Press.

Davidson, R. 1995. “Cerebral Asymmetry, Emotion, and Affective Style.” Pp. 361–387 in J. Davidson & K. Hugdahl (eds). *Brain Asymmetry*. MIT Press.

Davidson, R. & Irwin, W. 2002. “The Functional Anatomy of Emotion and Affective Style.” Pp. 473–490 in *Foundations in Social Neuroscience.* Ed. by Caccippo, J.; Bernstson, G.; Adolphs, R.; Carter, C. S.; Davidson, R.; McClintock, M.; McEwen, B.; Meaney, M.; Schacter, D.; Sternberg, E.; Suomi, S.; Shelley, E. The MIT Press.

Davidson, R., Scherer, K., Goldsmith, H. (eds). 2003. *Handbook of Affective Sciences*. Oxford University Press.

Davies, M; Davies, H; Davies, K. 1992. *Humankind the Gatherer-Hunter*. Kent, England: Myddle-Brockton Publishers.

Davis KL, Kahn RS, Ko G, Davidson M. 1991. “Dopamine in schizophrenia: a review and reconceptualization.” *Am J Psychiatry*. 148:1474–1486.

Davison G., & Neale J. 1994. *Abnormal psychology* (6th ed). John Wiley & Sons .

Dawkins, R; Krebs, J. 1978. “Animal signals: Information or manipulation?” In JR Krebs; ND Davies (eds) *Behavioural Ecology*. Oxford: Blackwell.

Dawkins, R. 1976. *The Selfish Gene*. Oxford University Press.

Dawkins, R. 1989. *The Extended Phenotype*. Oxford University Press.

Dawson, E. Praffit, E., Roberts, Q., Daniels, J. 1995a. "Linkage Studies of Bipolar Disorder in the Region of Darier's Disease Gene on Chromosome 12q23-24." *American Journal of Medical Genetics*. 60: 94–102.

Dawson, E., Gill, M., Curtis, D., Castle, D., Hunt, N., Murray, R., Powell, J. 1995b. "Genetic Association Between Alleles of Pancreatic Phospholipase A2 Gene and Bipolar Affective Disorder." *Psychiatric Genetics*. 5: 177–180.

Deacon, T. 1983. "Human brain evolution." In H Jerison; I Jerison (eds) *Intelligence and Evolutionary Biology*. Berlin: Springer Verlag.

Decina, P; Kestenbaum, CJ; Farber, S; Kron, L; Gargan, M; Sackheim, HA; Fieve, R. 1983. "Clinical and psychological assessment of children of bipolar probands." *Am J Psychiatry*. 140: 48–553.

Delay, J; Deniker, P; Harl, J. 1952. "Utilization therapeutique psychiatrique d'une phenothiazine d'action centrale elective." *Ann Med Psychol* (Paris). 110: 112–117.

Delbrouck, J. 1958. "Contribution ala' genetique de la sicklemie." *Annalles Societie Belge de Mediciene Tropique*. 38: 103–133.

De Long, GR; Aldershof, A. "Association of special abilities with juvenile manic-depressive illness." *Ann Neurol*. 14:362.

Del Veccio, M; Famiglietti, LA; Amati, A. 1979. "Subgroups of manic-depressive patients in relation to the ABO system." *Acta Neurologica*. 34:150.

DelZompo, M., Bocchetta, A., Goldin, L., Corsini, G. 1984. "Linkage Between X-Chromosome Markers and Manic-Depressive Illness. Two Sardinian Pedigrees." *Acta Psychiatrica Scandinavia*. 70: 282–287.

Dennett, D. 1998. "Reflections on Language and Mind." Pp. 284–293 in *Language and Thought*. Ed. by P. Carruthers & J. Boucher. Cambridge University Press.

Denny-Brown, D. 1951. "The frontal lobes and their functions." In A Feiling (ed) *Modern Trends in Neurology*. New York: Harper.

DePaulo, R: Personal Communication, 1996

De Ritter, D. 2000. "Social Status and Stress." Pp. 468–473 in *Encyclopedia of Stress*. V. 3. Ed. by G. Fink. San Diego, CA: Academic Press.

Descartes, R. 1977 (1641). "Meditations." In *Rene Descartes: The Essential Writings*. J Blom (ed and transl), New York: Harper and Row.

Desmond, A; Moore, J. 1991. *Darwin: The Life of a Tormented Evolutionist*. London: Penguin.

Detera-Wadleigh, SD; de Miguel, C; Berrettini, WH; DeLisi, LE; Goldin, LR; Gershon, E. 1987. "Neuropeptide gene polymorphisms in affective disorder and schizophrenia." *J Psychiatr Res*. 21: 581–587.

Detera-Wadleigh, S. 1995. "Adrenocorticotropin receptor/melanocortin receptor-2 maps within a reported susceptibility region for bipolar illness on chromosome 18." *Am J Med Genet*, 60: 317–321.

Detera-Wadleigh, S. "Affected-sib-pair analyses reveal support of prior evidence for a susceptibility locus for bipolar disorder on 21q." *Am J Hum Genet*. 58: 1279–1285.

Detera-Wadleigh, S; Badner, J; Goldin, L. 1996. "Analysis of linkage to bipolar illness on chromosome 21q." *Am J Hum Genet.* 58: 1279–1285.

Detera-Wadleigh, S. 1997. "Initial genome scan of the NIMH genetics initiative bipolar pedigrees: chromosomes 4, 7, 9, 18, 19, 20 and 21q." *Am J Med Genet*. 74: 254–262.

Detera-Wadleigh, S., Badner, J., Goldin, L., Berrittini, W., Sanders, A., et al. 1996. "Affected Sib-pair Analyses Reveal Support of Prior Evidence for a Susceptibility Locus for Bipolar Disorder, on 21q. *American Journal of Human Genetics*. 58: 1279–1285.

Detera-Wadleigh, S., Badner, J., Yoshikawa, T., Sanders, A., Goldin, L.,et al. 1997. "Initial Genome Scan of the NIMH Genetics Initiative Bipolar Pedigrees: Chromosomes 4, 7, 9, 18, 19, 20 and 21q."*American Journal of Medical Genetics*. 74: 254–262.

Deth, R. 2003. *Molecular Origins of Human Attention*. Boston, MA: Kluwer Academic Publishers.

Devinsky, O. & Luciano, D. 1993. "The Contributions of Cingulate Cortex to Human Behavior." Pp. 527–556 in *Neurobiology of Cingulate Cortex and Limbic Thalamus: A Comprehensive Handbook*. Ed. by Vogt, B. & Gabriel, M. Boston: Birkhauser.

DeVries, A., Glasper, E., Detillion, C. 2003 "Social Modulation of Stress Responses." *Physiology & Behavior*. V79, N3: 399–407.

de Vries, H. 1900. "Sur les unites des caracteres specifiques et leur application a l'etude des hybrides." *Revue Generale de Botanique*.

de Waal, FBM 1986. "Integration of dominance and social bonding in primates." *Quarterly Review of Biology*. 61: 459–479.

de Waal, F. 1988. "The Reconciled Hierarchy." Pp. 105–136 in *Social Fabrics of the Mind*, Ed. by M. Chance. Hillsdale, NJ: Erlbaum.

de Waal, F. 1989. "Dominance "style" and primate social organization." In V Standen and RA Foley (eds) *Comparative Socioecology: The Behavioural Ecology of Humans and Other Mammals*. Oxford: Blackwell.

de Waal, F. 1996. *Good Natured: The Origins of Right and Wrong in Humans and Other Animals*. Harvard University Press.

de Waal. F. 2002. "Evolutionary Psychology: The Wheat and the Chaff." Pp. 187–191 in *Current Directions in Psychological Science*. Vol. 11. N. 6.

Dittus, W. 1979. "The evolution of behaviours regulating density and age-specific sex ratios in a primate population." *Behaviour*. 69: 265–302.

Dobzhansky, T. 1962. *Mankind Evolving*. Yale University Press.

Donis-Keller, H; Green, P; Helms, C; Cartinhaur, S; Weiffenbach, B: Stephens, K: Keith, T; Bowden, D; Smith, D; Lander, E; Botstein, D; Akots, G; Rediker, K; Gravius, T; Brown, V; Rising, M; Parker, C; Powers, J; Watt, D; Kauffman, E; Bricker, A; Phipps, P; Muller-Kahle, H; Fulton, T; Ng, S; Schumm, J; Braman, J; Knowlton, R; Barker, D; Crooks, S; Lincoln, S; Daly, M; Abrahamson, J. 1987. "A genetic linkage map of the human genome." *Cell*. 51: 319–337.

Donnelly, E; Murphy, D; Goodwin, F; Waldman, I. 1982. "Intellectual function in primary affective disorder." *Br J Psychiatry*. 140: 633–636.

Doolittle, R; Feng, D; Tsang, S; Cho, G; Little, E. 1996. "Determining divergence times of the major kingdoms of living organisms with a protein clock." *Science*. Jan 26; 271 (5248): 470–7).

Drake, J. 1991. "A constant rate of spontaneous mutation in DNA-based microbes." *Proc Natl Acad Sci USA*. 88: 7160–7164.

Drake, J. 1993. "Rates of spontaneous mutation among RNA viruses." *Proc Natl Acad Sci USA*. 90: 4171–4175.

Dugatkin, L. 1995. "Formalizing Allee's Ideas on Dominance Hierarchies: An Intrademic Selection Model." *American Naturalist*. V. 146. N. 6: 954–960.

Duman, R; Heninger, G; Nestler, E. 1997. "A molecular and cellular theory of depression." *Arch Gen Psychiatry*. 54: 597–606.

Dunbar, R. 1988. *Primate Social Systems*. London: Croom Helm.

Dunner, D; Fleiss, J; Addonisio, G; Fieve, R. 1976a. "Assortative mating in primary affective disorder." *Biological Psychiatry*. 11: 43–51.

Dunner, D; Gershon, E; Goodwin, F. 1976b. "Heritable factors in the severity of affective illness." *Biol Psychiatry*. 11: 3142.

Dunner, D; Russek; FD; Russek, B; Fieve, R. 1982. "Classification of bipolar affective disorder subtypes." *Compr Psychiatry*, 23: 186–189.

Dunner, D; Jie, SQ; Ping, ZY; Dunner, P. 1984. "A study of primary affective disorder in the Peoples' Republic of China." *Biological Psychiatry*. 186–189.

Dupue, R.; Luciana, M.; Arbisi, P.; Collins, P.; & Leon, A. 2002. "Dopamine and the Structure of Personality: Relation of Agonist-Induced Dopamine to Positive Emotionality." Pp. 1071–1091 in *Foundations in Social Neuroscience*. Ed. by Caccippo, J.; Bernstson, G.; Adolphs, R.; Carter, C. S.; Davidson, R.; McClintock, M.; McEwen, B.; Meaney, M.; Schacter, D.; Sternberg, E.; Suomi, S.; Shelley, E. The MIT Press.

Durham, W. 1976. "Resource competition and human aggression." *Quarterly Rev Biology*. 51: 385–415.

During, M., Naegele, J., O'Malley, K., Geller, A. 1994. "Long-term Behavioral Recovery in Parkinsonisan Rats by an HSV vector Expressing Tyrosine Hydroxylase." *Science*. Nov 25; 266 (5225): 856–857.

During, M., Ryder, K., Spencer, D. 2002. "Hippocampal GABA Transporter Function in Temporal-Lobe Epilepsy." *Nature*. 376: 174–177.(13 July 2002).

Durkheim, E. 1951 (1897). *Suicide: A Study in Sociology*. Glencoe, IL: Free Press.

Dusser de Barenne, J. 1920. "Recherchers experimentales sur les functions du system nerveaux central, faites en particulier sur deux chat cont le neopallium avait ete enleve." *Arch Neerl Physiol*. 4: 31–123.

Eaton, W; Regier, DA; Locke, BZ; Taube, C. 1981. "The epidemiological catchment area program of the national institute of mental health." *Public Health Rep*. 96: 319–325.

Eaves, L. 1969. "The genetic analysis of continuous variation." *BR J Math Stat Psychology*, 22: 131–147.

Eaves, L. 1977. "Inferring the cause of human variation." *J Royal Stat Soc*. Series A (General), 140: 324–355.

Ebstein, R., & Auerbach, J. 2002. "Dopamine D4 Receptor and Serotonin Transporter Polymorphisms and Temperament in Early Childhood." Pp. 137–150 in *Molecular Genetics and the Human Personality*. Ed. by J. Benjamin, R. Ebstein, & R. Belmaker. Washington, D.C.: American Psychiatric Publishing.

Ebstein R. & Kotler, M. 2002. Personality, Substance Abuse, and Genes. Pp. 151–163 in *Molecular Genetics and the Human Personality*. Ed. by J. Benjamin, R. Ebstein, & R. Belmaker. Washington, D.C.: American Psychiatric Publishing.

Eccles, J. 1951. *The Neurophysiological Basis of Mind: The Principles of Neurophysiology*. Oxford: Oxford University Press.

Eckel, C. and Grossman, P. 1997. "Equity and Fairness in Economic Decisions: Evidence from Bargaining Experiments." Pp. 281 301 in *Advances in Economic Psychology*. (eds). G. Antonides and W. van Raaj. NY: Wiley.

Edenberg, H. 1997. "Initial genomic scan of the NIMH genetics initiative bipolar pedigrees: chromosomes 3, 5, 15, 16, 17 and 22." *Am J Med Genet*. 74: 238–246.

Edelman, G. 1992. *Bright Air, Brilliant Fire*. NY: Basic Books.

Edelman, G. and Tononi, G. 2000. *A Universe of Consciousness*. NY: Basic Books.

Edwards, J. 1973. "Extramarital involvement: Fact and theory." *J Sex Res*. 9: 210–224.

Edwards. R. 1993. "Neural Degeneration and the Transport of Neurotransmitters." *Ann Neurol*. Nov; 34 (5): 638–645.

Egeland J. 1983. "Bipolarity: The iceberg of affective disorders?" *Compr Psychiatry*, 24: 337–344.

Egeland, J; Gerhard, DS, Pauls, DL; Sussex, JN; Kidd, KK; Allen, CR; Hostetter, AM; Housman, D. 1987. "Bipolar affective disorders linked to DNA markers on chromosome 11." *Nature*. 325:783–787.

Ehrhardt, AA; Baker, S. 1974. "Fetal androgens, human central nervous system differentiation and behavior sex differences." In R. Friedman; R. Richart; R. Vande Wiele (eds). *Sex Differences in Behavior*. New York: Wiley.

Ehrinnger, H; Hornykiewicz, O. 1960. "Verteilung von noradrenilin und dopamine im gehirn des menschen und ihr verhalten bei erkrankungen des extrapyramidalen system." *Klin Wochenschr*. 38: 1236–1239.

Ehrmann, W. 1963. "Social determinants of human sexual behavior." In G Winokur (ed) *Determinants of Human Sexual Behavior*. Springfield, IL: CC Thomas.

Eibl-Eibesfeldt, I. 1989. *Human Ethology*. NY: Aldine de Gruyter.

Eibl-Eibesfeldt, I. 1972. *Love and Hate*. NY: Holt, Rinehart Winston.

Eibl-Eibesfeldt, I. 1974. "The myth of the agression-free hunter-gatherer society." In R. Hoolway (ed). *Primate Aggression, Territoriality and Xenophobia*. New York: Academic Press.

Eibl-Eibesfeldt, I. 1975. *Ethology: The Biology of Behavior*. New York: Holt, Rinehart and Winston.

Eisenberg, J. 1981. *The Mammalian Radiations: An Analysis of Trends in Evolution, Adaptation and Behavior*. Chicago: University of Chicago Press.

Eisenberg, L 1986. "Mindlessness and brainlessness in psychiatry." *Br J Psychiatry*. 148: 497–508.

Eisenberg, L. 1995. "The social construction of the human brain." *Am J Psychiatry*. 152: 1563–1575.

Eisenberg, N. 1994. "Empathy." Pp. 247–253 in *Encyclopedia of Human Behavior*, edited by V. S. Ramachandran. NY: Academic Press.

Eisler, R. & Levine, D. 2002. "Nurture, Nature, and Caring: We are not Prisoners of Our Genes." Pp. 9–52. In *Brain and Mind*. Vol. 3, No.1.

Ekman, P. 1973. "Cross cultural studies of facial expression." In P. Ekman (ed). *Darwin and Facial Expression*. New York: Academic Press.

Ekman, P; Friesen, W. 1974. "Nonverbal behavior in psychopathology." Pp 203-224 in K. Friedman; M. Katz, (eds). *The Psychology of Depression: Contemporary Theory and Research*. Washington, DC: Winston and Sons.

Ellenberger, D. 1970. *The Discovery of the Unconscious*. NY: Basic Books, 1970.

Ellenberger, H. 1958. "A clinical introduction to psychiatric phenomenology and existential analysis." In R. May, E. Angel, H. Ellenberger (eds). *Existence: A New Dimension in Psychiatry and Psychology*. New York: Basic Books.

Ellenwood, EH; Petrie, W. 1979. "Drug-induced psychoses." Pp 301–336 in R. Pickens; L. Heston (eds) *Psychiatric Factors in Drug Abuse*. NY: Grune and Stratton.

Ellis, A; Abarbanel, A (eds). 1961. *The Encyclopedia of Sexual Behavior*. New York: Hawthorn.

Ellis, A; Glantz, K; Pearce, J; Mayr, E, Wilson, DR. 1992. "The cognitive approach and evolutionary biology." *McLean Hospital Symposium*. Belmont, MA: McLean CME Videos.

Ellis, H. 1904. *A Study of British Genius*. London: Hurst and Blachett.

Ellis, H. 1926. *Studies in the Psychology of Sex*. Philadelphia: Davis.

Elston, RC; Stewart, J. 1971. "A general model for the genetic analysis of pedigree data." *Hum Hered*. 21: 523–542.

Elston, RC; Yelverton, K. 1975. "General models for segregation analysis." *Am J Hum Genetics*. 27: 31–45.

Emery, AEH; Mueller, R. 1992. *Elements of Medical Genetics*. 8th ed. Edinburgh: Churchill Livingstone.

Erdal, D. and Whiten, A. 1996. "Egalitarianism and Machiavellian Intelligence in Human Evolution." In *Modeling the Human Mind*. (eds). P. Mellars and K. Gibson. Cambridge: The MacDonald Institute.

Erlenmeyer-Kimling, L; Jarvik, L. 1963. "Genetics and intelligence: A review." *Science*. 142: 1477–1479.

Erlenmeyer-Kimling, L; Paradowski, W. 1966. "Selection and schizophrenia." *Am Naturalist*. 100: 651–665.

Esquirol, J. 1845. *Mental Maladies: A Treatise on Insanity*. Philadelphia: Lea and Blanchard.

Essen-Moller, E. 1985. "Twin research and psychiatry." *Acta Psychiatr Scand*. 39: 65–77.

Estroof, TW; Dackis, CA; Gold, MS; Pottash, A. 1985. "Drug abuse and bipolar disorders." *Int J Psychiatry Med*. 15: 37–40.

Ewald H, Mors O, Flint T. 1995. "A possible locus for manic-depressive illness on chromosome 16p13." *Psychiatr Genet*. 5: 71–81.

Ewald, H., Degn, B., Mors, O. 1998. "Significant Linkage Between Bipolar Affective Disorder and Chromosome 12q24." *Psychiatric Genetics*. 8: 131–140.

Ewald H, Flint T, Kruse TA, Mors O. 2002. "A genome-wide scan shows significant linkage between bipolar disorder and chromosome 12q24.3 and suggestive linkage to chromosomes 1p22-21, 4p16, 6q14-22, 10q26 and 16p13.3." *Mol Psychiatry*. 7(7):734–44.

Eysenck, H. 1977. *Crime and personality*. 3rd ed. Routledge & Kegan Paul Pub.

Eysenck, H. 1987. "The definition of personality disorders and the criteria appropriate for their description." *Journal of Personality Disorders*.1(3): 211–219.

Falconer, D. 1965. "The inheritance liability to certain diseases estimated from the incidence among relatives." *Ann Hum Genetics*. 29: 51–71.

Fallin M, Lasseter V, Wolyniec P, McGrath J, Nestadt G, Valle D, Liang K, Pulver A. 2004. "Genomewide Linkage Scan for Bipolar-Disorder Susceptibility Loci among Ashkenazi Jewish Families." *Am J Hum Genet*. Jun 18;75(2).

Faraone, S. 1996. "Diagnostic accuracy and confusability analyses: An application to the Diagnostic Interview for Genetic Studies." *Psychol Med*. 26: 401–410.

Farde, L; Wiesel, FA; Hall; H. 1987. "No D2 receptor increase in PET study of schizophrenia." *Arch Gen Psychiatry*. 44: 671–672.

Farde, L; Wiesel, FA; Halldin, C; Sedval, G. 1988. "Central D2 receptor occupancy in schizophrenic patients treated with antipsychotic drugs." *Arch Gen Psychiatry*. 45: 71–78.

Farde, L; Nordström, AL; Wiesel, F. 1992. "PET analysis of central D1 and D2 receptor occupancy in patients treated with classical neuroleptics and clozapine." *Arch Gen Psychiatry*. 49: 538–544.

Fehr. E. and Gachter, S. 2000. "Fairness and Retaliation: the Economics of Reciprocity." *Journal of Economic Perspectives*. V. 14. N. 3: 159–181.

Feldman, M.P. 1977. *Criminal behavior: A psychological analysis*. NY: John Wiley & Sons Pub.

Felsenstein, J. 2000. "From Population Genetics to Evolutionary Genetics: A View through the Trees." Pp. 609–627 in *Evolutionary Genetics*. Ed. by Singh, R. and Krimbas, C. Cambridge University Press.

Ferguson, J.; Young, L.; Insel, T.; 2002. "The Neuroendocrine Basis of Social Recognition." *Front Neuroendocrinology*. 23: 200–224.

Festinger, L. 1954. "A theory of social comparison processes." *Hum Rel*. 7: 117–140.

Fieve, R. 1973. *Moodswing: The Third Revolution in Psychiatry*, NY: Morrow.

Fieve, RR; Go, R; Dunner, DL; Elston, R. 1984. "Search for biological/genetic markers in a long-term epidemiological and morbid risk study of affective disorders." *J Psychiatr Res*. 18: 425–445

Firschein, L. 1961. "Population dynamics of the sickle-cell trait in the black Caribs of British Honduras." *American Journal of Human Genetics*. 13: 233–254.

Fisher, R. 1918. "The correlation between relatives on the supposition of Mendelian inheritance." *Trans R Soc Edin*, 52: 399–433.

Fisher, R. 1971-1974 (1928). "The social selection of human fertility." In J. Bennett (ed). *The Collected Papers of RA Fisher*. 5 Vols. London: Adelaide.

Fisher, R. 1930. *The Genetical Theory of Natural Selection*. Oxford: Clarendon Press.

Fleischer, S. & Slotnik, B. 1978. "Disruption of maternal behavior in rats with lesions of the septal area." *Physiological Behavior*. 21: 189–200.

Fleming, A., Morgan, H., & Walsh, C. 1996. "Experiential Factors in Postpartum Regulation of Maternal Care." Pp. 295–332 in *Parental Care: Evolution, Mechanisms, and Adaptive Intelligence.* Ed. by Rosenblatt, J. and Snowden, C. NY: Academic Press.

Flemenbaum, A; Larson, J. 1976. "ABO-Rh blood groups and psychiatric diagnosis: A critical review." *Dis Nerv Syst*. 37: 581–583.

Foley, R. 1992. "Studying human evolution by analogy." In S. Jones, R. Martin, D. Pilbeam (eds). *The Cambridge Encyclopedia of Human Evolution.* Cambridge: Cambridge University Press.

Ford, C. 1945. *A Comparative Study of Human Reproduction*. New Haven: Yale University Press.

Ford, C; Beach, F. 1951. *Patterns of Sexual Behavior*. New York: Harper and Row.

Ford, E. 1945. "Polymorphism." *Biological Review*. 20: 73–88.

Fowles, D. 1992. "Schizophrenia: Diathesis-stress revisited." *Annual Review of Psychology*. 43: 303–336.

Fox, R. 1986. "Fitness by any other name." *Beh Brain Sciences*. 9:1 92–193.

Friemer, NB; Reus, V; Escamilla, M; McInnes, L; Spesny, M; Leon, P; Service, S; Smith, L; Silva, S; Rojas, E; Gallegos, A; Meza, L; Fournier, E; Baharloo, S; Blankenship, K; Tyler, D; Batki, S; Vinogradov, S; Weissenbach, J; Barondes, S; Sandkuijl, L. 1996. "Genetic mapping using haplotype, association and linkage methodes suggests a locus for severe bipolar disorder (BPI) at 18q22-q23." *Nature Genetics*. 12: 436–441.

Freud, S. 1987. *Phylogentic Phantasy*. Harvard University Press.

Freud. S. 1917. Introductory Lectures on Psychoanalysis. S. E., 16: 243–463.

Freud, S. 1921. "Mourning and Melancholia." In J Strachey (translator), *The Standard Edition of the Complete Psychological Works of Sigmund Freud.* London: The Hogarth Press.

Freud, S. 1921. "Group Psychology and the Analysis of the Ego." *The Standard Edition of the Complete Psychological Works of Sigmund Freud.* London: The Hogarth Press.

Freud, S. 1930. *Civilization and Its Discontents*. NY: Norton.

Friddle, C., Koskela, R., Ranade, K., Herbert, J., Cargill, M., et al. 2000. "Full Genome Scan for Linkage in 50 Families Segregating the Bipolar Affective Disease Phenotype." *American Journal of Human Genetics.* 66: 205–215.

Friston, K. 2002. "Beyond Phrenology: What Can Neuroimaging Tell Us About Distributed Circuitry?" *Annual Review of Neuroscience*. 25: 221–250.

Frith, U. (1993) 1997. "Autism." Pp. 92–98 in *Scientific American: Mysteries of the Mind.* Spec. Issue V. 7, N.1.

Frith, U. 1989. *Autism: Explaining the Enigma*. Cambridge, MA: Basil Blackwell.

Fuchs, E. and Flügge, G. 2003. "Chronic Social Stress: Effects on Limbic Brain Structures." *Physiology & Behavior*. V79, N3: 417–427.

Fuster, J. 1999. "Cognitive Functions of the Frontal Lobes." Pp. 187–195 in *The Human Frontal Lobes.* Ed. by B. Miller & J. Cummings. NY: The Guilford Press.

Fuster, J. 1997. *The Prefrontal Cortex: Anatomy, Physiology, and Neuropsychology of the Frontal Lobe*. Third Edition. NY: Lippincott-Raven.

Gabor Miklos, G., Campbell, K. and Kankel, D. 1994. "The rapid emergence of bio-electronic novelty, neuronal architectures, and organismal performance." *Flexibility and Constraint in Behavioural Systems*. Ed. by R. J. Greenspan. NY: John Wiley and Sons: 269–293

Gainer, H. and Wray, S. 1992. "Oxytocin and Vasopressin: From Genes to Peptides." *Ann NY Acad Sci*. 652: 14–28.

Galton, F. 1865. "Inquiries into Human Faculty and Character." *Macmillan's Magazine*. 12: 157-166, 318–327.

Galton, F. 1884. "Measurement of character." *Fortnightly Rev*. 42: 179–185.

Gardner, R. 2002. "MacLean's Paradigm and Its Relevance for Psychiatry's Basic Science." Pp.85–105 in G. Cory & R. Gardner, eds. *The Evolutionary Neuroethology of Paul MacLean: Convergences and Frontiers*. Westport, CT: Praeger.

Gardner, R. 1982. "Mechanisms in Major Depressive Disorder: An Evolutionary Model." *Archives of General Psychiatry*. 39. 1436–1441.

Gardner, R. 1988. "Psychiatric syndromes as infrastructure for intraspecific communication." In M. Chance (ed*) Social Fabrics of the Mind.* London: Lawrence Erlbaum.

Gardner, R. 1995. "Sociobiology and its application to psychiatry." In H. Kaplan; B. Sadock (eds). *Comprehensive Textbook of Psychiatry*. 6th ed. Baltimore: Williams and Wilkins.

Garrod, A. 1902. "The incidence of alkaptonuria: A study of chemical individuality." *Lancet*. 2: 1616–1620.

Garrod, A. 1909. *Inborn Errors of Metabolism.* London: Oxford University Press.

Gawin, FA; Ellenwood, E. 1988. "Cocaine and other stimulants." *N Engl J Med.* 318: 1173–1182.

Gazzaniga, M. 1992. *Nature's Mind: The Impact of Darwinian Selection on Thinking, Emotions, Sexuality, Language and Intelligence*. New York: Basic Books.

Geertz, C. 1973. *The Interpretation of Cultures*. New York: Basic Books.

Gehring, W. 1998. *Master Control Genes in Development and Evolution: The Homeobox Story.* Yale University Press.

Gershon, E., Bander, J., Detera-Wadley, S., Ferraro, T., Berrittini, W. 1996. "Maternal Inheritance and Chromosome 18 Allele Sharing in Unilineal BP Illness Pedigrees." *American Journal of Medical Genetics*. 67: 202–207.

Gershon, ES; Liebowitz, J. 1975. "Sociocultural and demographic correlates of affective disorders in Jerusalem." *J Psychiatr Res*. 12: 37–50.

Gershon, E; Bunney, W; Leckman, J, Van Eerdewegh, M; De Bauche, B. 1976. "The inheritance of affective disorders: a review of data and of hypotheses." *Behav. Genet.* 6: 227–261.

Gershon, E: Hamovit, J. 1979. "Genetic methods and preventive psychiatry." *Prog Neuropsycho-pharmacol Biol Psychiatry.* 3: 565–573.

Gershon, E; Hamovit, J; Guroff, JJ; Dibble, E; Leckman, JF; Sceery, W; Targum, SD; Nurnberger, JI; Goldin, LR; Bunney, W. 1982. "A family study of schizo-affective, bipolar I, bipolar II, unipolar and normal control probands." *Arch Gen Psych.* 39: 1157–1167.

Gershon, E. 1983. "The genetics of affective disorders." Pp 434–538 in L. Grinspoon (ed). *Psychiatry Update.* Vol 2. Washington, DC: American Psychiatric Press.

Gershon, E; Goldin, L. 1984. "Clinical methods in psychiatric genetics, I. Robustness of genetic marker investigation strategies." *Arch Gen Psych.* 41: 173–180.

Gershon, E; McKnew, D; Cytryn, L; Hamovit, J; Schreiber, J; Hibbs, E; Pelligrini, D. 1985. "Diagnosis in school-aged children of bipolar affective disorder patients and controls." *J Aff Dis.* 8: 283–291.

Gershon, E; Weissman, M; Guroff, J: Prusoff, B; Leckman, J. 1986. "Validation of criteria for major depression through controlled family study." *J Aff Dis.* 11: 125–131.

Gershon, E; Hamovit, J; Guroff, J; Nurnberger, J. 1987. 'Birth-cohort changes in manic and depressive disorder in relatives of bipolar and schizoaffective patients." *Arch Gen Psychiatry.* 44: 314–319.

Gershon, E; DeLisi, L; Hamovit, J; Nurnberger, J; Maxwell, ME; Schreiber, J; Dauphiais, D; Dingman, CW; Guroff, J. 1988. "A controlled family study of chronic psychoses: Schizophrenia and schizoaffective disorder." *Arch Gen Psychiatry.* 45: 328–336.

Gershon, E. 1996a. "Maternal inheritance and chromosome 18 allele sharing in unilineal bipolar illness pedigrees." *Am J Med Genet.* 67: 202–207.

Gershon, E. 1996b. "Detection of linkage to affective disorders in the catalogued Amish pedigrees: A reply to Pauls, et al." *Am J Hum Genet.* 58: 1381–1385.

Geschwind, N: Personal Communication, 1985

Ghiselin, M. 1967. *The Economy of Nature and the Evolution of Sex.* Berkeley: University of California Press.

Gibson, RW; Cohen, M; Cohen, R. 1959. "On the dynamics of the manic-depressive personality." *Am J Psychiatry.* 115: 1101–1107.

Gibson, R. 1963. "Psychotherapy of manic-depressive states." *Psychiatr Res Rep.* 17: 91–102.

Gilbert, J: Personal Communication, 1992.

Gilbert, P. 1984. *Depression: From Psychology to Brain State.* London: Lawrence Erlbaum.

Gilbert, P. 1988. "Emotional disorders, brain state and psychosocial evolution." In W Dyden; P Trower (eds). *Developments in Cognitive Psychotherapy*. London: Sage.

Gilbert, P. 1989. *Human Nature and Suffering*. London: Lawrence Erlbaum.

Gilbert, P 1992. *Depression: The Evolution of Powerlessness*. London: Lawrence Erlbaum.

Gilbert, P; Price, J; Allen, S. 1995. "Social comparison, social attractiveness and evolution: How might they be related?" *New Ideas in Psychology*. 13: 149–165.

Gilbert, P. 1996. "Social comparison in affective disorders." *Proceedings of the Xth Congress, World Psychiatric Association*. Madrid.

Gilbert, P. & Bailey, K., (eds). 2000. *Genes on the Couch: Explorations in Evolutionary Psychotherapy*. Philadelphia, PA: Taylor & Francis.

Gingrich, JA; Caron, M. 1993. "Recent advances in the molecular biology of dopamine receptors." *Ann Rev Neurosci*. 16: 299–321.

Ginns, E; Ott, J; Egelund, J; Allen, C; Fann, C; Pauls, D; Weissenbach, J; Carulli, J; Falls, K; Keith, T; Paul, S. 1996. "A genome wide search for chromosomal loci linked to bipolar affective disorder in Old Order Amish." *Nature Genetics*. 12: 431–35.

Gintis, H. 2000. "Strong Reciprocity and Human Sociality." *Journal of Theoretical Biology*. 206: 169–79.

Glantz, K; Pearce, J. 1989. *Exiles from Eden: Psychotherapy from and Evolutionary Perspective*. New York: Norton.

Gloor, P. 1997. *The Temporal Lobes and the Limbic System*. Oxford University Press.

Godwin, J. & Crews, D. 2002. "Hormones, Brain, and Behavior in Reptiles." Pp. 545–585 in *Hormones, Brain, and Behavior*. Ed. by Pfaff, D. V. II. NY: Academic Press

Goertzel, MG; Goertzel, V; Goertzel, T. 1978. *Three Hundred Eminent Personalities*. San Francisco: Jossey-Bass.

Goldberg, E. 2001. *The Executive Brain: Frontal Lobes and the Civilized Mind*. Oxford University Press.

Goldin, LR; Gershon, ES; Lake, CR; Murphy, DL; McGinnis, M; Sparkes, RS. 1982. "Segregation and linkage studies of plasma dopamine-beta-hydroxylase (DBH), erythrocyte catechol-O-methyltransferase (COMT), and platelet monoamine oxidase (MAO): Possible linkage between the ABO locus and a gene controlling DBH activity." *Am J Hum Genetics*. 34: 250–262.

Goldin, LR; Gershon, ES; Targum, SD; Sparkes, PS; McGinnis, M. 1983. "Segregation analyses in families of patients with bipolar, unipolar and schizoaffective disorders." *Am J Hum Genetics*. 35: 274–287.

Goldstein, D. 1995, "Stress as a Scientific Idea: A Homeostatic Theory of Stress and Distress." *Homeostasis*. V. 36. N. 4: 177–215.

Goltz, F. 1892. "Der hund ohne grosshirn." *Pflug Arch Ges Physiol.* 51: 570–614.

Gonzalez-Mariscal, G. & Poindron, P. 2002. "Parental Care in Mammals: Immediate Internal and Sensory Factors of Control." Pp. 215–298 in *Hormones, Brain, and Behavior*. Ed. by Pfaff, D. V. I. NY: Academic Press.

Good, B. 1992. "Culture and psychopathology." In T. Schwartz, G. White; C. Lutz (eds). *New Directions in Psychological Anthropology*. Cambridge University Press.

Goodall, J. 1986. *The Chimpanzees of Gombe: Patterns of Change*. Belknap/Harvard University Press.

Goodall, J. 1990. *Through a Window*. Boston: Houghton Mifflin.

Goodall, J. 1986. *The Chimpanzees of Gombe: Patterns of Behavior*. Belknap/Harvard University Press.

Goodwin, F; Jamison, K. 2007. *Manic Depressive Illness*. 2nd Edition. Oxford University Press

Goodwin, F; Jamison, K. 1990. *Manic Depressive Illness*. Oxford University Press.

Gottesman, I. 1972. *Schizophrenia and Genetics: A Twin Study Vantage Point*. New York: Academic Press.

Gottesman, I; Carey, G. 1983. "Extracting meaning and direction from twin studies." *Psychiatr Dev.* 1: 35–50.

Gottesman, I. 2002. "Foreword: Genes, Temperament, and Personality: Past and Prospect." Pp. xiii–xx in *Molecular Genetics and the Human Personality*. Ed. by Benjamin, J. Ebstein, R., & Belmaker, R Washington, DC: American Psychiatric Publishing, Inc.

Gould, SJ; Lewontin, R. 1979. "The spandrels of San Marco and the panglossian paradigm. A critique of the adaptationist programme." *Proceedings of the Royal Society of London*. B 205: 581–598.

Gouldner, A. 1960. "The Norm of Reciprocity: a Preliminary Statement." *American Sociological Review*. 25: 161–178.

Graves, P. and Haystead, T. 2003. "A Functional Proteomics Approach to Signal Transduction." Pp. 1–24 in *Recent Progress in Hormone Research*. Ed. by A. Means. V. 58.(The Human Genome and Endocrinology). Chevy Chase, MD: The Endocrine Society.

Gray, J. 1987. *Fear and Stress*. 2nd Edition. Cambridge University Press.

Graybiel, A. 2000. "The basal ganglia." *Current Biology* 10: R509–R511.

Greden, J; Carroll, B. 1981. "Psychomotor function in affective disorders: An overview of new monitory techniques." *Am J Psychiatry*. 138: 1441–1448.

Greenberg, JS; Greenly, JR; Benedict, M. 1992. "Contributions of persons with serious mental illness to their families." *Hospital and Community Psychiatry*. 45: 475–480.

Greenberg, LS; Safran, J. 1984. "Integrating affect and cognition." *Cognitive Ther and Res*. 8: 559–578.

Greenberg, LS; Safran, J. 1987. *Emotion is Psychotherapy*. New York: Guilford.

Greenberg, N. 2002. "Adaptive Functions of the Corpus Striatum." Pp. 45–81 in G. Cory & R. Gardner, eds. *The Evolutionary Neuroethology of Paul MacLean: Convergences and Frontiers*. Westport, CT: Praeger.

Grossman, J., Carter, A., Volkmar, F. 1997. "Social Behavior in Autism." Pp. 440–454 in *The Integrative Neurobiology of Affiliation*. Ed. by C. S. Carter, I. Lederhendler & B. Kirkpatrick. NY: Annals of the NY Academy of Sciences (Vol. 809).

Gruzelier, J; Seymour, K; Wilson, L; Jolley, A; Hirsch, S. 1988. "Impairments on neuropsychiatric tests of temporohippocampal and frontohippocampal functions and word fluency in remitting schizophrenia and affective disorders." *Biol Psychiatry*. 21: 889–899.

Gurling, H. 1990. "Genetic linkage and psychiatric disease." *Nature*. 344: 22.

Gurling, H. 1998. "Chromosome 21 Workshop." *Psychiatric Genetics*. 8: 109–113.

Gurtman, M. 1986. "Depression and the response of others: Reevaluating the reevaluation." *J Abnormal Psychology*. 95: 99–101.

Gusella, JF; Wexler, NS; Conneally, PM; Naylor, SL; Anderson, MA; Tanzi, RE; Watkind, PC; Ottina, K; Wallace, MR; Sakaguchi, AY; Young, AM; Shoulson, I; Bonilla, E; Martin, J. 1983. "A polymorphic DNA marker genetically linked to Huntington's disease." *Nature*. 306: 234–238.

Gutnick, M. J. & Mody, I., (eds). 1995. *The Cortical Neuron*. Oxford University Press.

Haeckel, E. 1866. *Generelle Morphologie der Organismen*, Vol I, II. Berlin: Reimer.

Haldane, J. 1932. *The Causes of Evolution*. London: Longmans.

Haldane, J. 1955. "Population genetics." *New Biology*. 18: 34–51.

Haldipur, C. 1989. "Psychiatric nosology and taxonomy in ancient India." *Acta Psychiatrica Scand*. 80: 148-150.

Hales, RE; Yudofsky, S. 1987. *Textbook of Neuropsychiatry*. Washington: American Psychiatric Press.

Hall, E. 1968. "Human needs and inhuman cities." *Smithsonian Annual*. 11.

Hallipike, C. 1984. "The relevance of the theory of inclusive fitness to human society." *J Soc Biol Struct*. 7: 131–144.

Hamburg, B. 1971. "The biosocial basis of sex differences." In S. Washburn; E. McCown (eds). *Human Evolution: Biosocial Perspectives*. Menlo Park, CA: Benjamin/ Cummings.

Hamburg, D. 1974. "Ethological perspectives on human aggressive behavior." In N. White (ed). *Ethology and Psychiatry*. Toronto: University of Toronto Press.

Hamilton, W. 1964. "The Genetical Evolution of Social Behavior, I & II." Pp. 1–16, & 17–52 in *Journal of Theoretical Biology*. V. 7.

Hamilton, W. 1963."The Evolution of Altruistic Behavior." *American Naturalist* 97, 354–356.

Hamilton, W. 1967. "Extraordinary sex ratios." *Science*. 157: 477–488.

Hamilton, W. 1971. "Selection of selfish and altruistic behavior." In J. Eisenberg; W. Dillon (eds). *Man and Beast: Comparative Social Behavior*. Washington, DC: Smithsonian Institution Press.

Hamilton, W. "Innate aptitudes of man: An approach from evolutionary genetics." In R. Fox (ed). *Biosocial Anthropology*. London: Malaby.

Hammen, C; Peters, S. 1977. "Differential responses to male and female depressive reactions." *J Clin Consult Psychology*. 45: 994–1001.

Hammen, C; Peters, S. 1978. "Interpersonal consequences of depression: Responses to men and women enacting a depressed role." *J Abnormal Psychology*. 87: 322–332.

Hammerstein, P. 1996. "Darwinian adaptation, population genetics and the streetcar theory of evolution." *J Math Biol.* 34: 511–532.

Harbor, SN; Barchas, PR; Barchas, J. 1981. "A primate anologue of amphetamine induced behaviors in humans." *Biological Psychiatry*. 16: 181–196.

Harbor, SN; Barchas, PR; Barchas, J. 1984. "The regulatory effect of social rank on behavior after amphetamine administration." In P. Barchas (ed). *Social Hierarchies: Essays Toward a Sociophysiological Perspective*. Westport: Greenwood Press.

Hardy, G. 1908. "Mendelian proportions in a mixed population." *Science*. 28: 49–50.

Hare RD, Harpur TJ, Hakstian AR et al. 1990. "The Revised Psychopathy Checklist: descriptive statistics, reliability, and factor structure." *Psychological Assessment* 2: 338–341.

Hare, RD. 1993. *Without conscience: The disturbing world of the psychopaths among us*. Pocket Books.

Hare RD, 1998. *Without Conscience: The disturbing world of the psychopaths among us*. New York, NY: Guilford.

Hare, RD. 1970. *Psychopathy: Theory and research*. NY: John Wiley & Sons Pub.

Hare, E; Price, J; Slater, E. 1971. "Age distribution of schizophrenia and neurosis: Findings in a national sample." *Br J Psychiatry*. 119: 445–458.

Harlow, H. 1986. *From Learning to Love: The Selected Papers of H. F. Harlow*. Ed. by C. Harlow. NY: Praeger.

Harlow, H. & Harlow., M. 1965. "The Affectional Systems." Pp. 386–334 in *Behavior of Non-Human Primates*. Ed. by A. Schrier, H. Harlow, & F. Stollnitz. New York: Academic Press.

Harlow, H. 1958. *The evolution of learning*. In A. Roe; G. Simpson (eds). *Behavior and Evolution*. Yale University Press.

Harpending, H. & Sobus, J. 1987. “Sociopathy as an adaptation.” *Ethology and Sociobiology*.8(3s): 63s–72s.

Harris, J. 2002. “Empathy, Autism, and the Integration of the Triune Brain.” Pp. 155–166 in G. Cory & R. Gardner (eds). *The Evolutionary Neuroethology of Paul MacLean: Convergences and Frontiers*. Westport, CT: Praeger.

Harris, V. 1964. *The Life of the Rainbow Lizard*. London: Hutchinson.

Harrison, GA; Tanner, JM; Pilbeam, D; Baker, P. 1988. *Human Biology*, 3rd ed. Oxford University Press, pp. 214–215.

Hartmann, L. 1992. “Presidential address: Reflections on humane values and biopsychosocial integration.” *Am J Psychiatry*. 149: 1135–1147.

Harvald, B; Hauge, M. 1965. “Hereditary factors elucidated in twin studies.” In J. Neal,; M. Shaw; W. Shull (eds). *Genetics and the Epidemiology of Chronic Diseases*. Washington, DC: Public Health Service Publication 1163.

Hauge, M; Harvald, B; Fischer, M; Gotlieb-Jensen, K; Juel-Nielsen, N; Raeblid, I; Shapiro, R; Videbest, T. 1968. “The Danish twin register.” *Acta Geneticae Medicae et Gemellologiae*. 17: 315–332.

Hawkes, K. 1990. “Showing off: Tests of an hypothesis about men’s foraging goals.” *Ethology and Sociobiology*. 12: 29–54.

Hayek, F. 1952. *The Sensory Order: An Inquiry into the Foundations of Theoretical Psychology*. Chicago: University of Chicago Press.

Heller, W.; Nitschke, J. Miller, G. 1998. "Lateralization in Emotion and Emotional Disorders." Pp. 26-32 in *Current Directions in Psychological Science*. Vol. 7. N.1.

Hemingway, H. and Marmot, M. 1999. “Psychosocial Factors in the Aetiology and Prognosis of Coronary Heart Disease: Systematic Review of Prospective Cohort Studies.” *BMJ*. 318: 1460–1467.

Henry, GM; Weingartner, H; Murphy, D. 1971. “Idiosyncratic patterns of learning and word association during mania.” *Am J Psychiatry*. 128: 564–574.

Henry, GM; Weingartner, H; Murphy, D. “Influence of affective states and psychoactive drugs on verbal learning and memory.” *Am J Psychiatry*. 130: 966–971.

Henry, JP. & Wang, S. 1998. “Effects of Early Stress on Adult Affiliative Behavior.” *Psychoneuro-endocrinology*. V, 23. N.8: 863–875.

Henry, JP 1982. “The relation of social to biological processes in disease.” *Soc Sci Med*. 16: 369–380.

Henry, JP; Stephens, P. 1977. *Stress, Health and the Social Environment: A Sociobiologic Approach to Medicine*. New York: Springer Verlag.

Herbert, J. & Schulkin, J. 2002. “ Neurochemical Coding of Adaptive Responses in the Limbic System.” Pp. 659–689 in *Hormones, Brain, and Behavior*. Ed. by Pfaff, D. V. I. NY: Academic Press.

Herve, D.; Blanc, G.; Glowinski, J.; Tassin, J. 1982. "Reduction of dopamine utilization in the prefrontal cortex but not in the nucleus accumbens after selective destruction of noradrenergic fibers innervating the ventral tegmental area in the rat." *Brain Research.* 237: 510–516.

Heun, R. & Maier, W. 1993. "The Distinction of Bipolar II Disorder from Bipolar I and Recurrent Unipolar Depression: Results of a Controlled Family Study." *Acta Psychiatrica Scandinavia.* 87: 279–284.

Hill, D. 1968. "Depression: Disease, reaction or posture?" *Am J Psychiatry.* 125: 445–457.

Himmelhoch, J; Mulla, D; Neil, J; Detre, T; Kupfer, D. 1976. "Incidence and significance of mixed affective states in the bipolar population." *Arch Gen Psychiatry.* 33: 1062–1066.

Hinchliffe, M; Hooper, D; Roberts, F; Vaughn, P. 1975. "A study of the interaction between depressed patients and their spouses." *Br J Psychiatry.* 126: 164–172.

Hinde, RH. 1974. *Biological Bases of Human Behavior.* New York: McGraw-Hill.

Hinde, RH. 1987. *Individuals, Relationships and Culture: Links between Ethology and the Social Sciences.* Cambridge: Cambridge University Press.

Hinde, RS. 1989. "Relations between levels of complexity in behavioral sciences." *J Nerv Ment Dis.* 177: 655–657.

Hodgkinson, S; Sherrington, R; Gurling, H; Marchbanks, R; Reeders, S; Mallet, J; McInnes, M; Petursson, H; Brynjolfson, J. 1987. "Molecular evidence for heterogeneity in manic- depression." *Nature.* 325: 805–806.

Hoffman E., McCabe, K., & Smith, V. 1998. "Behavioral Foundations of Reciprocity: Experimental Economics and Evolutionary Psychology." Pp. 335–352 in *Economic Inquiry.* V. 36. N. 3.

Hoffman, M. 2000. *Empathy and Moral Development.* Cambridge University Press.

Hoffman, M. 1981. "Is Altruism Part of Human Nature?" Pp. 121–137 in *Journal of Personality and Social Psychology.* 40.

Holder, R: Personal Communication, 1989

Holzinger, K. 1929. "The relative effect of nature and nurture influences on twin differences." *J Edu Psychology.* 20: 241–248.

Hong KS, McInnes LA, Service SK, Song T, Lucas J, Silva S, Fournier E, Leon P, Molina J, Reus VI, Sandkuijl LA, Freimer N. 2004. "Genetic mapping using haplotype and model-free linkage analysis supports previous evidence for a locus predisposing to severe bipolar disorder at 5q31-33." *Am J Med Genet.* 2004 Feb 15;125B(1):83–6.

Hopkinson, G; Ley, P. 1969. "A genetic study of affective disorder." *Brit J Psychiatry.* 115: 917–922.

Hudson, J; Pope H. 1990. "Affective spectrum disorder." *Am J Psychiatry.* 47:1: 552–564.

Humphrey, N. 1976. "The Function of the Intellect." Pp. 303–317 in *Growing Points in Ethology*. Ed. by P. Bateson and R. Hinde Cambridge University Press.

Hume, D. 1978 (1739). *A Treatise of Human Nature*. Oxford: Oxford University Press.

Hundert, E. 1989. *Philosophy, Psychiatry and Neuroscience: Three Approaches to the Mind.* Oxford: Oxford University Press.

Hundert, EM: Personal Communication, 1990

Huxley, J. 1964. *Essays of a Humorist*. The Twentieth Century.

Huxley, T. 1914. "On the natural inequality of men." In Sir M. Foster; Sir R. Lankester (eds). *The Scientific Memoirs of Thomas Henry Huxley*. 5 Vols. Birmingham: Cornish Brothers.

Huxley, J; Mayr, E; Osmond, H; Hoffer, A. 1964. "Schizophrenia as a genetic morphism." *Nature*. 204:220–221.

Hyman, SE; Nestler, E. 1993. *The Molecular Foundations of Psychiatry*. Washington, DC: American Psychiatric Press.

Iidaka, T.; Terashima, S.; Yamashita, K.; Okada, T.; Sadato, N.; Yonekura, Y. 2003. "Dissociable Neural Responses in the Hippocampus to the Retrieval of Facial Identity and Emotion: An Event-Related fMRI Study." *Hippocampus*. 13: 429–436.

Ingram, V. 1956. "Molecular defect in sickle-cell anaemia." *Nature*. 178: 792–794.

Ionasescu, V. and Zellweger, H. 1983. *Genetics in Neurology*. NY: Raven Press.

Irvine, DG; Miyashita, H. 1965. "Blood types in relation to depressions and schizophrenia: A preliminary report." *Can Med Assoc J*. 92: 551–554.

Isaac, G. 1978. "The Food-sharing behavior of protohuman hominids." Pp. 90–108 in *Scientific American*. V. 238.

Itzkoff, S. 1990. The *Making of the Civilized Mind: the Evolution of Intelligence*. NY: Peter Lang.

Izard, CE; Blumberg, S. 1985. "Emotion theory and the role of emotions in anxiety in children and adults." In A. Tuma; J. Maser (eds). *Anxiety and Anxiety Disorders*. Hillsdale, NJ: Lawrence Erlbaum.

James, W. 1890. *Principles of Psychology*. New York: Holt.

James, W. 1902. *The Varieties of Religious Experience*. New York: Longmans.

Jamison, KR; Goodwin, F. 1983. "Psychotherapeutic treatment of manic-depressive patients on lithium." In M Greenhill; A Gralnick (eds). *The Interrelationship of Psychopharmacology and Psychotherapy*. New York: Macmillan.

Jamison, K. 1989. "Mood disorders and patterns of creativity in British writers and artists." *Psychiatry*. 52: 125-134.

Jamison, K. 1993. *Touched with Fire: Manic-Depressive Illness and the Artistic Temperment*. New York: Free Press.

Jamison, K. 1995. *An Unquiet Mind.* New York: Knopf.

Jamison, K: Personal Communication, 1996

Janowsky, D; Leff, M; Epstein, R. 1970. "Playing the manic game: Interpersonal maneuvers of the acutely manic patient." *Arch Gen Psychiatry.* 22: 252–261.

Janowsky, D; el-Yousef, MK; Davis, J. 1974. "Interpersonal maneuvers of manic patients." *Am J Psychiatry.* 131: 250–255.

Janowsky, D. & Overstreet, D. 2000. "Cholinergic Dysfunction in Mood Disorders." Pp. 95–120 in *Bipolar Disorder: Basic Mechanisms and Therapeutic Implications.* Ed. by Soares, J. & Gershon, S. NY: Marcel Dekker.

Janssen, P; Niemegeers, CJE; Awouters, F. 1988. "Pharmacology of risperidone (R64 766), a new antipsychotic with serotonin-S2 and dopamine-D2 antagonist properties." *J Pharm Exper Therap.* 244: 685–693.

Jaspers, K. 1963 (1913). *General Psychopathology.* (J Hoenig; MW Hamilton, transl). Manchester: University of Manchester Press.

Jencks, C. 1990 "Varieties of Altruism." pp. 53–67 in *Beyond Self-Interest.* Edited by J. Mansbridge. University of Chicago Press.

Jerison, H. 1973. *Evolution of the Brain and Intelligence.* New York: Academic Press.

Johannsen, W. 1909. *Elemente der exakten Erblichkeitslehre.* Gustav Fischer, Jena.

Johnson, G; Leeman, M. 1977. "Analysis of familial factors in bipolar affective illness." *Arch Gen Psychiatry.* 34: 1074–1083.

Johnson, G; Hunt, GE; Robertson, S; Doran, T. 1981. "A linkage study of manic-depressive disorders with HLA antigens, blood groups, serum proteins and red cell enzymes." *J Affective Disorders* 3: 43–58.

Jones, I., Jacobson, N., Green, E., et al. 2002. "Evidence for Familial Cosegregation of Major Affective Disorder and Genetic Markers Flanking the Gene for Darier's Disease." *Molecular Psychiatry.* 7 (4): 424–427.

Jones, I., Kent, L., Craddock, N. 2002. "Genetics of Affective Disorders." Pp. 211–245 in *Psychiatric Genetics and Genomics.* Ed. by McGuffin, P., Owen, M., Gottesman, I. Oxford University Press.

Kagan, J. 1982. "The emergence of self." *J Child Psychology Psychiatry.* 23: 363-381.

Kagan, J. 1984. *The Nature of the Child.* New York: Basic Books.

Kahneman, D., Diener, E., and Schwartz, N. (eds). 1999. *Well-Being: The Foundations of Hedonic Psychology.* NY: Russell Sage.

Kallmann, F. 1954. "Genetic Principles in Manic-Depressive Psychoses." In P Hoch; J Zubin (eds). *Depression.* New York: Grune and Stratton.

Kan, YW; Dozy, A. 1978. "Polymorphism of DNA sequence adjacent to human beta-globin structural gene: Relationship to sickle mutation." *Proc Natl Acad Sci USA.* 75: 5631–5635.

Kandel, E., Schwartz, J., and Jessell, T. 1995. *Essentials of Neural Science and Behavior*. Third Edition. Norwalk, CT: Appleton & Lange.

Kandel, E., Schwartz, J., and Jessell, T. 2000. *Essentials of Neural Science and Behavior*. Fourth Edition. NY: McGraw-Hill.

Kant, I. 1947 (1785). "Groundwork of the Metaphysic of Morals." In HJ Paton (trans). *The Moral Law*. London: Hutchinson.

Kant, I. 1790. *Kritik der Urteilskraft*. Berlin.

Kanyas K, Baur MP, Wienker TF, Maier W, Rietschel M, Propping P, Nothen M. 2001. "A genome screen for genes predisposing to bipolar affective disorder detects a new susceptibility locus on 8q." *Hum Mol Genet*. 2001 Dec 1;10(25): 2933–44. [with erratum in *Hum Mol Genet*. 2002 Jul 1;11(14):1685].

Karlsson, J. 1970. "Genetic association of giftedness and creativity with schizophrenia." *Hereditas*. 66: 177–182.

Kato, T. 2005. "Mitochondrial dysfunction in bipolar disorder: from ^{31}P-magnetic resonance spectroscopic findings to their molecular mechanisms." *Int Rev Neurobiol* 2005; 63: 21–40.

Kato, T.,and Kato, N. 2000. "Mitochondrial dysfunction in bipolar disorder." *Bipolar Disorder* 2000; 2: 180–190.

Kebabian, JW; Calne, D. 1997. "Multiple receptors for dopamine." *Nature*. 277: 93-96.

Keck, P: Personal Communication, 1997

Keenleyside, M. 1979. *Diversity and Adaptation in Fish Behaviour*. Berlin: Springer-Verlag.

Kegan, R. 1982. *The Evolving Self*. Cambridge, Mass: Harvard University Press.

Kelsoe, J; Ginns, E; Egeland, J; Gerhard, D; Goldstein, A; Bale, S; Pauls, D; Long, R; Kidd, K; Conte, G; Housman, D; Paul, S. 1989. "Reevaluation of the linkage relationship between chromosome 11p loci and the gene for bipolar affective disorder in the old order Amish." *Nature*. 342: 238 243, 1989.

Kelsoe, J. 1992. "The search for genes for psychiatric illness." *Neuropsychopharmacology*. 6: 215–217.

Kelsoe, J: Personal Communication, 1996

Kemper, T. 1978. *A Social Interaction Theory of Emotions*. New York: Wiley.

Kemper, T. 1990. *Social Structure and Testosterone*. New Brunswick: Rutgers University Press.

Kendell, R. 1975. "The concept of disease and its implications for psychiatry." *Br J Psychiatry*, 127: 305–315.

Kendler, K; Eaves, L. 1986. "Models for joint effects of genotype and environment on liability to psychiatric illness." *Am J Psychiatry*, 143:279–289.

Kendler, K; Heath, AC; Martin, N. 1987. "Symptoms of anxiety and symptoms of depression: Same genes, different environments?" *Arch Gen Psychiatry.* 44: 451–457.

Kendler, K; Neale, M; Kessler, R; Heath, AC; Eaves, L. 1993. "Major depression and generalized anxiety disorder: Same genes, (partly) different environments?" *Arch Gen Psychiatry.* 49: 716–722.

Kendler, K; Pedersen, N; Neale, M; Mathe, A. 1995. "A pilot study of affective illness including hospital- and population-ascertained subsamples: Results of model fitting." *Behavioral Genetics.* 25: 217–232.

Kendler, K. 2002. "Hierarchy and heritability: the role of diagnosis and modeling in Psychiatric Genetics." *American Journal of Psychiatry.* 159: 515–518.

Kendler, K. & Gardner, C. 1997. "The risk for psychiatric disorders in relatives of schizophrenic and control probands: A comparison of three independent studies." *Psychological Medicine.* 27: 411–419.

Kendler, K., Pedersen, N., Johnson, L., Neale, M., Mathe, A. 1993. "A pilot Swedish twin study of affective illness, including hospital and population-ascertained subsamples." *Archives of General Psychiatry.* 50: 699–700.

Kendler, K., Pedersen, N., Neale, M., Mathe, A. 1995. "A pilot Swedish twin study of affective illness, including hospital and population-ascertained subsamples: Results of model fitting." *Behavior Genetics.* 25: 217–232.

Kendrew, J. 1978. "Introduction." In R Duncan; M Smith (eds). *The Encyclopedia of Ignorance.* NY: Pocket Books.

Kendrick, K. 2002. "Electrophysiological effects of androgens." Pp. 511–525 in *Hormones, Brain, and Behavior.* Ed. by Pfaff, D. V. III. NY: Academic Press

Kennedy JL, Farrer LA, Andreasen NC, Mayeux R, St George-Hyslop P. 2003. "The genetics of adult-onset neuropsychiatric disease: complexities and conundra?" *Science.* Oct 31; 302(5646):822–826.

Kenrick, D.T., Dantchik, A., & MacFarlane, S. 1983. "Personality, environment, and criminal behavior: An evolutionary perspective." In: *Personality theory, moral development, and criminal behavior.* (eds). W.S. Laufer & J.M. Day. Lexington Books.

Kerbeshian, J; Burd, L; Klug, M. 1995. "Comorbid Tourette's disorder and bipolar disorder: An etiologic perspective." *Am J Psychiatry.* 152: 1646–1651.

Kernberg, O. 1980. *Internal World and External Reality.* New York: Jason Aronson.

Kettlewell, H. 1956. "Further selection experiments on industrial melanism in the Lepidoptera." *Heredity,* 10:287–302.

Kettlewell, H. 1958. "A survey of frequencies of Biston betularia and its melanic forms in Great Britain." *Heredity.* 12: 51–72.

Kettlewell, H. 1961. "The phenomenon of industrial melanism in Lepidoptera." *Ann Rev Entomol.* 6: 245–262.

Kettlewell, H. 1965. "Insect survival and selection for pattern." *Science*. 148: 1290–1296.

Kidd, K; Matthysse, S. 1978. "Research designs for the study of gene-environment interactions in psychiatric disorders: Report of an FFRP panel." *Arch Gen Psychiatry*. 3: 925.

Kidd, K; Egelund, JA; Molthan, L; Pauls, DL; Kruger, SD; Messner, K. 1984. "Amish study IV: Genetic linkage study of bipolar pedigrees." *Am J Psychiatry*. 41: 1042–1048.

Kidd, J; Egelund, JA; Pastist, A; Castiglione, C; Pletcher, B; Morton, L; Kidd, K. 1987. "Searching for a major genetic locus for affective disorder in the Old Order Amish." *J Psychiatric Research*. 21: 577–588.

Kimura, M. 1955. "Stochastic processes and distribution of gene frequencies in natural selection." *Cold Spring Harbor Symposium Quant Biol*. 22: 33–53.

Kimura, M. 1967. "On the evolutionary adjustment of spontaneous mutation rates." *Genet Res*. 9: 23–34.

Kimura, M. 1968. "Evolutionary rate at the molecular level." *Nature*. 217: 624–626.

Khantzian, E. 1985. "The self-medication hypothesis of addictive disorders." *Am J Psychotherapy*. 28: 59–70.

Kleinman, A. 1982. "Neurasthenia and depression." *Culture, Medicine and Psychiatry*. 6: 117–190.

Kleinman, A. 1985. "Some uses of the social sciences in medicine." In D Fiske; R Shweder (eds). *Metatheory in Social Science*. Chicago: University of Chicago Press.

Kleinman, A. 1987. "Anthropology and psychiatry." *Br J Psychiatry*. 151: 447–454.

Kleinman, A. 1988. *Rethinking Psychiatry*. London: Collin Macmillan.

Kleppner, S. & Tobin, A. 2002. "GABA." *Encyclopedia of the Human Brain*. V. 2, Pp. 353–367 in Ed. by Ramachandran, V. NY: Academic Press.

Kluver, H; Bucy, P. 1937. "'Psychic Blindness' and other symptoms following bilateral temporal lobectomy in rhesus monkeys." *Amer J Physiol*. 119: 352–353.

Knauft, B. 1991. "Violence and Sociality in Human Evolution." *Current Anthropology*, 32: 391–428.

Knauft, B. 1994. "Culture and Cooperation in Human Evolution." Pp. 37–67 in L. Sponsel and T. Gregor, eds. *The Anthropology of Peace and Nonviolence*. Boulder, CO: Lynne Rienner.

Koch, C. 1999. *Biophysics of Computation: Information Processing in Single Neurons*. Oxford University Press.

Koestler, A. 1967. *The Ghost in the Machine*. London: Hutchinson.

Kofoed, L. & MacMillan, J. 1986. "Alcoholism and antisocial personality: The sociobiology of an addiction." *The Journal of Nervous and Mental Disease.* 174(6): 332–335.

Kohlberg, L. 1984. *The Psychology of Moral Development.* Vol. 2. San Francisco: Harper & Row.

Kohler, W. 1925. *The Mentality of Apes.* New York: Vintage.

Kolb, B. 1995. *Brain Plasticity and Behavior.* Mahwah, NJ: Erlbaum.

Kolb, B.; Gibb, R. and Robinson, T. 2003. Brain Plasticity and Behavior." *Current Directions in Psychological Science.* V. 12, N1 (February): 1–5.

Kolb, B.; Forgie, M.; Gibb, R.; Gorny, G. & Rowntree, S. 1998. "Age, Experience, and the Changing Brain." *Neuroscience and Biobehavioral Reviews.* 22: 143–159.

Konner, M. 1983. *The Tangled Wing: Biological Constraints on the Human Spirit.* New York: Harper Colophon Books.

Kosfeld, M., Heinrichs, M., Zak, P., Fischbacher, U., & Fehr, E. 2005. Oxytocin increases trust in humans. *Nature.* V.435/2 (June):673–676.

Kosten, T. 1993. "Pharmacotherapies for cocaine abuse: Neurobiological abnormalities reversed with drug intervention." *Psychiatr Times.* 10: 25.

Kotler, M. 1997. "Excess dopamine D_4 receptor (D4DR) exon III seven repeat allele in opioid-dependent subjects." *Molecular Psychiatry.* 2: 251–254.

Kovacs, M; Pollock, M. 1995. "Bipolar disorder and comorbid conduct disorder in childhood and adolescence." *J Am Acad Child Adol Psychiatry.* 34: 715–723.

Kraepelin, E. 1904. *Lectures on Clinical Psychiatry.* London: Balliere.

Kraepelin, E. 1921. *Manic-Depressive Insanity and Paranoia.* R. Barclay (Trans). Edinburgh: Livingstone.

Krafft-Ebbing, R. 1965 (1880). *Psychopathia Sexualis* (F. Kraf, transl). New York: Bell.

Kramer, D. 1990. "Conceptualizing wisdom: The primacy of affect-cognition relations." In R Steinberg (ed). *Wisdom: Its Nature, Origins and Development.* Cambridge: Cambridge University Press.

Krasnegor, N. and Bridges, R., eds. 1990. *Mammalian Parenting: Biochemical, Neurobiological, and Behavioral Determinants.* Oxford University Press.

Krebs, JR; Davies, N. 1981. *An Introduction to Behavioural Ecology.* Oxford: Blackwell.

Kreitman, M. and Aguadé, M. 1986. "Genetic Uniformity in Two Populations of *Drosophila Melanogaster* as Revealed by Filter Hybridization of Four-Nucleotide-Recognizing-Restriction Enzyme Digests." *Proceedings of the National Academy of Sciences USA* V. 83: 3562–3566.

Kroeber, A. 1952. 'The superorganic.' In AL Kroeber (ed). *Anthropology.* Chicago: University of Chicago Press.

Kropotkin, P 1972 (1902). *Mutual Aid: A Factor in Evolution*. New York: New York University Press.

Kruger, SD; Turner, WS; Kidd, K. 1982. "The effects of requisite assumptions of linkage analysis of manic-depressive illness with HLA." *Biol Psychiatry*. 17: 720–724.

Kruger, S; Cooke, RG; Hasey, GM; Jorna, T; Persad, E. 1995. "Comorbidity of obsessive compulsive disorder in bipolar disorder." *J Affective Disorders*. 34: 117–120.

Kuhn, T. 1970. *The Structure of Scientific Revolutions*. Chicago: University of Chicago Press.

Kukleta, M. 2001. "Psychosocial Stress and Health." *Homeostasis*. V41, N 1–2: 35–40.

Kupfer, D; Pickar, D; Himmelhoch, J; Detre, T. 1975. "Are there two types of unipolar depression?" *Arch Gen Psychiatry*. 32: 866–871.

LaBerge, D. 1995. *Attentional Processing: The Brain's Art of Mindfulness*. Harvard University Press.

Lack, D. 1954. *The Natural Regulation of Animal Numbers*. Oxford University Press.

Lackman, H; Papolos, D. 1995. "A molecular model for bipolar affective disorder." *Med Hypothesies*. 45:3: 255–264.

Laing, R. 1970. *Divided Self*. Penguin Books.

Lalouel, J; Morton, N. 1981. "Complex segregation analysis with pointers." *Hum Hered*. 31:312–321.

Lalouel, J., Rao, D., Morton, N., Elston, R. 1983. "A Unified Model for Complex Segregation Analysis." *American Journal of Human Genetics*. 35: 816–826.

Lamarck, J-B: *Philosophy Zoologique*. Paris, 1809

Lambotte-Legrandes, J; Lambotte-Legrande, C. 1958. "Sicklemie et Malaria." *Annalles Societie de Mediciene Tropique*. 38: 45–54.

Lander, E; Botstein, D. 1986. "Strategies for studying heterogeneous genetic traits in humans using a linkage map of restriction fragment polymorphisms." *Proc Nat Acad Sci USA*. 83: 7353–7357.

Landis, C; Page, J. 1938. *Modern Society and Mental Disease*. New York: Farrar and Rinehart.

Lapeyre, P. and Lledo, P-M. 1994. "Homeostasis." Pp. 517–528 in *The Encyclopedia of Human Behavior*. Ed. by V. Ramachandran. San Diego, CA: Academic Press.

Lathrop, GM; Lalouel, JM; Julier, C; Ott, J. 1984. "Strategies for multilocus linkage analysis in humans." *Proc Nat Acad Sci USA*. 81: 3443–3446.

Laviola, G. & Andersen, S. 2003. "Brain Development, sex differences and stress: implications for psychopathology." *Neuroscience and Biobehavioral Reviews*. 27: 1–2.

Lavori, P; Keller, M; Roth, S. 1984. "Affective disorders and ABO blood groups: New data and a reanalysis of the literature using the logistic transformation of proportions." *J Psychiatric Res.* 18: 119–129.

Lazarus, R. 1966. *Psychological Stress and the Coping Processes.* New York: McGraw Hill.

Lazarus, R. 1982. "Thoughts on the relation between emotion and cognition." *Am Psychologist.* 37: 1019–1024.

Leckman, JF; Gershon, ES; McGinniss, MH; Caruso, KA; Merikangas, KR; Pauls, DL; Kidd, K. 1979. "Subtypes of depression: Family study perspectives." *Arch Gen Psychiatry.* 41: 833–838.

Lerner, I. 1954. *Genetic Homeostasis.* New York: Wiley.

Lesch, K.; Greenberg, B.; Higley, J.; Bennett, A.; Murphy, D. 2002. "Serotonin Transporter, Personality, and Behavior. Pp. 109–135 in *Molecular Genetics and the Human Personality.* Ed. by J. Benjamin, R. Ebstein, & R. Belmaker. Washington, DC: American Psychiatric Publishing.

Levine, D. 1986. "A Neural Network Theory of Frontal Lobe Function." Pp, 716–727 in *Proceedings of the Eighth Annual Conference of the Cognitive Science Society.* Hillsdale, NJ: Erlbaum Associates.

Levine, D. and Jani, N. 2002. "Toward a Neural Network Theory of the Triune Brain. In Cory, G. and Gardner, R., eds. *The Evolutionary Neuroethology of Paul MacLean: Convergences and Frontiers.* Westport, CT: Praeger.

Levi-Strauss, C. 1944. "The social and psychological aspects of chieftainship in a primitive tribe: The Nambikuara of northwestern Mato Grosso." *Transactions of the New York Academy of Sciences.* 7: 16–32.

Levitan, M; Montagu, A. 1971. *Textbook of Human Genetics.* New York: Oxford University Press.

Lewis, A. 1933. "Inheritance of mental disorders." *Eugenics Reviews.* 25: 79–84.

Lewis, A. 1934. "Melancholia: A clinical survey of depressive states." *J Ment Sci.* 80: 277–378.

Lewis, A. 1936. "Prognosis in the manic-depressive psychosis." *Lancet.* 2: 997–999.

Libet, JM; Lewinshon, P. 1973. "Concept of social skill with special reference to the behavior of depressed persons." *J Consult Clin Psychology.* 40: 304–312.

Lichter, J.; Barr, C.; Kennedy, J.; Van Tol, H.; Kidd, K.; Livak, K. 1993. "A hypervariable segment in the human dopamine receptor D4 (DRD4) Gene." *Human Molecular Genetics.* 2: 767–773.

Lieberman, P. 2000. *Human Language and Our Reptilian Brain.* Harvard University Press.

Lilionfeld, SO; Marino, L. 1995. "Mental disorder as a Roschian concept: A critique of Wakefield's 'harmful dysfunction' analysis." *Journal of Abnormal Psychology.* 104: 411–420.

Lin, P.; McInnis, M.; Potash, J.; et al. 2005. "Assessment of the effect of age at onset on linkage to bipolar disorder: Evidence on chromosomes 18p and 21q." *Am. J. of Hum. Genet.* 77: 545–555.

Linnaeus, C. 1781. *Politica Naturae.* (F. Brand, transl) London.

Lishman, W. 1978. *Organic Psychiatry.* Oxford: Blackwell.

Livingstone, F. 1958. "Anthropological implications of sickle cell gene distribution in West Africa." *American Anthropologist.* 60: 533–562.

Livingstone, F. 1964. "The distribution of the abnormal hemoglobin genes and their significance for human evolution." *Evolution.* 18: 685–699.

Livingstone, F. 1967. *Abnormal Hemoglobins in Human Populations.* Chicago: Aldine.

Livingstone, F. 1969. "An analysis of the ABO blood group clines in Europe." *American Journal of Physical Anthropology.* 31: 1–10.

Lloyd Morgan, C. 1909. "Mental factors in evolution." In A. Seward (ed) *Darwin and Modern Science.* Cambridge University Press.

Locke, J. 1975 (1690). *An Essay Concerning Human Understanding.* Oxford University Press.

Loeber, R. 1982. "The stability of antisocial and delinquent child behavior: A review." *Child Development.* 53: 1431–1446.

Loehlin, J. 1992. *Genes and Environment in Personality Development.* Newbury Park, CA: Sage.

Loewenstein, W. 1999. *The Touchstone of Life.* Oxford University Press.

Lorenz, K. 1963. *On Aggression.* London: Methuen.

Lorenz, K. 1965. *Evolution and Modifications of Behavior.* University of Chicago Press.

Lorenz, K. 1981. *The Foundations of Ethology.* New York: Touchstone.

Lorenz. K. 1950. "The comparative method in studying innate behaviour patterns." *Sym. Soc. Exp. Biology* 4: 221–268.

Loye, D. 2002a. *Darwin's Lost Theory.* NY: Seven Stories Press.

Loye, D. 2002b. "The Moral Brain." *Brain and Mind.* V.3. N.2: 133–150.

Lucotte, G; Landoulsi, A; Berriche, S. 1992. "Manic depressive illness linked to factor IX in a French pedigree." *Ann Gen.* 35: 93–95.

Luddens, H., Korpi, E., Seeburg, P. 1995. "GABA/benzodiazepine receptor heterogeneity: Neurophysiological Implications. *Neuropharmacology.* 34: 245–254.

Ludwig, A. 1992. "Creative achievenment and psychopathology: Comparisons among professions." *Am J Psychiatry.* 46: 330–356.

Lund, R. 1978. *Development and Plasticity of the Brain.* Oxford University Press.

Lundblad, N: Personal Communication, 1995.

Lundblad, S: Personal Communication, 1997.

Luria, A. 1976. R. *Cognitive Development: Its Cultural and Social Foundations.* M Lopez- Morillas; L Solotaroff (trans). Harvard University Press.

Luxenberger, H. 1930. "Psychiatrisch-neurologischhe zwillingspathologie." *Zentralblatt fur die gesamte Neurologie und Psychiatrie*. 56: 145–181.

Lykken, DT; Tellegen, A; Ianoco, W . 1982. "EEG spectra in twins: Evidence for a neglected mechanism of genetic determination." *Physiological Medicine*. 10: 759–776.

Lykken, DT; Bouchard, TJ; McGue, M; Tellegen, D. 1991. "The Minnesota Twin Family Registry: Some initial findings." *Acta Genet Med Gemellol*. 39: 35–70.

MacArthur, R. 1965. "Patterns of species diversity." *Biological Review*. 40: 510–533.

Machiavelli, N. 1979 (1531). "The Discourses." In P Bondanella; M Musa (eds and transl). *The Portable Machiavelli*. NY: Penguin.

MacLean, P. 1949. "Psychosomatic disease and the 'visceral brain': Recent developments bearing on the Papez theory of emotion." *Psychosom Med*. 11: 338–353.

MacLean, P. 1952. "Some psychiatric implications of physiological studies on the frontotemporal portion of the limbic system (visceral brain)." *Electroencephalography, Clinical Neurophysiology*. 4: 407–418.

MacLean, P. 1974. "Discussion." *Res Publ Assoc Res Nerv Ment Dis*. 52: 148.

MacLean, P. 1985. "Evolutionary psychiatry and the triune brain." *Psychological Medicine*. 15: 219–221.

MacLean, P. 1987. "The midline frontolimbic cortex and the evolution of crying and laughter." In E Perecman (ed). *The Frontal Lobes Revisited*. New York: IRBN Press.

MacLean, P. 1993. "Human Nature: Duality or Triality." Pp. 107–112 in *Politics and the Life Sciences*. V. 12. N.2.

MacLean, P. 1992. "Obtaining Knowledge of the Subjective Brain ('Epistemics')." p. 57–70 in *So Human a Brain*. Ed. by A. Harrington. Boston: Birkhauser.

MacLean, P. 1990. *The Triune Brain in Evolution: Role in Paleocerebral Functions*. NY: Plenum.

MacMillan, J. & Kofoed, L. 1984. "Sociobiology and antisocial personality: An alternative perspective." *Journal of Nervous and Mental Disease*. 172(12): 701–706.

Maier, W., Lichtermann, D., Minges, J., Hallmayer, J., Heun, R., Benkert, O., Levinson, D. 1993. "Continuity and discontinuity of affective disorders and schizophrenia. Results of a controlled family study. *Archives of General Psychiatry*. 50: 871–883.

Maier, W; Lichterman, D; Minges, J; Delmo, C; Huen, R. 1995. "The relationship between bipolar disorder and alcoholism: A controlled family study." *Psychological Medicine*. 25: 787–796.

Mallet, J; Meloni, R; Laurent, P. 1994. "Catecholamine metabolism and psychiatric or behavioral disorders." *Curr Opin Genet Dev*. 4: 419–426.

Malthus, T. 1798. *An Essay on the Principle of Population, as It Affects the Future Improvement of Society*. London: J Johnson.

Mandler, G. 1975. *Mind and Emotion*. New York: Wiley.

Marin, O.; Smeets, W.; & Gonzalez, A. 1998. "Evolution of the basal ganglia in tetrapods: A new perspective based on recent studies in amphibians." *Trends in Neuroscience*. 21(11): 487–494.

Marks, I. 1987. *Fears, Phobias and Rituals: Panic, Anxiety and their Disorders*. Oxford University Press.

Marks, I; Nesse, R. 1994. "Fear and fitness: An evolutionary analysis of anxiety disorders." *Ethology and Sociobiology*. 15: 247–261.

Martindale C. 1972. "Father's absence, psychopathology, and poetic eminence." *Psychol Rep*. 31: 843–847.

Marzuk, P; Barchas, J. 1995. "Psychiatry." *JAMA*. 273: 1715–1716.

Mascie-Taylor, CGN; Boldsen, J. 1988. "Assortative mating, differential fertility and abnormal pregnancy." *Ann Hum Biology*. 15: 223–228.

Maslow, A. 1936. "The role of dominance in the social and sexual behavior of infra-human primates, I. Observations at Vilas Park Zoo." *J Genetical Psychology*. 48: 261–277.

Maslow, A. 1937. "Dominance feeling, behavior and status." *Psychological Review*. 44: 404–429.

Maslow, A. 1971. *The Further Researches of Human Nature*. NY: Viking.

Mason, C. 1956. "Pre-illness intelligence of mental hospital patients." *J Consult Psychology*. 20: 297–300.

Master, A. 1967. "The distribution of ABO blood groups in psychiatric illness." *Brit J Psychiatry*. 113: 1309.

Masters, R. 1975. "Politics as a Biological Phenomenon." *Social Sci Info*. 14: 7–63.

Massat, I.; Souery, D.; Mendlewicz, J.; & Papadimitriou, G. 2000. "The GABAergic Hypothesis of Mood Disorders." Pp. 143–160 in in *Bipolar Disorder: Basic Mechanisms and Therapeutic Implications*. Ed. by Soares, J. & Gershon, S. NY: Marcel Dekker.

Mathews C., Reus V. 2003. "Genetic linkage in bipolar disorder." *CNS Spectr*. Dec: 8 (12): 891–904.

Matthysse, S; Kidd, K. 1981. "Estimating the genetic contribution to schizophrenia." *Am J Psychiatry*/ 133: 185–191.

Maudsley, H. 1886. *Natural Causes and Supernatural Seemings*. London: Keegan, Paul.

de Maupertius, PLM. 1746. *Venus Physique*. Paris: Haye et Husson.

Mayfield, D; Coleman, L. 1968. "Alcohol use and affective disorder." *Dis Nerv Syst*. 29: 467–474.

Maynard Smith, J. 1972. "Game theory and the evolution of fighting." In J Maynard Smith (ed) *On Evolution*. Edinburgh: Edinburgh University Press.

Maynard Smith, J. 1974. "The theory of games and the evolution of animal conflict." *J Theoretical Biol*. 47: 209–221.

Maynard Smith, J. 1976. "Evolution and the theory of games." *American Scientist*. 64: 41–45.

Maynard Smith, J. 1982. *Evolution and the Theory of Games*. Cambridge University Press.

Maynard Smith, J; Price, G. 1973. "The logic of animal conflict." *Nature*, 246: 15–18.

Maynard Smith, J; Parker, G. 1976. "The logic of assymetric contests." *Nature*. 246: 15–18.

Maynard Smith, J. 1982. "The Evolution of Social Behavior - a Classification of Models." Pp. 28–44 in *Current Problems in Sociobiology*. Ed by. King's College Sociobiology Group. Cambridge University Press.

Maynard Smith, J. 2002. "Equations of Life." Pp. 193–211 in *It Must be Beautiful: Great Equations of Modern Science*. Ed. by G. Farmelo. London: Granta Books.

Mayr, E. 1950. "Taxonomic categories in fossil hominids." *Cold Spring Harbor Symposium in Quantitative Biology*,. 15: 109–118.

Mayr, E. 1961. "Cause and effect in biology." *Science*. 134:1501–1506.

Mayr, E. 1963. *Populations, Species and Evolution*. Belknap/Harvard University Press.

Mayr, E. 1972. "Sexual selection and natural selection." In D Campbell (ed) *Sexual Selection and the Descent of Man 1871-1971*.

Mayr, E. 1977. "Behavioral programs and evolutionary strategies." *American Scientist*. 62: 650–659.

Mayr, E. 1988. *Toward a New Philosophy of Biology*. Cambridge, Mass: Belknap/Harvard University Press.

Mayr, E. 1992. *One Long Argument*. London: Penguin.

Mayr, E. 1997. *This is Biology*. Cambridge, Mass: Belknap/Harvard University Press.

Mazure, A; Booth, A; Dabbs, J. 1992. "Testosterone and chess competition." *Social Psychology Quarterly*. 55: 70–77.

McAllister, A. 1999. "Neurotrophins and Synaptic Plasticity." *Ann Rev of Neuroscience*. V. 22: 295–318.

McClelland, D. 1985. *Human Motivation*. Dallas: Scott Foresman.

McCord, J. 1983. "Personality, moral development, and criminal behavior." In: *Personality theory, moral development, and criminal behavior*. (eds). W. Laufer & J. Day. Lexington Books.

McElroy, S; Keck, P; Hudson, J; Pope, H. 1995. "Disorders of impulse control." In E Hollander; D Stein (eds) *Impulsivity and Aggression*. Chichester: Wiley.

McEwen, B. 2003. "Protective and Damaging Effects of Stress Mediators." Pp. 1127–1140 in *Foundations in Social Neuroscience*. Ed. by J. Cappacio, G. Berntson. The MIT Press.

McEwen, B. & Seeman, T. 2003. "Stress and Affect: Applicability of the Concepts of Allostasis and Allostatic Load." Pp. 1117–1138 in *Handbook of Affective Sciences*. Ed. by R. Davidson, K. Scherer, & H. Goldsmith. Oxford University Press.

McGuffin, P; Katz, R. 1989. "The genetics of depression and manic-depressive disorder." *Br J Psychiatry*. 155: 294–304.

McGuffin, P; Katz, R; Rutherford, J. 1991. "Nature, nurture and depression: A twin study." *Psychological Medicine*. 21: 329–335.

McGuffin, P; Katz, R; Rutherford, J. 1993. "Twin studies as vital indicators of phenotypes in molecular research." In T. Bouchard; P Propping (eds). *Twins as a Tool of Behavioral Genetics*. NY: Wiley.

McGuire, M; Raleigh, M. 1985. "Serotonin-behaviour interactions in Vervet monkeys." *Psychopharmacology Bull*. 21: 458–463.

McGuire, M; Raleigh, M; Johnson, C. 1983. "Social dominance in adult male vervet monkeys: General considerations." *Social Science Information*. 22: 89–123.

McGuire, M; Marks, I; Nesse, R; Troisi, A. 1992. "Evolutionary biology: A basic science for psychiatry?" *Acta Psychiatr Scand*. 86: 89–96.

McGuire, M. and Troisi, A. 1998. *Darwinian Psychiatry*. Oxford University Press.

McHugh, P; Slavney, P. 1986. *The Perspectives of Psychiatry*. Baltimore: The Johns Hopkins University Press.

McInnes, M; McMahon, F; Chase, G. 1993. "Anticipation in bipolar affective disorder." *Am J Hum Genetics*. 53: 385–390.

McKusick, M. 1990. *Mendelian Inheritance in Man: Catologs of Autosomal Dominant, Autosomal Recessive and X linked Phenotypes*. 9th ed., Baltimore, The Johns Hopkins University Press, p.258.

McNeil, T. 1971. "Prebirth and postbirth influence on the relationship between creative ability and recorded mental illness." *J Pers*. 39: 391–406.

McQueen, M.; Devlin, B.; Faraone, S.; et al. 2005. "Combined analysis from eleven linkage studies of bipolar disorder provides strong evidence of susceptibility loci on chromosomes 6q and 8q." *Am. J. of Hum. Genet*. 77: 582–595.

Mealey L. 1995. "The sociobiology of sociopathy: An integrated evolutionary model." *Behav. Brain Sci*. 18: 523–541.

Mednick, S., Kirkegaard-Sorense, L., Hutchings, B., et al. 1977. "An example of biosocial interaction research: The interplay of socioenvironmental and individual factors in the etiology of criminal behavior." In: *Biosocial bases of criminal behavior*. (eds). S.. Mednick & K.. Christiansen. Gardner Press.

Mednick, S.A., Gabrielli, W.F., Jr. & Hutchings, B. 1987. "Genetic factors in the etiology of criminal behavior." In: *The causes of crime: New biological approaches.* (eds). S.A. Mednick, Terrie E. Moffitt, & S.A. Stack. Cambridge University Press.

Meloni, R, Leboyer, M; Bellivier, F. 1995. "Association of manic-depressive illness with tyrosine hydroxylase microsatellite marker." *Lancet.* 345: 932.

Meltzer, H. 1992. "The importance of serotonin-dopamine interactions in the action of clozapine." *Br J Psychiatry.* 160(suppl 170): 22–29.

Mendel, G. 1866. *Versuche uber Phlanzenhybriden.* Verhandlungen des Naturforschended Vereines in Brunn, 4: 3–47, 1866

Mendelwicz, J; Fleiss, J; Fieve, R. 1972. "Evidence for X-linkage in the transmission of manic-depressive illness." *JAMA.* 222: 1624–1627.

Mendelwicz, J; Rainer, J. 1974. "Morbidity risk and genetic transmission in manic-depressive illness." *Am J Hum Gen.* 26: 692–701.

Mendelwicz, J; Rainer, J. 1977. "Adoption study supporting genetic transmission in manic-depressive illness." *Nature.* 268: 327–329.

Mendelwicz, J; Linkowski, P; Guroff, JJ; Van Praag, H. 1979. "Color blindness linkage to bipolar manic-depressive illness: New evidence." *Arch Gen Psychiatry.* 36: 1442–1447.

Mendelwicz, J; Linkowski, P; Wilmotte, J. 1980. "Linkage between glucose-6-phosphate dehydrogenase deficiency and manic-depressive psychosis." *Br J Psychiatry.*

Mendelwicz, J; Simon, P; Sevy, S; Charon, F; Brocas, H, Legros, S; Vassart, G. 1987. "Polymorphic DNA marker on X chromosome and manic-depression." *Lancet.* I: 1230–1232.

Mendlewicz, J; Souery, D. 1995. "Heredite et psychose maniaco-depressive." *Bulletin Academie Nationale Mediciene.* 179: 4:755–766.

Mendelwicz, J. & Rainer, J. 1977. "Adoption study supporting genetic transmission in manic-depressive Illness." *Nature.* 268: 326–329.

Mendelwicz, J., Linkowski, P., Guroff, J., Van Praag, H. 1979. "Color Blindness linkage to bipolar manic-depressive illness. New evidence." *Archives of General Psychiatry.* 36: 1442–1447.

Mendelwicz, J., Linkowski, P., Wilmotte, J. 1980. "Linkage between glucose-6-phosphate dehydrogenase deficiency and manic-depressive Psychosis." *British Journal of Psychiatry.* 137: 337–342.

Mendelwicz, J., Simon, P., Sevy, S., Charon, F., Brocas, H., et al. 1987. "Polymorphic DNA marker on X chromosome and manic depression." *Lancet.* 1: 1230–1232.

Middleton F, Pato M, Gentile K, Morley C, Zhao X, Eisener A, Brown A, Petryshen T, Kirby A, Medeiros H, Carvalho C, Macedo A, Dourado A, Coelho I, Valente J, Soares M, Ferreira CP, Lei M, Azevedo MH, Kennedy JL, Daly MJ, Sklar P, Pato C. 2004. "Genomewide linkage analysis of bipolar disorder by use of a high-density single-nucleotide-polymorphism (SNP) genotyping assay: a comparison with microsatellite marker assays and finding of significant linkage to chromosome 6q22." *Am J Hum Genet.* 2004 May;74(5): 886–897. Epub 2004 Apr 01.

Mill, JS. 1843. *A System of Logic.* London.

Miller, B. & Cummings, J. (eds). 1999. *The Human Frontal Lobes.* NY: The Guilford Press.

Miller, WR. Psychological deficit in depression. *Psychol Bull,* 82: 238–260, 1975

Mineka, S; Suomi, S. 1978. "Social separation in monkeys." *Psychological Bull.* 85: 1376-1400.

Mink, J. 1996. "The basal ganglia: Focused selection and inhibition of competing motor programs." *Progress in Neurobiology.* 50(4): 381–425.

Mithen, S. 1996. *Prehistory of the Mind.* London: Thames and Hudson.

Mitterauer, B. 2000. "Clock genes, feedback loops, and their possible role in the etiology of bipolar disorders: An integrative model." *Medical Hypotheses.* 55: 155–159.

Moffitt, T. 1987. "Parental mental disorder and offspring criminal behavior: An adoption study." *Psychiatry.* 50: 346–360.

Moldin, S; Reich, T; Rice, J 1991. "Current perspectives on the genetics of unipolar depression." *Behavioral Genetics.* 21: 211–242.

Moldin, S; Gottesman, I. 1995. "Population Genetics in Psychiatry." In H Kaplan and B Sadock (eds) *Comprehensive Textbook of Psychiatry.* 6th ed. Baltimore: Williams and Wilkins.

Mollica, R 1988. "Mood disorders: Epidemiology." In H Kaplan; B Sadock (eds) *Comprehensive Textbook of Psychiatry.* 5th ed. Baltimore: William and Wilkins.

Monat, R. 1992. "Werner syndrome: Molecular genetics and mechanistic hypotheses." *Exp. Gerontol.* 27 (4) : 447–453.

Monnat, R; Chiaverotti, T; Hackmann, A; Maresh, G. 1992. "Molecular structure and genetic stability of human hypoxanthine phosphoribosyltransferase (HPRT) gene duplications." *Genomics.* 13: 788–796.

Morgan, T. 1910. "Chance or purpose in the origin and evolution of adaptation." *Science.* 21: 201–210.

Morgane, P., Galler, J. and Mokler, D. 2005. "A review of systems and networks of the limbic forebrain/limbic midbrain." *Progress in Neurobiology.* V. 75: 143–160.

Morissette, J., Villeneuve, A., Bordeleau, L., et al. 1999. "Genome-wide search for linkage of bipolar affective disorders in a very large pedigree derived from a homogenous population in Quebec points to a locus of major effect on chromosome 12q23-q24." *American Journal of Medical Genetics (Neuropsychiatric Genetics).* 88: 567–587.

Morris, D. 1977. *Manwatching.* London, Triad-Granada.

Morrison, J. 1974. "Bipolar affective disorder and alcoholism." *Am J Psychiatry* 131: 1130–1133.

Morrison, J. 1975. "The family histories of manic-depressive patients with and without alcoholism." *J Nerv Ment Dis.* 160: 227–229.

Morton, N. 1992. *Outline of Genetic Epidemiology*. NY: Karger.

Morton, N. 1990. "Genetic linkage and complex diseases: A comment." *Genetic Epidemiology*. 7: 33–34.

Morton, N. and MacLean, C. 1974. "Analysis of family resemblance. III Complex segregation of quantitative traits." *American Journal of Human Genetics*. 26: 489–503.

Müller, M.; Keck, M.; Stechler, T.; Hoslboer, F. 2002. "Genetics of endocrine-behavior interactions." Pp. 263–301 in *Hormones, Brain, and Behavior*. Ed. by Pfaff, D. V. 5. NY: Academic Press.

Murdock, G. 1934. *Our Primitive Contemporaries*. Toronto: Macmillan.

Murdock, G. 1972. "Anthropology's mythology." *Proceedings of the Royal Anthropological Institute of Great Britain and Ireland for 1971*:17–24.

Murdoch, G. 1967. *Ethnographic Atlas*. Pittsburg: University of Pittsburg Press.

Murphy, G. 1923. "Types of word association in dementia praecox, manic-depressives and normal persons." *Am J Psychiatry*. 79: 539–571.

Murphy, D; Beigel, A; Weingartner, H; Bunney, W. 1974. "The quantitation of manic behavior." *Mod Probl Pharmacopsychiatry*. 131: 1389–1391.

Murray, C.; Lopez, A. 1996. *The Global Burden of Disease*. World Health Organization, Geneva, Switzerland.

Mury, M; Verdoux, H; Bourgeois, M. 1995. "Comorbidity of bipolar and eating disorders: Epidemiologic and therapeutic aspects." *Encephale*. 21: 545–553.

Myerson, A; Boyle, RD. 1941. "The incidence of manic-depressive psychosis in certain socially important families: Preliminary report." *Am J Psychiatry*. 98: 11–21, p. 20.

Nakanishi, S., Axel, R., Schneider, N. 1992. "Molecular diversity of glutamate receptors and implications for brain function." *Science*. Oct. 23; 258 (5082): 597–603.

Neel, JV. 1951. "The population genetics of two inherited blood dyscrasias in man." *Cold Spring Harbor Symposium on Quantitative Biology*. 13:141–158.

Neel, JV. 1956. "The genetics of human hemoglobin differences." *Annals Human Genetics*. 21: 1–30.

Neel, JV; Schull, WJ. 1954. "The study of natural selection in primitive and civilized human populations." *Human Biology* 30: 43–72.

Nemeroff, C; Schatzberg, AF (eds). 1995. *Psychopharmacology*. Washington: American Psychiatric Press.

Nesse, R. 1991. "What good is feeling bad?" *The Sciences*. 31: 30–37.

Nesse, R. 1990. "Evolutionary explanation of the emotions." *Human Nature*. 1: 261–289.

Nesse, R. & Young, E. 2000. "Evolutionary Origins and Functions of the Stress Response." Pp. 79–84 in *Encyclopedia of Stress*. V.2. Ed. by G. Fink. San Diego, CA: Academic Press.

Nelson, R. 2000. *An Introduction to Behavioral Endocrinology*. Second Edition. Sunderland, MA: Sinauer.

New, A.; Gelernter, J.; & Mitropoulou, V. 1999. "Serotonin Related Genotype and Impulsive Aggression." *Biological Psychiatry*. 45: 120S.

Newton, I. 1975 (1687). *Mathematical Principles of Natural Philosophy*. New York: Dover Books.

Nietzsche, F. 1989 (1887). *On the Geneology of Morals*. E Kaufmann (Ed and trans). New York, Vintage Books.

Ninio, J. 1991. "Transient mutators: A semiquantitative analysis of the influence of translation and transcription errors on mutation rates." *Genetics*. 129: 957–962.

Niznik, NB; Van Tol, HH. 1992. "Dopamine receptor genes: new tools for molecular psychiatry." *J Psychiatr Neurosci*. 17: 158–180.

Nothen, M., Cichon, S., Rohleder, H., Hemmer, S., Franzek, E., et al. 1999. "Evaluation of Linkage of BPD to Chromosome 18 in a Sample of 57 German Families." *Molecular Psychiatry*. 4: 76–84.

NIMH Genetics Initiative Bipolar Group. 1997. "Genomic survey of bipolar illness in the NIMH genetics initiative pedigrees: A preliminary report." *American Journal of Medical Genetics*. 74: 227–237.

Numan, M. 1994. ""Maternal Behavior." Pp. 221–302 in *Physiology of Reproduction*. 2nd ed Vol. 2. Ed. by Knobil, E. and Neill, J. NY: Raven Press.

Numan, M. and Sheehan, T. 1997. "Neuroanatomical Circuitry for Mammalian Maternal Behavior." Pp. 101–125 in *The Integrative Neurobiology of Affiliation*. Ed. by C. S. Carter, I. Lederhendler, and B. Kirkpatrick. NY: New York Academy of Sciences, Vol. 807.

Numan, M. & Insel, T. 2003. *The Neurobiology of Parental Care*. NY: Springer-Verlag.

Nurnberger, J; Guroff, J; Hamovit, J; Berrettini, W; Gershon, ES: A family study of rapid-cycling bipolar illness. J Aff Dis, 15:87–91, 1988

Odegaard, O. 1952. "The excess mortality of the insane." *Acta Psychiatr Scand*. 27: 353–367.

O'Dowd, B. 1993. "Structure of dopamine receptors." *J Neurochem*. 60: 810.

Oesterreicher, W. 1951. "Peculiarities of Indonesian psychiatry." *Folia Psychiatr Neurol Neurochir Neerlandica*. 54: 431–437.

Ohman, A. 1986. "Face the beast and fear the face: Animal and social fears as prototypes for evolutionary analyses of emotion." *Psychophysiology*. 23: 123–144.

Ohmark, D; Strayer, FF; Freedman, DG. 1980. *Dominance Relations: An Ethological View of Human Conflict and Social Interaction*. NY: Garland.

Onigini, E; Longo, V. 1989. "Dopamine receptor subtypes and arousal." *Int Rev Neurobiol.* 31: 239–255.

Oquendo, M. & Mann, J. 2000. "Serotonergic Dysfunction in Mood Disorders." Pp. 121–142 in *Bipolar Disorder: Basic Mechanisms and Therapeutic Implications.* Ed. by Soares, J. & Gershon, S. NY: Marcel Dekker.

O'Rourke, D; McGuffin, P; Reich T. 1983. "Genetic analysis of manic-depressive illness." *Am J Phys Anthropol.* 62: 51–59.

Ornstein, R. 1986. *Multimind: A New Way of Looking at Human Beings.* New York: Macmillan.

Owen, MJ. 1992. "Will schizophrenia become a graveyard for molecular geneticists?" *Psychological Medicine.* 22: 289–293.

Owen, MJ; Risch, N. 1995. "Atypical Antipsychotics." In C Nemeroff; A Schatzberg (eds) *Psychopharmacology.* Washington, DC: American Psychiatric Press.

Owen, M., McGuffin, P., Gottesman, I. 2002. "The Future and Post-Genomic Psychiatry." Pp. 445–460 in *Psychiatric Genetics and Genomics.* Ed. By P. McGuffin, M. Owen, & I. Gottesman. Oxford University Press.

Packard, M. and Knowlton, B. 2002. "Learning and Memory Functions of the Basal Ganglia." Pp. 563–593 in *Annual Review of Neuroscience.* 25.

Paglia, C. 1990. *Sexual Personae.* Yale University Press.

Paglia, C. 1994. *Vamps and Tramps.* NY: Vintage.

Palomino, H; Cerda-Flores, R; Blanco, R; Palomino, HM; Barton, SA; de Andrade, M; Chakraborty, R. 1997. "Nonsyndromic cleft lip with or without cleft palate." *J Craniofac Genet Dev Biol.* 17: 57–61.

Panksepp, Jaak. 1998. *Affective Neuroscience.* Oxford University Press.

Panksepp, Jaak. 2002. "Foreword." In *The Evolutionary Neuroethology of Paul MacLean: Convergences and Frontiers.* Ed. by G. Cory and R. Gardner. Westport, CT: Praeger.

Panksepp, Jaak (ed.). 2004. *Textbook of Biological Psychiatry.* Hoboken, NJ: Wiley-Liss.

Panksepp, Jaak and Panksepp, Jules. B. 2000. "The Seven Sins of Evolutionary Psychology." *Evolution and Cognition,* 6, 108–131.

Panksepp, Jaak and Burgdorf, Jeff. 2003. "'Laughing' Rats and the Evolutionary Antecedents of Joy." *Physiology and Behavior.* 79: 533–547.

Papez, J: 1937. "A proposed mechanism of emotions." *Archives Neurology and Psychiatry,* 79: 217–224.

Paradiso, S.; Johnson, D.; Andreasen, N.; O'Leary, D. 1999. "Cerebral blood flow changes associated with attribution of emotional valence to pleasant, unpleasant, and neutral visual stimuli in a PET study of normal subjects." *American Journal of Psychiatry.* 156(10): 1618–1629.

Parker, J; Theillie, A; Speilberger, D. 1961. "Frequency of blood types in an homogeneous group of manic-depressive patients." *J Ment Sci.* 107: 936–942.

Parker, G. 1974. "Assessment strategy and the evolution of fighting behavior." *J Theoretical Biol.* 47: 223–243.

Parker, G. 1984. "Evolutionarily stable strategies." In J. Krebs; N. Davies (eds) *Behavioural Ecology: An Evolutionary Approach.* 2nd edition. Oxford: Blackwell.

Parsons, T. 1966. *Societies: Evolutionary and Comparative Perspectives.* New York: Prentice Hall.

Pato C., Pato MT, Kirby A, Petryshen TL, Medeiros H, Carvalho C, Macedo A, Dourado A, Coelho I, Valente J, Soares MJ, Ferreira CP, Lei M, Verner A, Hudson TJ, Morley CP, Kennedy JL, Azevedo MH, Daly MJ, Sklar P. 2004. "Genome-wide scan in Portuguese Island families implicates multiple loci in bipolar disorder: fine mapping adds support on chromosomes 6 and 11." *Am J Med Genet.* 2004 May 15;127B(1): 30–34.

Patterson, S. and Aebersold, R. 2003. "Proteomics: the First Decade and Beyond." *Nature Genetics Supplement.* V23: 311–323.

Patterson, G.R., DeBaryshe, B.D. & Ramsey, E. 1989. "A developmental perspective on antisocial behavior." *American Psychologist.* 44(2): 329–335.

Pattison, E. & Kahan, J. 1983. "The deliberate self-harm syndrome." *American Journal of Psychiatry.* 140: 867–872.

Pauls, D., Morton, L., Egeland, J. 1992. "Risks of affective illness among first degree relatives of bipolar I old-order Amish probands." *Archives of General Psychiatry.* 49: 703–708.

Pauls, D. 1987. "Editorial." *Am Journal of Medical Genetics.* 60: 783–787.

Pauls, D. 1993. Behavioural disorders: Lessons in linkage. *Nature Genetics.* 3: 4–5.

Pavlov, I. 1929. *Lectures on Conditioned Reflexes.* (W Gantt, transl). London.

Paykel, E. 1978. "Contribution of life events to the causation of psychiatric illness." *Psychological Medicine.* 8: 245–253.

Paykel, E. 1982. "Life events and early environment." In E. Paykel (ed) *Handbook of Affective Disorders.* New York: Churchill Livingstone.

Paykel, E; Weissmann, M. 1973. "Social adjustment and depression: A longitudinal study." *Arch Gen Psychiatry.* 28: 659–663.

Pearson, K. 1904. "On the laws of inheritance in man, II." *Biometrica.* 3: 131:190.

Pebusque, M-J.; Coulier, F.; Birnbaum, D.; and Pontarotti, P. 1998. "Ancient large-scale genome duplications: Phylogenetic and linkage analyses shed light on chordate genome evolution." *Mol. Biol. Evol.* 15(9): 1145–1159.

Peirce, C. [1899] 1931. *Collected Papers of Charles Sanders Peirce.* Ed. by Hartshorne, C. and Weiss, P. Harvard University Press.

Pekkarinen, P., Terwilliger, J., Bredbacka, P., Lonkvist, J., Peltonen. L. 1995. "Evidence of a predisposing locus to BPD on Xq24-27.1 in an extended Finnish fpdigree." *Genome Research.* 5: 105–115.

Penfield, W. 1938. "The cerebral cortex and conciousness." *Arch Neurol Psychiatry*. 40: 417–442.

Pershonok, D: Personal Communication, 1993

Peroutka S. and Howell T. 1994. "The molecular evolution of G protein-coupled receptors: focus on 5-hydroxytryptamine receptors." *Neuropharmacology* Mar-Apr;33 (3-4): 319–24.

Petronis, A. 2003. "Epigenetics and Bipolar Disorder: New opportunities and challenges." *American Journal of Medical Genetics*. Part C. 123C: 65–75.

Petronis, A; Kennedy, J. 1995. "Unstable genes--unstable minds?" *Am J Psychiatry*.152: 164–172.

Piaget, J. [1932] 1965. *The Moral Judgement of the Child*. NY: Free Press.

Piaget, J. 1971. *Biology and Knowledge: An Essay on the Relations between Organ Regulations and Cognitive Processes*. Edinburgh University Press.

Pianka, E. 1970. "On *r*- and *K*- selection." *American Naturalist*. 104: 592-597.

Pichot, P. 1994. "Nosological Models in Psychiatry." *British Journal of Psychiatry*. 164: 232–240.

Pinker, S. 1997. *How the Mind Works*. NY: Norton.

Plato. In E. Hamilton; H. Cairns (eds). 1963. *Plato: The Collected Dialogues including Letters*. Princeton: Bollingen/Princeton University Press.

Plomin, R. 1988. "The nature and nurture of cognitive abilities." In R. Sternberg (ed). *Advances in the Psychology of Human Intelligence*. London: Lawrence Erlbaum.

Plomin, R. 1990. "The role of inheritance in behavior." *Science*. 248: 183–188.

Plomin, R; McLearn, G; Owen, M; McGuffin, P. 1991. "Quantitative genetics, molecular genetics and dimensions of normal variation." *Behavior and Brain Science*. 2.

Plomin, R; Owen, M; McGuffin, P. 1994. "The genetic basis of complex human behaviors." *Science*. 264: 1733–1739.

Ploog, D. 2002. personal communication to coauthor Cory.

Plutchik, R. 1984. "Emotions: A general psychoevolutionary theory." In K. Scherer; P. Ekman (eds). *Approaches to Emotion*. Hillsdale, NJ: Lawrence Erlbaum.

Pope, H; Lipinski, J. 1978. "Diagnosis in schizophrenia and manic-depression: A reassessment of the specificity of 'schizophrenic' symptoms in light of current research." *Arch Gen Psychiatry*. 35: 811–828.

Pope, HG Jr; Yurgelun-Todd D. 1990. "Schizophrenic individuals with bipolar first-degree relatives: Analysis of two pedigrees." *J Clin Psychiatry*. 51: 97–101.

Pope, H. 1983. "Distinguishing bipolar disorder from schizophrenia in clinical practice: Guidelines and case reports." *Hosp Community Psychiatry*. 34: 322–328.

Popper, K; Eccles, J. 1977. *The Self and Its Brain*. NY: Springer.

Popp, J; De Vore, I. 1979. "Aggressive competition and social dominance theory: Synopsis." In D. Hamburg; E. McCown (eds). *The Great Apes*. Menlo Park, CA: Benjamin Cummings.

Pons, L; Nurnberger, J; Murphy, D. 1981. "Mood-independent abberancies in associative processes in bipolar affective disorder." *J Nerv Ment Dis*. 167: 56–65.

Porges, S. 1998. "Love: An emergent property of the mammalian autonomic nervous system." *Psychoneuroendocrinology*. 23: 837–862.

Porzgen, P; Park. S.; Hirsh, J; Sonders, M; Amara, S. 2001. "The antidepressant-sensitive dopamine transporter in Drosophila melanogaster: a primordial carrier for catecholamine. *Molecular Pharmacology*. Jan; 59(1): 83–95.

Pospisil, L. 1972. *Anthropology of Law*. New Haven: HRAF Press.

Post, RM; Goodwin, F. 1978. "Approaches to brain amines in psychiatric patients: A reevaluation of cerebrospinal fluid studies." In L. Iversen; S. Iversen, S. Snyder (eds). *Handbook of Psychopharmacology*. Vol 13, Biology of Mood and Anti-anxiety Drugs. New York: Plenum.

Povinelli, D. 1987. "Monkeys, apes, mirrors and minds. The evolution of self-awareness in primates." *Human Evolution*. 2: 493–507.

Premack, D. and Woodruff, G. 1978. "Does the Chimpanzee have a Theory of Mind?" Pp. 515–526 in *Behavioral Brain Science*. 1.

Prescott, T. and C. Ibbotson, C. 1997. "A robot trace-maker: early invertebrate behavior." *Artificial Life* 3: 289–306.).

Pribram, K. 1954. "Functions of the 'olfactory brain.'" *Ann NY Acad Sci*. 58: 109–138.

Pribram, K. 1958. "Comparative neurology and the evolution of behavior." In A. Roe; G. Simpson (eds). *Behavior and Evolution*. New Haven: Yale University Press.

Pribram, K. 1994. "Brain and the Structure of Narrative." Pp. 375–415 in *Neural Networks for Knowledge Representation and Inference*. Ed. by Levine, D. & Aparicio, M. Hillsdale, NJ: Lawrence Erlbaum Associates.

Pribram, K. 1973. "The Primate Frontal Cortex—Executive of the Brain." Pp. 293–314 in *Psychophysiology of the Frontal Lobes*. Ed. by Pribram, K. & Luria, A. NY: Academic Press.

Price, D. 1971. *Science Since Babylon*. New Haven: Yale University Press.

Price, J. 1967. "The dominance heirarchy and the evolution of mental illness." *Lancet*. 2: 243.

Price, J. 1968. "The genetics of depressive behaviour." In A. Coppen; A. Wilk (eds). *Recent Developments in Affective Disorders*. London: Royal Medico-Psychological Association.

Price, J. 1972. "Genetic and phylogenetic aspects of mood variation." *International J Mental Health*. 1: 123-144.

Price, J. 1988. "Alternative channels for negotiating asymmetry in social relationships." In M. Chance (ed) *Social Fabrics of the Mind.* London: Lawrence Erlbaum.

Price, J. "The effect of social stress on the behaviour and physiology of monkeys." In K. Davison; A. Kerr (eds*). Contemporary Themes in Psychiatry.* London: Gaskell Press.

Price, J. 1992. "Accentuate the positive, eliminate the negative: The role of boosting and putting-down signals in mental health." In D. Trent (ed) *The Promotion of Mental Health.* Vol I. Avebury: Aldershot.

Price, J. 1993. "Mania and depression as alternative strategies in social competition." *Proceedings of the Royal College of Psychiatrists.* Scarborough.

Price, J. 2002. "The Triune Brain, Escalation De-escalation Strategies, and Mood Disorders." Pp. 107–117 in G. Cory & R. Gardner, eds. *The Evolutionary Neuroethology of Paul MacLean: Convergences and Frontiers.* Westport, CT: Praeger.

Price, J. 1967. "Hypothesis: The dominance hierarchy and the evolution of mental illness." *Lancet* ii, 243–246.

Price, J.; Sloman, L.; Gardner, R.; & Rohde, P. 1994. "The social competition hypothesis of depression." *British Journal of Psychiatry.* 164, 309–315.

Price, J. 2002. "Limbic System." Pp. 695–702 in *Encyclopedia of the Human Brain.* V. 2. NY: Elsevier Science.

Price, J; Gardner, R. 1995. "The paradoxical power of the depressed patient." *British Medical Journal.* 164: 309–315.

Price, J. 1996. "Social competition in the treatment of bipolar disorder." *Proceedings of the X^TH^ Congress, World Psychiatric Association.* Madrid.

Price J.; Sloman, L. 1982 "The evolutionary model of psychiatric disorder." *Arch Gen Psychiatry.* 41: 211.

Price, J; Sloman, L. 1984. "The evolutionary model of psychiatric disorder." *Arch Gen Psychiatry.* 41: 211.

Price, J; Sloman, L. 1987. "Depression as yielding behavior: An animal model based on Schelderup-Ebbe's pecking order." *Ethology and Sociobiology.* 8: 85–95.

Price, J: Sloman, L; Gardner, R; Gilbert, P; Rhode, P. 1994. "The social communication hypothesis of depression." *Br J Psychiatry.* 164: 309–315.

Prolo, P. & Licinio, J. 2002. "DRD4 and Novelty Seeking. Pp. 91–108 in *Molecular Genetics and the Human Personality.* Ed. by J. Benjamin, R. Ebstein, & R. Belmaker. Washington, DC: American Psychiatric Publishing.

Provine, WB. 1971. *The Origins of Theoretical Population Genetics.* Chicago: University of Chicago Press.

Przewlocki, R. 2002. Stress, Opioid Peptides, and Their Receptors." Pp. 691–734 in *Hormones, Brain, and Behavior.* Ed. by Pfaff, D. V. I. NY: Academic Press.

Puig-Antich, J; Goetz, D; Davies, M; Kaplan, T; Davies, S; Ostrow, D; Asnis, L; Twomey, J; Iyergar, S; Ryan, ND.1989. "A controlled family history study of prepuberal major depressive disorder." *Arch Gen Psychiatry*. 46: 406–418.

Purves, D; Lichtman, JW. 1980. "Elimination of synapses in the developing nervous system." *Science*, 210: 153–157.

Rado, S. 1928. "The problem of melancholia." *Int J Psychoanal*, 9: 420–438.

Raleigh, M; McGuire, M; Brammer, G; Yuwiler, A. 1984. "Social status and whole blood serotonin in vervets." *Arch Gen Psychiatry.* 41:4: 405–410.

Raleigh, M; McGuire, M; Brammer, G; Yuwiler, A 1991. "Serotonergic mechanisms promote dominance acquisition in adult male vervet monkeys." *Brain Res*, 181–190.

Ranchor, A. & Sanderman, R. 2000. "Education Levels and Stress." Pp. 15–19 in *Encyclopedia of Stress*. V. 2. Ed. by G. Fink. San Diego, CA: Academic Press.

Rand, A. 1967. *Capitalism: The Unknown Ideal*. New American Library.

Rand, A. 1964. *The Virtue of Selfishness*. New American Library.

Rao, D. and Gu, C. 2002. "Principles and Methods in the Study of Complex Phenotypes." Pp. 1–32 in *Molecular Genetics and the Human Personality*. Ed. by Benjamin, J., Ebstein, R., & Belmaker, R. Washington, D. C.: American Psychiatric Publishing, Inc.

Rao, DC; Morton, NE: 1976. "Resolution of cultural and biological inheritance by path analysis." *Am J Hum Genetics*. 28: 228.

Ray, JC; Sapolsky, RM. 1992. "Styles of social behavior and their endocrine correlates among high-ranking wild baboons." *American Journal of Primatology*, 28: 231–250.

Reich, T; Clayton, PJ; Winokur, G. 1969. "Family history studies, V. The genetics of mania." *Am J Psychiatry*. 125: 1358–1369.

Reich, T: Personal Communication, 1996

Reich, T; James, JW; Morris, CA. 1972. "The use of multiple thresholds in determining the mode of transmission of semi-continuous traits." *Annals Hum Genetics*, 36: 163.

Reichert, H. 2005. "A tripartite organization of the urbilaterian brain: developmental genetic evidence from Drosophila. " *Brain Res Bull*. Sep 15; 66(4–6):491–494.

Reite, M; Field, T. 1985. *The Psychobiology of Attachment and Separation*. New York: Academic Press.

Rendel, JM. 1967. *Canalization and Gene Control*. New York: Academic Press.

Renieris, P; Stefanis, C; Lykoura, E; Varsou, E. 1979. "Subtypes of schizophrenia and ABO blood groups." *Neuropsychobiology*, 8: 57:59.

Retezeanu, A; Christodorescu, D. 1978. "ABO blood groups in affective and in schizophrenic psychosis". *Rev Roum Med, Neurol Psychiat.* 16: 271.

Rice, DP: Personal Communication, 1997

Rice, JP; Cloninger, CR; Reich, T. 1978 "Multifactorial inheritance with cultural transmission and assortative mating, I. Description and basic properties of unitary models." *Am J Hum Genetics*. 30:618.

Rice, JP; Reich, T; Andreasen, NC; Endicott, J; Van Eerdewegh, M; Fishman, R; Hirschfeld, RMA; Klerman, GL. 1987a. "The familial tansmission of bipolar illness." *Arch Gen Psychiatry*. 44: 441–447.

Rice, JP; Endicott, J; Knesevich, MA; Rochberg, N. 1987b. "The estimation of diagnostic sensitivity using stability data: An application to the affective disorders." *Arch Gen Psychiatry*. 49: 824–830.

Rice, JP. 1997. "Initial genome scan of the NIMH genetics initiative bipolar pedigrees: chromosomes 1, 6, 8, 10, and 12." *Am J Med Genet.* 74: 247–253.

Richards, R. 1987. *Darwin and the Emergence of Evolutionary Theories of Mind and Behavior*. Chicago: University of Chicago Press.

Richards, R. 1981. "Relationships between creativity and psychopathology: An evaluation and interpretation of the evidence." *Genet Psychol Monogr.* 103: 261–324.

Richards, RL; Kinney, D; Lunde, I; Benet, M; Merzel, AP. 1988. "Creativity in manic- depressives, cyclothymes, their normal relatives and control subjects." *J Abnormal Psychology*,. 97: 281–288.

Ridley, M. 1996. *The Red Queen: Sex and the Evolution of Human Nature*. New York: Macmillan.

Rihmer, Z; Arato, M. 1981. "ABO blood groups in manic-depressive patients." *J Affective Dis*. 3: 1–7.

Risch, N. 1990. "Genetic linkage and complex diseases, with special reference to psychiatric disorders." *Genetic Epidemiology*. 7: 3–7.

Risch, N. & Botstein, D. 1996. "A Manic Depressive History." *Nat. Genet.* 12:351–353.

Risch, N; Baron, M. 1982. "X-linkage and genetic heterogeneity in bipolar-related major affective illness: Reanalysis of linkage data." *Ann Hum Genet*. 46: 153–166.

Risch, N; Baron, M; Mendelwicz, J. 1986. "Assessing the role of X-linked inheritance in major affective disorder." *J Psychiatric Res*. 20: 275–288.

Risch, N; Botstein, D. 1996. "Editorial." *Nature Genetics*. 12: 351–353.

Ritvo, L. 1990. *Darwin's Influence on Freud: A Tale of Two Sciences*. London: Yale University Press.

Roberts, E. 1987. "Gamma-Aminobutyric Acid (GABA)." V. 1 Pp. 441–445 in *Encyclopedia of Neuroscience.* Ed. by Adelman, G. Boston: Birkhauser.

Robertson, G; Taylor, PJ. 1985. "Some cognitive correlates of affective disorders." *Psychol Med*. 15: 297–309.

Robins, LN. 1978. "Psychiatric epidemiology." *Arch Gen Psychiatry*. 35: 697–702.

Robins, L.., Tipp, J. & Przybeck, T. 1991. "Antisocial personality." In: *Psychiatric disorders in America* (eds) L.N. Robins & D.A. Regier. Free Press.

Robins, L; Helzer, J; Weissman, M; Orvaschel, H; Gruenberg, E; Burke, JD; Regier, D. 1984. "Lifetime prevalence of specific psychiatric disorders in three sites." *Arch Gen Psychiatry*. 41: 949–958.

Romanes, G. 1884. *Mental Evolution in Animals*. London: Keegan, Paul, Trench & Co.

Romanes, G. 1886. *Darwin and After Darwin*, Vol I-III. Chicago: Open Court.

Ronai, Z.; Guttman, A; Nemoda, Z.; Staub, M.; Kalasz, H.; Sasvari-Szekely, M. 2000. "Rapid and sensitive genotyping of dopamine D4 receptor tandem repeats by automated ultrathin-layer gel electrophoresis." *Electrophoresis*. 21: 2058–2061.

Rosanoff, A; Handy, L; Plesset, I. 1935. "The etiology of manic-depressive syndromes with special reference to their occurence in twins." *Am J Psychiatry*. 91: 725–762.

Rose, G. 1980. "Neuroepidemiology: Some general principles." In F Rose (ed) *Clinical Neuroepidemiology*. Turnbridge Wells, UK.

Rosecan, J: Personal Communication, 1996

Rosenberg, R. 1997. "Molecular neurogenetics: The genome is settling the issue." *JAMA*. 278: 1282–1283.

Rosenblatt, J. and Snowden, C. 1996. *Parental Care: Evolution, Mechanisms, and Adaptive Intelligence*. NY: Academic Press.

Rosenthal, D. 1974. *Genetic Theory and Abnormal Behavior*. New York: McGraw Hill.

Rousseau, J. 1965 (1755). *Discours sur l'origine et les fondements de l'ine`galite' parmi les hommes*. Paris: Gallimard,

Rothstein, M., Mehlman. M., Andrews, L. 2002. *Andrews' Genetics: Ethics, Law and Policy*. Thompson West.

Rowe, D. 1990. "Inherited dispositions toward learning delinquent and criminal behavior: New evidence." In: *Crime in biological, social, and moral contexts*, eds. L. Ellis & H. Hoffman. Praeger Pub.

Ruse, M. 1986. *Taking Darwin Seriously: A Naturalistic Approach to Philosophy*. Oxford: Blackwell.Rushton JP. 1990. "Race, brain size and intelligence: A reply to Chernovsky." *Psychol Rep*. 66: 659–666.

Ruzickova, M., Turecki, G., Alda, M. 2003. "Pharmacogenetics and mood stabilization in bipolar disorder." *American Journal of Medical Genetics, Part C*. 123C: 18–25.

Ryan, F. 1955. "Spontaneous mutation in non-dividing bacteria." *Genetics*. 40: 726–738.

Ryff, C. and Singer, B. 1998. "The concept of positive human health." *Psychol. Inquires*. 9: 1–10.

Ryle, G. 1949. *The Concept of Mind.* London: Hutchinson.

Saarela, M., Hlushchuk, Y., Williams, A., Schürmann, M., Kalso, E., and Hari, R. 2006. "The Compassionate Brain: Humans Detect Intensity of Pain from Another's face." *Cerebral Cortex*, doi:10.1093/cercor/bhj141.

Sackheim, HA; Steif, BL. 1988. "Neuropsychology of depression and mania." In A Georgotas; R Cancro (eds). *Depression and Mania.* New York: Elsevier.

Salamone, J. 1994. "The involvement of nucleus accumbens dopamine in appetitive and aversive motivation." *Behavioral Brain Research.* 61: 117–133.

Sapir, E. 1966 (1932). "Cultural anthropology and psychiatry." In D Mandelbaum (ed) *Culture, Language and Personality: Selected Essays.* Berkeley: University of California Press.

Sapolsky, R. 1989. "Hypercortisolism among socially subordinate wild baboons originates at the CNS level." *Arch Gen Psychiatry.* 46: 1047–1051.

Sapolsky, R. 1990a. "Adrenocortial function, social rank and personality among wild baboons." *Biol Psychiatry.* 28: 862–878.

Sapolsky, R. 1990b. "Stress in the Wild." *Scientific American.* 1: 106–113.

Sapolsky, R. 1993. "Endocrine alfresco: Psychoendocrine studies of wild baboons." *Recent Progress in Hormone Research*, 48: 437–468.

Sapolsky, R. 1994. "Individual differences in the stress response." *Seminars in the Neurosciences.* 6: 261–269.

Sapolsky, R. 1999. "The Physiology and Pathophysiology of Unhappiness." Pp. 453–469 in D. Kahneman, E. Deiner, & N. Schwarz (eds.). Well-Being: *The Foundations of Hedonic Psychology.* NY: Russell Sage Foundation.

Sapolsky, R. 1992. "Neuroendocrinology of the Stress Response." Pp. 287–324. in J. Becker, S. Breedlove, & D. Crews (eds.) *Behavioral Endocrinology.* The MIT Press.

Sarnat, HB; Netsky, M. 1981. *Evolution of the Nervous System.* 2nd edition. Oxford: Oxford University Press.

Sartorius, N; Davidian, H; Kernberg, G. 1983. *Depressive Disorders in Different Cultures: Report on the WHO Collaborative Study on Standardized Assessment of Depressive Disorders.* Geneva: World Health Organization, 1983

Sasaki, T; Billett, E; Petronis, A; Ying, D; Parsons, T; Macciardi, FM; Meltzer, H: Lieberman, J; Joffe, R; Ross, CA; McInnes, M; Li, SH; Kennedy, JL. 1996. "Psychosis and genes with trinucleotide repeat polymorphism." *Human Genetics.* 97: 2: 244–246.

Say, J-B. 1855 *A Treatise on Political Economy.* Trans. By C. Prinsep. Philadelphia: Lippincott, Grambo & Co.

Schaeffer, S. and Aguadé, M. 2000. "Evidence for Balancing, Directional, and Background Selection in Molecular Evolution." Pp. 124–142 in *Evolutionary Genetics.* Ed. By Singh, R. & Krimbas, C. Cambridge University Press.

Schar, G.; Price-Stover E.; Clemons K.; Feldman D. & Stevens D. 1986. "Progesterone binding and inhibition of growth in *Trichophyton mentagrophytes*." *Infection and Immunity* 52: 763–767.

Scherer, K; Ekman, P (eds). *Approaches to Emotion*. Hillsdale, NJ: Lawrence Erlbaum

Schjelderupp-Ebbe, T. 1922. "Beitrage zur sozialpsychologie des haushuns." *Zeitschrift fur Psychologie*. 88: 225–252.

Schmalhausen, I. 1949. *Factors of Evolution: The Theory of Stabilizing Selection*. Philadelphia: Blakiston.

Schnider, A. and Gutbrod, K. 1999. "Traumatic Brain Injury." Pp. 487–505 in B. Miller & J. Cummings (eds.). *The Human Frontal Lobes*. NY: The Guilford Press.

Schore, AN. 1994. *Affect Regulation and the Origin of the Self: The Neurobiology of Emotional Development*. London: Lawrence Erlbaum.

Schulkin, J. 2002. "Hormonal Modulation of Central Motivational States." Pp. 633–657 in *Hormones, Brain, and Behavior*. Ed. by Pfaff, D. V. I. NY: Academic Press.

Schulsinger, F; Kety, SS; Rosenthal, D; Wender, P. 1979. "A family study of suicide." In M Schou; E Stromgren (Eds) *Origin, Prevention and Treatment of Affective Disorders*. London: Academic Press, 1979

Schulze, T. and McMahon, F. 2003. "Genetic Linkage and Association Studies in Bipolar Affective Disorder: A Time for Optimism." *Journal of American Medical Genetics, Part C*. 123C: 36–47.

Schulze TG, Buervenich S, Badner JA, Steele CJ, Detera-Wadleigh SD, Dick D, Foroud T, Cox NJ, MacKinnon DF, Potash JB, Berrettini WH, Byerley W, Coryell W, DePaulo JR Jr, Gershon ES, Kelsoe JR, McInnis MG, Murphy DL, Reich T, Scheftner W, Nurnberger JI Jr, McMahon FJ. 2004. "Loci on chromosomes 6q and 6p interact to increase susceptibility to bipolar affective disorder in the national institute of mental health genetics initiative pedigrees." *Biol Psychiatry*. 2004 Jul 1; 56(1): 18–23.

Seeman, P; Guan, HC; Van Tol, HHM. 1993. "Dopamine D_4 receptors elevated in schizophrenia." *Nature*. 365: 441–445.

Segman, R. & Lerer, B. 2000. "Genetic and Causal Factors of Bipolar Disorder." Pp. 31–48 in *Bipolar Disorders: Basic Mechanisms and Therapeutic Implications*. Ed. by Soares, J. & Gershon, S. NY: Marcel Dekker, Inc.

Seguin, E. 1866. *Idiocy: And its Treatment by the Physiological Method*. NY: William Wood.

Segurado R, Detera-Wadleigh SD, Levinson DF, Lewis CM, Gill M, Nurnberger JI Jr, Craddock N, DePaulo JR, Baron M, Gershon ES, Ekholm J, Cichon S, Turecki G, Claes S, Kelsoe JR, Schofield PR, Badenhop RF, Morissette J, Coon H, Blackwood D, McInnes LA, Foroud T, Edenberg HJ, Reich T, Rice JP, Goate A, McInnis MG, McMahon FJ, Badner JA, Goldin LR, Bennett P,

Willour VL, Zandi PP, Liu J, Gilliam C, Juo SH, Berrettini WH, Yoshikawa T, Peltonen L, Lonnqvist J, Nothen MM, Schumacher J, Windemuth C, Rietschel M, Propping P, Maier W, Alda M, Grof P, Rouleau GA, Del-Favero J, Van Broeckhoven C, Mendlewicz J, Adolfsson R, Spence MA, Luebbert H, Adams LJ, Donald JA, Mitchell PB, Barden N, Shink E, Byerley W, Muir W, Visscher PM, Macgregor S, Gurling H, Kalsi G, McQuillin A, Escamilla MA, Reus VI, Leon P, Freimer NB, Ewald H, Kruse TA, Mors O, Radhakrishna U, Blouin JL, Antonarakis SE, Akarsu N. 2003. "Genome scan meta-analysis of schizophrenia and bipolar disorder, part III: Bipolar disorder." *Am J Hum Genet.* 2003 Jul; 73(1): 49–62. Epub 2003 Jun 11.

Sejnowski, T. 1999. Review of Allman, J. 1999. *Evolving Brains.* NY: Scientific American Library, in *Science*, V 283, N 5405, 1999, p 1121).

Selye, H. 1978. *The Stress of Life.* NY: McGraw-Hill.

Selye, H. 1936. "A Syndrome Produced by Diverse Nocuous Agents." *Nature.* 138: 32–33.

Sevcikova, L.; Jurkovicova, J.; Stefanikova, Z.; Bottilikova, S.; Sabolova, M.; Machacova, E; Aghova, L. 2003. "Factors of Perceived Stress in School Children." *Homeostasis.* V42, N5:193–95.

Sham, P. 1996. "Genetic epidemiology." *Br Med Bull.* 52: 408–433.

Shapiro, RW; Ryder, JP; Svejgaard, A; Rafaelson, O. 1977. "HLA antigens and manic-depressive disorders: Further evidence of an association." *Psychological Medicine.* 7: 387–396.

Shaw, C. & McEachern, J. 2001. *Toward a Theory of Neuroplasticity.* NY: Taylor and Francis.

Sherman, S. and Guillery, R. 2001. *Exploring the Thalamus.* NY: Academic Press.

Sherrington, C. 1906. *The Integration of Action of the Nervous System.* New York: Scribner.

Sherrington, R; Brynjolfsson, J; Petursson, H. 1988. "Localization of a susceptibility locus for schizophrenia on chromosome 5." *Nature.* 336: 164–267.

Shmukler, Y. and Buznikov, G. 1998. "Functional Coupling of Neurotransmitters with Second Messengers during Cleavage Divisions: Facts and Hypotheses." *Perspectives on Developmental Neurobiology.* Vol. 5. pp. 469–480;

Silberman, EK; Weingartner, H; Post, R. 1988. "Thinking disorder in depression: Logic and strategy in abstract reasoning task." *Arch Gen Psychiatry.* 40: 775–780.

Simon, N. 2002. "Hormonal Processes in the Development and Expression of Aggressive Behavior." Pp. 339–392 in *Hormones, Brain, and Behavior.* Ed. by Pfaff, D. V. I. NY: Academic Press.

Simonton, D. 1984. *Genius, Creativity and Leadership: Historiometric Inquiries.* London: Harvard University Press.

Simpson, G., McMahon, F. McInnis, M., MacKinnon, D. Edwin, D., Folstein, S., DePaulo. J. 2002. "Diagnostic Reliability of Bipolar II Disorder." *Archives of General Psychiatry*. 59: 736–740.

Simpson, SG; Folstein, SE; Meyers, DA; McMahon, FJ; Brusco, DM; DePaulo, JR Jr. 1993. "Bipolar II: The most common bipolar phenotype?" *Am J Psychiatry*. 150: 901–903.

Sklar P, Pato MT, Kirby A, Petryshen TL, Medeiros H, Carvalho C, Macedo A, Dourado A, Coelho I, Valente J, Soares MJ, Ferreira CP, Lei M, Verner A, Hudson TJ, Morley CP, Kennedy JL, Azevedo MH, Lander E, Daly MJ, Pato CN. 2004. "Genome-wide scan in Portuguese Island families identifies 5q31-5q35 as a susceptibility locus for schizophrenia and psychosis." *Mol Psychiatry*. 2004 Feb;9(2): 213–218.

Slater, E, 1953. *Psychotic and Neurotic Illnesses in Twins*. London: Her Majesty's Stationery Office.

Slavin, MO; Kriegman, D. 1992. *The Adaptive Design of the Human Psyche*. New York: Guilford Press.

Sloman, L. 1976. "The role of neurosis in phylogenetic adaptation with particular reference to early man." *Am J Psychiatry*. 133: 543–547.

Sloman, L. 2002. "Involuntary Defeat Strategy as Backdrop for Depression." Pp. 120–132 in G. Cory & R. Gardner (eds.). *The Evolutionary Neuroethology of Paul MacLean: Convergences and Frontiers*. Westport, CT: Praeger.

Smeraldi, E; Negri, F; Melica, AM; Scorze-Smeraldi, R. 1978. "HLA system and affective disorders: A sibship genetic study." *Tissue Antigens*. 12: 270–274.

Smeraldi, E; Negri, F; Heimbuch, RC; Kidd, KK. 1981. "Familial patterns and possible modes of inheritance of primary affective disorders." *J Aff Disorders*. 3: 173–182.

Smith, S; Kimberling, W; Pennington, B; Lubs, H. 1983. "Specific reading disability: Identification of an inherited form through linkage analysis." *Science*. 219: 1345–1347.

Smoller, W. & Finn, C. 2003. "Family, Twin, and Adoption Studies of Bipolar Disorder." *American Journal of Medical Genetics*. 123C: 48–58.

Smyth, C., Kalsi, G., Curtis, D., et al. 1997. "Two-Locus Admixture Linkage Analysis of Bipolar and Unipolar Affective Disorder Supports the Presence of Susceptibility Loci on Chromosomes 11p15 and 21q22. *Genomics*. 39: 271–278.

Soares, J. & Gershon, S. (eds.) 2000. *Bipolar Disorder: Basic Mechanisms and Therapeutic Implications*. NY: Marcel Dekker.

Sokoloff, P; Levesque, D; Martres, MP. 1992. "The dopamine D3 receptor: From cloning to characterization of the native receptor in the brain." *Proc 7TH International Catecholamine Symposium*.

Spalt, L. 1975. "Sexual behavior and affective disorders." *Dis Nerv Syst*. 36: 974–997.

Spence, MA; Bishop, DT; Boehnke, M; Elston, RC; Falk, C; Hodge, SE; Ott, J; Rice, J; Merikangas, K; Kupfer, D. 1992. "Methodological issues in linkage analyses for psychiatric disorders: Secular trends, assortative mating, bilineal pedigrees." "Report of the MacArthur Foundation, Network 1 Task Force on Methodological Issues." *Human Heredity*. 1992

Spence, MA; Flodman, PL; Sadovnick, AD; Bailey-Wilson, JE; Ameli, H; Remick, RA. 1995. "Bipolar disorder: Evidence for a major locus." *Am J Med Genet.* 60: 5.

Spencer, H. 1851. *Social Statics: Or the Conditions Essential to Human Happiness Specified.* London: John Chapman.

Sperry, RW. 1964. *Problems Outstanding in the Evolution of Brain Function. James Arthur Lecture.* New York: American Museum of Natural History, 1964

Spitz, R. 1965. *The First Year of Life*. NY: International Universities Press.

Spitzer, RL; Endicott, J; Robins, E. 1978. "Research diagnostic criteria: Rationale and reliability." *Arch Gen Psychiatry*. 35: 773–782.

Spuhler, JN. 1959. "Somatic paths to culture." *Human Biology*. 31: 1–13.

Staley, K., Smith, R., Schaack, J., Wilcox, C., Jentsch, T. 1996. "Alteration of GABA A receptor function following gene transfer of the CLC-2 chloride channel." *Neuron.* Sep; 17 (3): 543–551.

Stamm, J. 1955. "The function of the medial cerebral cortex in maternal behavior in rats." *Journal of Comparative Physiological Psychology.* 48: 347–356.

Stanwood, G. & Zigmond, M. 2000. "Dopamine, Central." Pp. 739–746 in *Encyclopedia of Stress*. V. 1. Ed. by G. Fink. San Diego, CA: Academic Press.

Stensted, A. 1952. "A study of manic-depressive psychoses: Clinical, social, and genetic investigations." *Acta Psychiatr et Neurol* (Suppl 79): 1–111.

Sterling, P. and Eyer, J. 1981. "Allostasis: A New Paradigm to Explain Arousal Pathology." Pp. 629–652 in *Handbook of Life Stress, Cognition and Health.* Ed. by S. Fisher & J. Reason. NY: Wiley.

Stern, C. and Passingham, R. 1996. "The nucleus accumbens in monkeys (Macaca fascicularis): II. Emotion and motivation." *Behavioral Brain Research.* 75: 179–193.

Sternberg, E. 2000. "Does Stress Make You Sick and Belief Make You Well." Pp. 1–3 in *Neuroimmumodulation: Perspectives at the New Millenium.* Ed. by Conti, A; Maestroni, G. NY: The New York Academy of Sciences.

Stevens, A; Price, J. 1996. *Evolutionary Psychiatry: A New Beginning.* London: Routledge.

Stevens, A; Price, J. 2000. *Evolutionary Psychiatry:A New Beginning.* 2nd Ed. London: Routledge.

Stine, OC; Xu, JF; Koskela, R. 1995. "Evidence for linkage of bipolar disorder to chromosome 18 with a parent-of-origin effect." *Am J Hum Gen* 56: 1384–1394.

Stine, O. 1997. "Initial genome screen for bipolar disorder in the NIMH genetics initiative pedigrees: chromosomes 2, 11, 13, 14, and X." *Am J Med Genet.* 74: 263–269.

Stine, O., Xu, J., Koskela, R., McMahon, F., Geswend, M., Friddle, C., et al. 1995. "Evidence of linkage of BPD to chromosome 18 with a parent of origin effect." *American Journal of Human Genetics.* 57: 1384–1394.

Stoka, A. 1999. "Phylogeny and evolution of chemical communication: an endocrine approach." *Journal of Molecular Endocrinology.* 22: 207–225.

Storey, A.; Walsh, C.; Quinton, R.; Wynne-Edwards, K. 2000. "Hormones of paternal responsiveness in new and expectant fathers." *Evol. Hum.* I: 95.

Strauss, C. & Lahey, B. 1984. "Behavior disorders of children." In: *Comprehensive handbook of psychopathology,* (eds). Adams, H.E. & Sutker, P.B. Plenum Press.

Straub, R., Lehner, T., Luo, Y., Loth, J., Shao, W., Sharpe, L., et al. 1994. "A possible vulnerability locus for bipolar affective disorder on chromosome 21q22.3." *Nature Genetics.* 8: 291–296.

Strober, M. 1984. "Familial aspects of depressive disorder in early adolescence." In E Weller; R Weller (eds). *Current Perspectives on Major Depressive Disorders in Children.* Washington, DC: American Psychiatric Press.

Strober, M; Morrel, W; Burroughs, J; Lampert, C; Danforth, H; Freeman, R. 1988. "A family study of bipolar I disorder in adolescence: Early onset of symptoms linked to increased familial loading and lithium resistance." *J Aff Dis.* 15: 255–268.

Stuss, D. and Knight, R. 2002. *Principles of Frontal Lobe Function.* Oxford University Press.

Suarez, B; Croughan, J. 1982. "Is the major histocompatibility complex linked to genes that increase susceptibility to affective disorders? A critical appraisal." *Psychiatr Res.* 7: 19–27.

Suarez, B; Reich, T. "HLA and major affective disorders." *Arch Gen Psychiatry.* 41: 22-27.

Suarez, BK; Reich, T; Trost, J. 1985. "Limits of the two allele single locus model with incomplete penetrance." *Annals Hum Genetics.* 42: 87–94.

Sulloway, F. 1979. *Freud, Biologist of Mind: Behind the Psychoanalytic Legend.* New York: Basic Books.

Sulloway, F. 1996. *Born to Rebel: Birth Order, Family Dynamics and Creative Lives.* New York: Pantheon.

Sutton, WS. 1902. "Chromosomes." *Biol Bull.* 4: 231–248.

Suzuki, DT; Griffiths, AJF; Miller, J; Lewontin, RC. 1989. *An Introduction to Genetic Analysis* 4th ed. San Fransisco: W.H. Freeman, 1989, p.704.

Swallow, SR; Kuiper, NA. 1988. "Social comparison and negative self evaluation: An application to depression." *Clin Psychology Rev.* 8: 55–76.
Swedlund, A. 1974. "The use of ecological hypotheses in australopithecine taxonomy." *American Anthropologist.* V. 76 (3) 515–529.
Swedlund, A. 1978. "Historical demography as population ecology." *Annual Review of Anthropology.* 7: 137–173.
Swedlund, A. and Armelagos, G. (eds). 1990. *Disease in Populations in Transition: Anthropological and Epidemiological Perspectives.* South Hadley: Bergin and Garvey.
Symons, D. 1979. *The Evolution of Human Sexuality.* Oxford University Press.
Szasz, T. 1974. *The Myth of Mental Illness.* NY: Harper & Row.
Tankersley, D., Stowe, C., Huettel, S. 2007. "Altruism is associated with an increased neural response to agency." *Nature Neuroscience.* V10 N2 (Feb 2007): 150–151.
Tanna, V; Winokur, G. 1968. "A study of association and linkage of ABO blood types and primary affective disorders." *Br J Psychiatry.* 114: 117–1181.
Tanna, V; Go, RCP; Winokur, G; Elston, RC. 1977. "Possible linkage between group-specific component (Gc protein) and pure depressive disease." *Acta Psychiatr Scand.* 55: 111–115.
Targum, SD; Gershon, ES; Van Eerdewegh, M; Rogentine, N. 1979. "Human HLA system not closely linked to or associated with bipolar manic-depressive illness." *Biological Psychiatry.* 14: 615–636.
Tarter, RE; Hegedus, AM; Gavaler, JS. 1985. "Hyperactivity in sons of alcoholics." *J Studies Alcohol.* 46: 259–261.
Taylor, MA; Abrams, R. 1980. "Reassessing the bipolar-unipolar dichotomy." *J Aff Disorders.* 2: 195–217 .
Taylor, MA; Abrams, R. 1981. "Early- and late-onset bipolar illness." *Arch Gen Psychiatry.* 38: 51–68.
Taylor, M. Berenbaum, S., Jampala, V., Cloninger, C. 1993. "Are schizophrenia and affective disorder related? Preliminary data from a family study." *American Journal of Psychiatry.* 150: 278–285.
Taylor, S. 2002. *The Tending Instinct.* NY: Time Books.
Teicher, M., Andersen, S., Polcari, A., Anderson, C., Navalta, C., Kim, D. 2003. "The neurobiological consequences of early stress and childhood maltreatment." *Neuroscience and Biobehavioral Reviews.* 27: 33–44.
Thapar, A; McGuffin, P. 1996. "Genetic influences on life events in childhood." *Psychological Medicine.* 26: 813–820.
Thierry, A.; P. Tassin, J.; Blanc, G; Glowinski, J. 1976. "Selective activation of the mesocortical DA system by stress." *Nature* 263: 242–244.
Thomas, E. 1995. *The Very Best Men.* New York: Touchstone.
Thompson, D'A. *On Growth and Form.* New York: Macmillan.

Thomson, G. 1981. "A review of theoretical aspects of HLA and disease associations." *Theor Pop Biol.* 20: 168–208.

Tickell, C. 1993. "The human species: A suicidal success." Pp. 219–226 in *Geographical Journal.* Vol. 159.

Tiger, L. 1975. "Somatic factors and social behaviour." In R. Fox (ed) *Biosocial Anthropology*. New York: Wiley.

Tiitinen, H.; Sinkkonen, J.; Reinikainen, K.; Alho, K.; Lavikainen, J.; Naatanen, R. 1993. "Selective attention enhances the auditory 40-Hz transient response in humans." *Nature*. 364: 59–60.

Tillich, P. 1950 *The Courage to Be*. Yale University Press.

Tinbergen, N. 1951. *The Study of Instinct*. Oxford University Press.

Tooby, J. and Cosmides, L. 1989. "Evolutionary psychology and the generation of culture, Part I." Pp. 29–49 in *Ethology and Sociobiology*. V. 10.

Tooby, J. and DeVore, I. 1987 "The Reconstruction of Hominid Behavioral Evolution through Strategic Modeling." in *Primate Primate Models for the Origin of Human Behavior*. Ed. by W. Kinsey. New York: SUNY Press.

Tooby, J; Cosmides, L. 1990a. "On the universality of human nature and the uniqueness of the individual: The role of genetics and adaptation." *J Personality*. 58: 1:17–67.

Tooby, J; Cosmides, L. 1990b. "The past explains the present: Emotional adaptations and the structure of ancestral environments." *Ethology and Sociobiology*. 11: 375–421.

Traub, R.; Spruston, N.; Soltesz, I.; Konnerth, A.; Whittington, M.; Jefferys, G. 1998. "Gamma-frequency oscillations: A neuronal population phenomenon, regulated by synaptic and intrinsic cellular processes, and inducing synaptic plasticity." *Progress in Neurobiology*. 55: 563–575.

Trimble, M; Mendez, M; Cummings, J. 1997. "Neuropsychiatric symptoms from the temporolimbic lobes." *J Neuropsychiatry Clin Neurosci*. 9: 429–438.

Trivers, R. 1981. "*Sociobiology and Politics*." Pp. 1–44 in *Sociobiology and Human Politics*, Ed. by E. White. Lexington, MA: D. C. Heath & Company.

Trivers, R. 1971a. "The evolution of reciprocal altruism." Pp. 35–57 in *The Quarterly Review of Biology*. V. 46.

Trivers, RL. 1971b. "Parental investment and sexual selection." In B Campbell (ed) *Sexual Selection and the Descent of Man 1871—1971*. Chicago: Aldine.

Trivers, RL. 1974. "Parent-offspring conflict." *American Zoologist*. 14:249–264.

Trivers, RL. 1985. *Social Evolution*. Menlo Park, CA: Benjamin/Cummings.

Trower, P; Gilbert, P. 1989. "New theoretical conceptions of social anxiety and social phobia." *Clin Psychology Rev*. 9: 19–35.

Trudeau, V. 1997. "Neuroendocrine regulation of gonadotropin-II release and gonadal growth in goldfish, carrassius auratus." *Rev Reprod*. 2. 55–68.

Tsuang, M; Faraone, SV; Fleming, J. "Familial transmission of major affective disorders: Is there evidence supporting the distinction between unipolar and bipolar disorders?" *Br J Psychiatry*. 146: 268–271.

Tsuang, M; Faraone, SV; Lyons, MJ. 1993. "Identification of the phenotype in psychiatric genetics." *European Arch Psychiatry Neurol Sci*. 682: 1–12.

Tsuang, M; Faraone, S. 1990. *The Genetics of Mood Disorders*. Baltimore: The Johns Hopkins University Press.

Tsuang, M. and Faraone, S. 2003. "Heterogeneity and the genetics of bipolar disorder." *American Journal of Medical Genetics. Part C.* V123C. N1:1–9.

Tsuang, M., Winokur, G., Crowe, R. 1980. "Morbidity risks of schizophrenia and affective disorders among first degree relatives of patients with schizophrenia, mania, depression and surgical conditions." *British Journal of Psychiatry*. 137: 497–504.

Tucker, D; Luu, P.; and Pribram, K. 1995. "Social and emotional self-regulation." Pp. 213–239 in *Annals of the New York Academy of Sciences*. Vol. 769.

Tuke, DH. 1878. *Insanity: In Ancient and Modern Life*. London: Macmillan.

Turner, WJ; King, S. 1983. "BPD-2 an autosomal form of bipolar affective disorder." *Biological Psychiatr.*, 18: 63–88.

Valles. V., Van Os, J., Guillamat, R., Gutierrez, B., Campillo, M., Gento, P., Fananas, L. 2000. "Increased morbid risk for schizophrenia in families of in-patients with bipolar illness."*Schizophrenia Research*. 42: 83–90.

van Baal, J. 1975. *Reciprocity and the Position of Women*. Amsterdam: Van Gorcum, Assen.

van den Berghe, PL; Barash, DP.1977. "Inclusive fitness and human family structure." *American Anthropologist*. 79: 809-823.

van der Dennen, JGM. 1986. "Ethnocentrism and in-group/out-group differentiation: A review and interpretation of the literature." In V Reynolds; V Falger; I Vine (eds) *The Sociobiology of Ethnocentrism: Evolutionary Dimensions of Xenophobia, Discrimination, Racism and Nationalism*. London: Croom Helm.

Van Tol, HHM; Bunzow, JR; Guan, HC. 1991. "Cloning of the gene for a human dopamine D4 receptor with high affinity for the antipsychotic clozapine." *Nature*. 350: 610–614.

Van Wulfften-Palthe, PM. "Psychiatry and neurology in the tropics." In C deLangen, A Lichtenstien (eds) *A Clinical Textbook of Tropical Medicine*. Batavia: Kloff.

Venken, T.; Claes, S.; Sluijs, S.; Paterson, A.; van Duijn, C.; Adolfsson, R.; Del-Favero, J.; Van Broeckhoven, C. 2005. "Genomewide scan for affective disorder susceptibility loci in families of a northern Swedish isolated population." *American Journal of Human Genetics*. 76: 237–248.

Verdoux H, Bourgeois M. 1995. "Social class in unipolar and bipolar probands and relatives." *J Affect Disord*. 33: 181–187.

Vico, G. 1725. *Scienza Nuova*. Naples.

Vincent, J.; Cardinaud, B.; Vernier, P. 1998. "Evolution of monoamine receptors and the origin of motivational and emotional systems in vertebrates. *Bull Acad Natl Med.* 182(7): 1505–14; discussion 1515–6)

Virkkunen, M. 1976. "Reactive hypoglycemic tendency among habitually violent offenders." *Nutritional Review* 44 (suppl): 94–103.

Vogel, F. 1981. "Neurobiological approaches in human behavioral genetics." *Behavior Genetics*. 11: 87–102.

Vogel, F; chalt, E; Kruger, J. 1979a. "The electroencephalogram (EEG) as a research tool in human behavior genetics: Psychological examinations in healthy males with various inherited EEG variant, I. Rationale of the study, material, methods, heritability of test parameters." *Hum Genetics.* 47: 1–45.

Vogel, F; Schalt, E; Kruger, J. 1979b. "The electroencephalogram (EEG) as a research tool in human behavior genetics: Psychological examinations in healthy males with various inherited EEG variant, II. Results." *Hum Genetics*, 47: 47–80.

Vogel, F; Schalt, E; Kruger, J. 1979c "The electroencephalogram (EEG) as a research tool in human behavior genetics: Psychological examinations in healthy males with inherited EEG variant, III. Interpretation of the results." *Hum Genetics*. 47: 81–111.

Von Knorring, AL; Cloninger, CR; Bohman, M; Sigvardsson, S. 1983. "An adoption study of depressive disorders and substance abuse." *Arch Gen Psychiatry*. 40: 943–950.

von Neumann, J. 1961–1963 (1928). *The Theory of Games*. In A Taub (ed) *John von Neumann: Complete Works*. 1903–1957, 6 Vol. Oxford: Pergamon,

von Neumann, J; Morgenstern, O. 1944. *The Theory of Games and Economic Behavior*. Princeton University Press.

von Tschermak, P. 1900. "Mendel's law concerning the behavior of progeny of varietal hybrids." *Berichte der Deutchen Botanischen Gesellschaft* 18: 232–239.

Waddington, CH. 1957. *The Strategy of the Genes*. London: Allen and Unwin.

Wakefield, JC. 1992. "Disorder as harmful dysfunction: A conceptual critique of DSM-III-R's definition of mental disorder." *Psychological Review*. 99: 232–247.

Wallace, AR. 1891. "On the law which has regulated the introduction of new species." *Ann Mag Nat Hist.* 16: 184–196, 1855. Rptd in *Natural Selection and Tropical Nature*. London: Macmillan, 1891.

Weatherall, DJ; Clegg, JB. 1981. *The Thalassaemia Syndromes*. Oxford: Blackwell.

Weber, M. 1947. "The theory of social and economic organization." In HH Gerth; CW Mills (eds) *From Max Weber*. Oxford: Oxford University Press.

Weinberg, W; Brumback, RA. 1990. "Mania in childhood." *Am J Diseases of Child.*130: 380–385.

Weinberger, D. 1987. "Implications of normal brain development for the pathogenesis of schizophrenia." *Arch Gen Psychiatry.* 44: 660.

Weinrich, JD 1977. "Human sociobiology: Pair-bonding and resource predictability (effects of social class and race)." *Behav Ecology Sociobiology.* 2: 91–118.

Weisfeld. G. 2002. "Neural and Functional Aspects of Pride and Shame." Pp. 193–214 in G. Cory and R. Gardner (eds). The Evolutionary Neuroethology of Paul MacLean: Convergences and Frontiers. Westport, CT: Greenwood/Praeger.

Weismann, A. 1909. *Die Selektionstheorie: Eine Untersuchung.* Jena: Fischer.

Weiss, P. 1950. *Genetic Neurology.* Chicago: University of Chicago Press.

Weiss, JM; Glazer, H; Pohorecky, LA; Bailey, WH; Schneider, LH. 1979. "Coping behavior and stress-induced behavioral depression: Studies of the role of brain catecholamines." In R Depue (ed) *The Psychobiology of Depressive Disorders.* New York: Academic Press.

Weiss, JM; Simpson, PG. 1985. "Neurochemical mechanisms underlying stress-induced depression." In T Field; P McCabe; N Schneiderman (eds) *Stress and Coping.* London: Lawrence Erlbaum.

Weiss, RD; Collins, DA. 1992. "Substance abuse and psychiatric illness: The dually diagnosed patient." *Am J Addiction.* 1: 93.

Weissing, FJ. 1996. "Genetic versus phenotypic models of selection: Can genetics be neglected in a long-term perspective?" *J Math Biol.* 34: 533–555.

Weissman, MM; Prusoff, B; Merikangas, KR. 1984a. "Is delusional depression related to bipolar disorder?" *Am J Psychiatry,.* 141: 892–893.

Weissman, MM; Gershon, ES; Kidd, KK; Prusoff, BA; Leckman, JF; Dibble, E; Hamovit, J; Thompson, WD; Pauls, DL, Guroff, JJ. 1984b. "Psychiatric disorders in the relatives of probands with affective disorders: The Yale-NIMH collaborative family study." *Arch Gen Psychiatry,* 41: 13–21.

Weissman, MM; Leaf, P; Tischler, G; Blazer, D; Karno, M; Bruce, M; Florio, L. 1988. "Affective disorders in five United States communities." *Psychol Med.* 18: 141–153.

Weissman, M., Gershon, E., Kidd, K., Prusoff, B. et al. 1984. "Psychiatric disorders in the relatives of probands with affective disorders: The Yale University—NIMH Collaborative Study." *Archives of General Psychiatry.* 41: 13–21.

Weitkamp, LR; Purdue, LH; Huntzinger, RS. 1980. "Genetic marker studies in a family with unipolar depression." *Arch Gen Psychiatry.* 37: 1187–1192.

Weitkamp, LR; Stancer, HC; Persad, E; Flood, C; Guttormsen, S. 1981. "Depressive disorders and HLA: A gene on chromosome six that can affect behavior." *NEJM.* 305: 1301–1306.

Weitz, S. 1977. *Sex Roles: Biological, Psychological and Social Foundations.* New York: Oxford University Press.

Weitz, S. 1979. "Proximity behaviors." In S Weitz (ed) *Nonverbal Communications*, 2nd ed. New York: Oxford University Press.
Welner, Z; Rice, J. 1988. "School-aged children of depressed parents: A blind and controlled study." *J Affective Disorders*. 15: 291–302.
Weller, E; Weller, R. 1988. "Neuroendocrine changes in affectively ill children and adolescents." *Endocrinol Metab Clin North Am*. 17: 41–53.
Wender, P; Kety, S; Rosenthal, D; Schulsinger, F; Ortmann, J; Lunde, I. 1986. "Psychiatric disorders in the biological and adoptive families of individuals with affective disorders." *Arch Gen Psychiatry*. 43: 923–929.
Wenegrat, B. 1984. *Sociobiology and Mental Disorder: A New View*. Menlo Park, CA: Addison Wesley.
West-Eberhard, M. 2003. *Developmental Plasticity and Evolution*. Oxford University Press.
West-Eberhard, M. 2005. *Developmental plasticity and the origin of species differences*. Proceedings of the National Academy of Science. V. 102. Suppl. 1 6543–6549.
Westermarck, E. 1891. *The History of Human Marriage*. New York: Macmillan.
Wickelgren, I. 1998. "Getting the brain's attention." *Science*. 278: 35–37.
Widlocher, DJ. 1983. "Psychomotor retardation: Clinical, theoretical and psychometric aspects." *Psychiatr Clin North Am*. 6: 27–40.
Widom, C.S. 1976. "Interpersonal and personal construct systems in psychopaths." *Journal of consulting and clinical psychology* 44(4): 614–623.
Williams, J; Gibbon, M; First, M; Spitzer, R; Davies, M; Borus, J; Howes, M; Kane, J; Pope, H; Rounsaville, B. 1988. "Structured clinical interview for DSM-III-R (SCID), II. Multisite test-retest reliability." *Arch Gen Psychiatry*. 49: 630–636.
Williams, G. 1966. *Adaptation and Natural Selection*. Princeton: Princeton University Press.
Williams, G. 1975. *Sex and Evolution*. Princeton: Princeton University Press.
Williams, G; Nesse, R. 1991. "The dawn of Darwinian medicine." *Quarterly Review of Biology*. 66: 1–22.
Wilson, D; Sober, E. 1983. "Reintroducing group selection to the human behavioral sciences." *Behavioral and brain sciences*. 17: 159-187.
Wilson, D. R; Widmer, R; Cadoret, R; Judiesch, K. 1987. "Anxiety in family practice." *J Affective Dis*. 12: 179–183.
Wilson, D. R. 1991. "A healthy approach to the end of life!" *Proceedings of the Andover-Newton Theological Seminary Annual Colloquium*. Boston.
Wilson, D. R. 1992. "Evolutionary epidemiology." *Acta Biotheoretica*. 40: 87–90.
Wilson, D. R. 1993. "Evolutionary epidemiology: Darwinian theory in the service of medicine and psychiatry." *Acta Biotheoretica*. 41: 205–218.
Wilson, D. R. 1994. "The Darwinian roots of human neurosis." *Acta Biotheoretica*. 42: 49–62.

Wilson, D. R; Stanton, S; Wilson, SL. 1995. "Serotonin and dopamine in the service of kinship selection." *Presented at the European Sociobiological Society 18TH Annual Meeting.* Christ's College, Cambridge.

Wilson, D. R. "Darwinian psychiatry." *Proceedings of the XTH Congress, World Psychiatric Association.* Madrid.

Wilson, D. R. 1997. "Phylogenetics of neurotransmitters." *Proceedings of the All-Ohio Institute on Psychiatry.* Columbus: ODMH.

Wilson, D. R. 2002. "The Evolved Basis of Mood and Thought Disorders." In Cory, G. and Gardner, R., eds. *The Evolutionary Neuroethology of Paul MacLean: Convergences and Frontiers.* Westport, CT: Praeger.

Wilson, E. O. 1975. *Sociobiology: The New Synthesis.* Belknap/Harvard University Press.

Wilson, E. O; Bossert, W. 1971. *A Primer of Population Biology.* Sunderland, Mass: Sinauer Associates.

Wimsatt, W. 1989. "Evolutionary entrenchment." (Proceedings of the Human Behavior and Evolution Society Second Annual Meeting), *Ethology and Sociobiology.* 10: 5.

Winokur, G; Tanna, VL. 1969. "Possible role of X-linked dominant factor in manic-depressive disease." *Dis Nerv Syst.* 30: 89–94.

Winokur, G. 1980. "Is there a common genetic factor in bipolar disorder and unipolar affective disorder?" *Compr Psychiatry.* 21: 460-468.

Winokur, G; Clayton, P. 1986. *The Medical Basis of Psychiatry.* Philadephia: Saunders.

Winokur, G., Clayton, P., Reich, T. 1969. *Manic-Depressive Illness.* Pp. 112–125. St. Louis: C.V. Mosby.

Wise, R. A., M. A. Bozarth. 1984. "Brain reward circuitry: four circuit elements 'wired' in apparent series." *Brain Research Bulletin.* 12: 203–208.

Witter, M. and Groenewegen, H. 1992. "Organizational Principles of Hippocampal Connections." Pp. 37–60 in *The Temporal Lobes and the Limbic System.* Ed. By Trimble, M. and Bolwig, T. Petersfield, UK: Wrightson Biomedical Publishing

Wolf, P. 1987. "Definitions of antisocial behavior in biosocial research." In: *The causes of crime: New biological approaches.* Eds. S.A. Mednick, T. Moffitt, & S. Stack. Cambridge University Press.

Wong, D.; Wagner, H.; Tune, L. 1986. "Positron emission tomography reveals elevated D_2 dopamine receptors in drug-naive schizophrenics." *Science.* 234: 1558.

Wood, J. 1989. "Theory and research concerning social comparisons of personal attributes." *Psychological Bull.* 106:2 31–238.

World Health Organization. 1992. *Manual of the International Classification of Diseases, Injuries and Causes of Death*, 10th Ed. Geneva: World Health Organization.

Wright, S. 1968. *Evolution and the Genetics of Populations*. Chicago: University of Chicago Press.

Wyatt, R.; Karoum, F; Suddath, R; Hitri, A. 1988. "The role of cocaine in dopamine use and abuse." *Psychiatr Ann.* 18: 531–534.

Wynne-Edwards, K. 2001. "Hormonal changes in mammalian fathers." *Horm. Behav*. 40: 139–145.

Wynne-Edwards, V. 1962. *Animal Dispersion in Relation to Social Behavior*. Edinburgh: Oliver and Boyd.

Wynne-Edwards, V. 1986. *Evolution Through Group Selection*. Oxford: Blackwell.

Yakovlev, P. 1948. "Motility, behavior, and the brain." *Journal of Nervous and Mental Disease*, 107: 313–335.

Yerkes, R. 1925. *Almost Human*. London: Cape.

Young, L. 2007. "Is bipolar disorder a mitochondrial disease? *J. Psychiatry Neurosci* 2007; 32(3): 160–161.

Youngren, M; Lewinshon, PM. 1980. "The functional relationship between depression and problematic interpersonal behavior." *J Abnormal Psychol.* 89: 333–341.

Zerbin-Rudin, E. 1982. "Genetic research and the theory of schizophrenia." *International Journal Ment H.* 1: 42–62.

Zingg, H. 2002. "Oxytocin." Pp. 779–802 in *Hormones, Brain, and Behavior*. Ed. by Pfaff, D. V. III. NY: Academic Press.

Zuckerman, S. 1932. *The Social Life of Monkeys and Apes*. London: Butler and Tanner.

Zuckerman, M. 2002. "Genetics of Sensation Seeking." Pp. 193–209 in *Molecular Genetics and the Human Personality*. Ed. by J. Benjamin, R. Ebstein, & R. Belmaker. Washington, D.C.: American Psychiatric Publishing.

NAME INDEX

SUBJECT INDEX